Resource Management in Satellite Networks

Optimization and Cross-Layer Design

Resource Management in Satellite Networks

Optimization and Cross-Layer Design

Giovanni Giambene
Università degli Studi di Siena

Springer

Giovanni Giambene
Dipartimento di Ingegneria Dell'Informazione
Università degli Studi di Siena
Via Roma, 56
53100 Siena
ITALY

ISBN 978-1-4419-4235-7 e-ISBN 978-0-387-53991-1

Printed on acid-free paper.

9 8 7 6 5 4 3 2 1

springer.com

Acknowledgements

The volume Editor and the authors would like to acknowledge the FP6 EU Network of Excellence SatNEx II (IST-027393) and, in particular, activities ja2230&2330 for the research work as the basis of this book.

Preface

Nowadays, satellites are used for a variety of purposes, including sensors and data collection, weather, maritime navigation and timing, Earth observation, and communications. In particular, satellite transmissions have an important role in telephone communications, television broadcasting, computer communications as well as navigation.

The use of satellites for communications was a brilliant idea of Arthur C. Clarke who wrote a famous article in October 1945 in the Wireless World journal, entitled "Extra Terrestrial Relays - Can Rocket Stations Give Worldwide Coverage?" that described the use of *manned* satellites in orbits at 35,800 km altitude, thus having synchronous motion with respect to a point on the Earth. This article was the basis for the use of *GEOstationary* (GEO) satellites for telecommunications. Subsequently, he also proved the usefulness of satellites as compared to transatlantic telephone cables.

Satellite communications deserve the special merit to allow connecting people at great distances by using the same (homogeneous) communication system and technology. Other very significant advantages of the satellite approach are: (*i*) easy fruition of both broadcast and multicast high bit-rate multimedia services; (*ii*) provision of backup communication services for users on a global scale (this feature is very important for emergency scenarios and disaster relief activities); (*iii*) provision of services in areas that could not be reached by terrestrial infrastructures; (*iv*) support of high-mobility users.

Three broad areas where satellites can be employed are: fixed satellite service, broadcast satellite service, and mobile satellite service. Particularly relevant is the significant global success of broadcast satellite services for both analogue and digital audio/TV by exploiting the inherent wide coverage area of GEO satellites. At the beginning of the 21st century more than 70 million European homes watch TV programs through direct satellite reception or through cable distribution systems.

New satellite system architectures are being envisaged to be fully IP-based and support digital video broadcasting and return channel protocols, such as DVB-S, DVB-S2 and DVB-RCS. Trends in telecommunications indicate that four growing market areas are messaging and navigation services, mobility services, video delivery services, and interactive multimedia services. In addition to this, interesting areas for investigation with big potential markets are: the extension of the DVB-S2/-RCS standard for mobile usage, satellite IP networks interconnected with terrestrial wireless systems, and the convergence of satellite communications and remote sensing for Earth observation.

Satellite resources (i.e., radio spectrum and transmission power) are costly and satellite communications impose special constraints with respect to terrestrial systems in terms of path loss, propagation delay, fading, etc. These are critical factors for supporting user service level agreements and *Quality of Service* (QoS).

The ISO/OSI reference model and the Internet protocol suite are based on a layered protocol stack. Protocols are designed such that a higher-layer protocol only makes use of the services provided by the lower layer and is not concerned with the details of how the service is being provided; protocols at the different layers are independently designed. However, there is tight interdependence between layers in IP-based next-generation satellite communication systems. For instance, transport layer protocols need to take into account large propagation delays, link impairments, and bandwidth asymmetry. In addition to this, error correction schemes are implemented at physical, link and (in some cases) transport layers, thus entailing some inefficiencies and redundancies. Hence, strict modularity and layer independence of the layered protocol model may lead to a non-optimal performance.

Satellite resources are costly and must be efficiently utilized in order to provide suitable revenue to operators. Users, however, do not care about the platform technology adopted and employed resource management scheme, but need QoS provision. Unfortunately, resource utilization efficiency and QoS support are conflicting needs: typically, the best utilization is achieved in the presence of a congested system, where QoS can difficulty be guaranteed. A new possible approach addressing both these issues is represented by the *cross-layer design* of the air interface, where the interdependency of protocols at different layers is exploited with the aim to perform a joint optimization or a dynamic adaptation. The innovation of this approach relies on the fact that it introduces direct interactions event between non-adjacent protocol layers with the aim to improve system performance.

The main aim of this book is to address the novel research area of cross-layer air interface design for satellite systems and provide a complete description of available methods, showing the possible efficiency improvements. A particular interest has been addressed here to the protocol stack defined by the ETSI TC-SES/BSM (*Satellite Earth Stations and Systems / Broadband Satellite Multimedia*) working group for IP-based satellite networks. In this framework, a protocol stack architecture has been identified, where lower layers depend on satellite system implementation (*satellite-dependent layers*) and higher layers are those typical of the Internet protocol stack (*satellite-independent layers*). These two blocks of stacked protocols are interconnected through the SI-SAP (*Satellite-Independent - Service Access Point*) interface that has acquired a crucial importance for the definition of cross-layer interactions and signaling.

This book has been conceived in the framework of the SatNEx Network of Excellence (www.satnex.org, project IST-507052, 2004–2006) that has made possible a tight cooperation of many European partners. Since the beginning (January 2004), SatNEx devoted the sub-work-package 2430, namely joint activity 2430 (ja2430), to the investigation of cross-layer issues that were soon considered as an original research field. Such activity attracted the interest of more than 14 SatNEx partners. In particular, research groups at the following European Universities or research Institutions contributed to ja2430:

- AUTh - Aristotle University of Thessaloniki, Greece
- CNIT - Consorzio Nazionale Interuniversitario per le Telecomunicazioni, Italy
- DLR - Deutsches Zentrum für Luft- und Raumfahrt e.V., Germany
- FhI - Fraunhofer Institute for Open Communication Systems, Germany
- ISTI - National Research Council (CNR), ISTI Institute, Italy
- RWTH - Rheinisch-Westfälische Technische Hochschule Aachen, COMNETS, Germany
- TéSA - France
- TUG - Graz University of Technology, Austria
- UAB - Universidad Autónoma de Barcelona, Spain
- UC3M - Universidad Carlos III de Madrid, Spain
- UoA - University of Aberdeen, UK
- UniS - University of Surrey, Centre for Communication Systems Research, UK
- UToV - University of Rome "Tor Vergata", Department of Electronic Engineering, Italy
- UVI - Universidad de Vigo, Departamento de Ingeniería Telemática, Spain.

I had the pleasure to coordinate the ja2430 activities, organizing 4 periodical meetings (plus *ad hoc* meetings dedicated to the coordination of this book activity), where objectives (organized according to *Focus Topics*, FTs), common scenarios and strategies were identified. In particular, the FTs below were defined, thus contributing to the different parts of this book:

- FT 1: QoS for multimedia traffic
- FT 2: Radio resource management
- FT 3: Protocol integration.

The main objective of ja2430 has been the study of novel radio resource management schemes able to support multimedia traffic with QoS guarantee in future satellite communication systems. Our aim has been to propose modifications to the ISO/OSI standard protocol stack by considering interactions and even new interfaces among non-adjacent protocol layers. Such approach can be particularly important in order to optimize the performance (i.e., efficiency) of resource management protocols.

After more than one year of SatNEx ja2430 activities, it was decided in September 2005 to organize the results obtained in a book. With the end of SatNEx activities in March 2006, the work of this book continued in SatNEx II (IST-027393, 2006–2009) in the two new sub-work-packages deriving from ja2430, that is ja2330 (entitled: "Radio Resource Allocation and Adaptation") and ja2230 (entitled: "Cross-Layer Protocol Design").

The activity carried out for this book has been a very good opportunity for the SatNEx community to integrate the competencies of different partners considering all the parts of the system design (i.e., propagation issues, resource management techniques, link design, QoS, transport protocols, etc.) and especially because SatNEx is unique in that its expertise covers both broadband (fixed) and mobile satellite systems. This has been an ideal condition for the study of mechanisms that involve interactions among several protocol layers.

Besides Part I of this book that is aimed to introduce satellite communications (Chapter 1), resource management techniques (Chapter 2), QoS issues (Chapter 3) and cross-layer design methods (Chapter 4), the two following parts are conceived according to the ETSI SES/BSM protocol stack, thus distinguishing cross-layer issues involving satellite-dependent layers (Part II, Chapters 5, 6 and 7) from those of satellite-independent layers (Part III, Chapters 8, 9 and 10).

Before concluding this preface, I would like to say that I feel honored to have coordinated this book work first in the framework of ja2430 and then in ja2230&ja2330. I take this opportunity to thank SatNEx for the economical support received and all the SatNEx Colleagues who have provided a continuous support to this initiative. Finally, a very special thank is for my Collaborator, Dr. Ing. Paolo Chini, for his significant support in helping me during these years of hard work on the book. Many thanks also to my Collaborator, Dr. Ing. Ivano Alocci, for his kind support.

Giovanni Giambene
CNIT - University of Siena
Via Roma, 56 - 53100 Siena, Italy
Phone: +39 0577 234603
Fax: +39 0577 233602
E-mail: giambene@unisi.it

Curriculum Vitae
Dr. Giovanni Giambene

Giovanni Giambene was born in Florence, Italy, in 1966. He received the Dr. Ing. degree in Electronics from the University of Florence, Italy, in 1993 and the Ph.D. degree in Telecommunications and Informatics from the University of Florence, Italy, in 1997. From 1994 to 1997, he was with the Electronic Engineering Department of the University of Florence, Italy. He was Technical External Secretary of the European Community COST 227 Action, entitled "Integrated Space/Terrestrial Mobile Networks". He also contributed to the Resource Management activity of the Working Group 3000 within the RACE Project, called "Satellite Integration in the Future Mobile Network" (SAINT, RACE 2117). From 1997 to 1998, he was with OTE of the Marconi Group, Florence, Italy, where he was involved in a GSM development program. In the same period he also contributed to the COST 252 Action ("Evolution of Satellite Personal Communications from Second to Future Generation Systems") research activities by studying the performance of Packet Reservation Multiple Access (PRMA) protocols suitable for supporting voice and data transmissions in low earth orbit mobile satellite systems. In 1999 he joined the Information Engineering Department of the University of Siena, Italy, first as research associate and then as assistant professor. He teaches the advanced course of Telecommunication Networks at the University of Siena. From 1999 to 2003 he participated to the project "*Multimedialità*", financed by the *Italian National Research Council* (CNR). From 2000 to 2003, he contributed to the activities of the "Personalised Access to Local Information and services for tOurists" (PALIO) IST Project within the fifth Research Framework of the European Commission (www.palio.dii.unisi.it). At present, he is involved in the SatNEx network of excellence of the FP6 programme in the satellite field, as work package leader of two groups on radio access techniques and cross-layer air interface design (www.satnex.org). He is also vice-Chair of the COST 290 Action (www.cost290.org), entitled "Traffic and QoS Management in Wireless Multimedia Networks" (Wi-QoST).

Contents

Acknowledgements v

Preface vii

Contents xiii

List of Contributors xix

List of Acronyms and Abbreviations xxiii

Part I Resource Management Framework for Satellite Communications

1 **INTRODUCTION TO SATELLITE COMMUNICATIONS AND RESOURCE MANAGEMENT** 3
- 1.1 Satellite communications 3
- 1.2 Basic issues in the design of satellite communication systems 10
- 1.3 Multiple access techniques 12
- 1.4 Radio interfaces considered and scenarios 15
  - 1.4.1 S-UMTS 15
  - 1.4.2 DVB-S standard 16
  - 1.4.3 DVB-RCS standard 17
  - 1.4.4 DVB-S2 standard 23
  - 1.4.5 Numerical details on the selected scenarios for performance evaluations 27
- 1.5 Satellite networks 28
  - 1.5.1 SI-SAP interface overview 31
- 1.6 Novel approaches for satellite networks 34
  - 1.6.1 Horizontal approach 34

1.6.2 Vertical approach ... 34
1.7 Conclusions ... 37

References ... 39

2 ACTIVITY IN SATELLITE RESOURCE MANAGEMENT ... 43
2.1 Introduction ... 43
2.2 Frequency/time/space resource allocation schemes ... 46
2.3 Power allocation and control schemes ... 50
2.4 CAC and handover algorithms ... 51
2.4.1 Handover algorithms ... 53
2.5 RRM modeling and simulation ... 54
2.6 Related projects in Europe ... 55
2.6.1 TWISTER: Terrestrial Wireless Infrastructure integrated with Satellite Telecommunications for E-Rural applications ... 56
2.6.2 MAESTRO: Mobile Applications & sErvices based on Satellite & Terrestrial inteRwOrking ... 56
2.6.3 SatNEx: Satellite Network of Excellence ... 57
2.6.4 NEWCOM: Network of Excellence in Wireless COMmunications ... 57
2.6.5 VIRTUOUS: Virtual Home UMTS on Satellite ... 58
2.6.6 COST Actions ... 58
2.6.7 The ISI Initiative ... 59
2.7 Conclusions ... 60

References ... 61

3 QoS REQUIREMENTS FOR MULTIMEDIA SERVICES ... 67
3.1 Introduction ... 67
3.2 Services QoS requirements ... 68
3.2.1 Performance requirements for conversational services ... 70
3.2.2 Performance requirements for interactive services ... 73
3.2.3 Performance requirements for streaming services ... 74
3.2.4 Performance requirements for background services-applications ... 76
3.3 IP QoS frameworks/models ... 76
3.4 Broadcast and multicast services ... 80
3.4.1 Delayed real-time service over GEO satellite distribution systems ... 83
3.4.2 Scenario characterization and results ... 85
3.5 Experimental results on QoS ... 89
3.6 Conclusions ... 92

References ... 93

4 CROSS-LAYER APPROACHES FOR RESOURCE MANAGEMENT 95
4.1 Introduction 95
4.2 Literature survey on cross-layer methods 96
4.3 The need of a cross-layer air interface design 102
4.4 Cross-layer design: requirements depending on the satellite scenario 105
4.4.1 Broadband satellite scenario requirements (DVB-S/S2) 105
4.4.2 Mobile satellite scenario requirements (S-UMTS) 108
4.4.3 LEO satellite scenario requirements 108
4.5 Conclusions 111

References 113

Part II Cross-Layer Techniques for Satellite-Dependent Layers

5 ACCESS SCHEMES AND PACKET SCHEDULING TECHNIQUES 119
5.1 Introduction 119
5.2 Uplink: access schemes 120
5.2.1 Random access in UMTS and application to S-UMTS . 121
5.2.2 The Packet Reservation Multiple Access (PRMA) protocol 129
5.2.3 Adopting PRMA-like schemes in S-UMTS 131
5.2.4 Stability analysis of access protocols 132
5.3 Downlink: scheduling techniques 134
5.3.1 Survey of scheduling techniques 134
5.3.2 Scheduling techniques for HSDPA via satellite 139
5.3.3 Scheduling techniques for broadcast and multicast services in S-UMTS 152
5.3.4 Packet scheduling with cross-layer approach 164
5.4 Conclusions 170

References 173

6 CALL ADMISSION CONTROL 177
6.1 Introduction to Call Admission Control 177
6.2 CAC and QoS management 179
6.3 CAC algorithms for GEO satellite systems 184
6.3.1 CAC schemes for MF-TDMA networks 184
6.3.2 CAC schemes for CDMA networks 188
6.4 Handover and CAC algorithms for non-GEO satellite systems 189
6.4.1 Intra-satellite handover and CAC schemes 191
6.4.2 Inter-satellite handover and CAC schemes 194

6.5 Directions for further research 199
6.6 Conclusions 200

References 201

7 DYNAMIC BANDWIDTH ALLOCATION 207
7.1 Dynamic bandwidth allocation: problem definition 207
7.1.1 Survey of allocation approaches 209
7.2 DBA schemes for DVB-RCS scenarios 211
7.3 Recent developments on DBA techniques 213
7.3.1 DVB-RCS dynamic channel allocation using control-theoretic approaches 213
7.3.2 Dynamic bandwidth de-allocation 214
7.3.3 Dynamic bandwidth allocation with cross-layer issues . 214
7.3.4 Joint timeslot optimization and fair dynamic bandwidth allocation in a system employing adaptive coding 218
7.3.5 Dynamic bandwidth allocation for handover calls 233
7.4 Conclusions 234

References 237

Part III Cross-Layer Techniques for Satellite-Independent Layers

8 RESOURCE MANAGEMENT AND NETWORK LAYER 243
8.1 Introduction 243
8.2 Overview IP QoS framework 244
8.2.1 Integrated services 244
8.2.2 Differentiated services 246
8.2.3 Multiprotocol Label Switching (MPLS) 247
8.3 Resource management for IP QoS 248
8.3.1 Relative DiffServ by MAC Scheduling 249
8.4 QoS mapping over satellite-independent service access point . 256
8.4.1 Model-based techniques for QoS mapping and support 257
8.4.2 A measurement-based approach for QoS mapping and support 258
8.4.3 Performance evaluation and discussion 262
8.5 QoS provisioning for terminals supporting dual network access - satellite and terrestrial 264
8.6 Switched Ethernet over LEO satellite: implicit cross-layer design exploiting VLANs 270
8.6.1 Protocol harmonization and implicit cross-layer design via IEEE VLAN 272
8.6.2 Performance evaluation 273

8.7 Conclusions 282

References 285

9 RESOURCE MANAGEMENT AND TRANSPORT LAYER 289
9.1 Introduction 289
9.2 Overview of TCP over satellite 290
9.2.1 TCP standard mechanisms 291
9.2.2 Criticalities of TCP on satellite links 292
9.2.3 Survey of proposed solutions 293
9.3 Cross-layer interaction between TCP and physical layer 294
9.4 Cross-layer interaction between TCP and MAC 298
9.4.1 A novel TCP-driven dynamic resource allocation scheme 299
9.5 Overview of UDP-based multimedia over satellite 305
9.5.1 Cross-layer methods for UDP 307
9.6 Conclusions 307

References 309

10 CROSS-LAYER METHODS AND STANDARDIZATION ISSUES 313
10.1 Introduction 313
10.2 Cross-layer design and Internet protocol stack 314
10.3 Cross-layer methodologies for satellite systems 314
10.3.1 Implicit and explicit cross-layer design methodologies 315
10.3.2 Cross-layer techniques categorized in terms of the direction of information flow 315
10.4 Potential cross-layer optimizations for satellite systems 317
10.4.1 Optimizations aiming at QoS harmonization across layers 317
10.4.2 Optimization of the Radio Resource Management 318
10.4.3 Optimizations combining higher and lower layers 319
10.5 Cross-layer signaling for satellite systems 320
10.6 Standardization issues 322
10.6.1 Standardization bodies and groups 323
10.6.2 European Conference of Postal and Telecommunications Administrations 323
10.6.3 ETSI 323
10.6.4 DVB 326
10.6.5 International Telecommunication Union 330
10.7 Conclusions 330

References 333

Index 335

List of Contributors

Rafael Asorey Cacheda
UVI - Universidad de Vigo,
Dep. Ingeniería Telemática, ETSI
Telecomunicación, Campus, 36200
Vigo, Spain
rasorey@det.uvigo.es

Kostantinos Avgeropoulos
AUTh - Aristotle University of
Thessaloniki, Thessaloniki,
Panepistimioupolis, 54124, Greece
k.avgeropoulos@gmail.com

Paolo Barsocchi
CNR-ISTI - National Research
Council (CNR), ISTI
Institute, Via G. Moruzzi, 1,
San Cataldo, 56124 Pisa, Italy
paolo.barsocchi@isti.cnr.it

Ulla Birnbacher
TUG - Graz University of
Technology, Inst. Comm. Net. and
Satellite Comm., Inffeldgasse 12,
A-8010 Graz, Austria
ulla.birnbacher@tugraz.at

Daniel Castro García
INFOGLOBAL, Spain

Nedo Celandroni
CNR-ISTI - National Research
Council (CNR), ISTI
Institute, Via G. Moruzzi, 1,
San Cataldo, 56124 Pisa, Italy
nedo.celandroni@isti.cnr.it

Wei Koong Chai
UniS - University of Surrey, CCSR,
Centre for Communication Systems
Research, Guildford,
Surrey GU2 7XH, UK
W.Chai@surrey.ac.uk

Paolo Chini
CNIT - University of Siena
Research Unit, Via Roma, 56,
53100, Siena, Italy
chini7@unisi.it

Antonio Cuevas
UC3M - Universidad Carlos III de
Madrid,
Avda. Universidad 30, 28911
Leganés, Spain
acuevas@it.uc3m.es

Franco Davoli
CNIT - University of Genoa
Research Unit, Via Opera Pia, 13,
16145, Genova, Italy
franco.davoli@cnit.it

Gorry Fairhurst
UoA - University of Aberdeen,
Department of Engineering,
Fraser Noble Building,
Aberdeen AB24 3UE, UK
gorry@erg.abdn.ac.uk

Erina Ferro
CNR-ISTI - National Research
Council (CNR), ISTI
Institute, Via G. Moruzzi, 1,
San Cataldo, 56124 Pisa, Italy
erina.ferro@isti.cnr.it

Giovanni Giambene
CNIT - University of Siena
Research Unit, Via Roma, 56,
53100, Siena, Italy
giambene@unisi.it

Samuele Giannetti
CNIT - University of Siena
Research Unit, Via Roma, 56,
53100, Siena, Italy
giannetti13@unisi.it

Francisco Javier González Castaño
UVI - Universidad de Vigo,
Dep. Ingeniería Telemática, ETSI
Telecomunicación, Campus, 36200
Vigo, Spain
javier@det.uvigo.es

Alberto Gotta
CNR-ISTI - National Research
Council (CNR), ISTI
Institute, Via G. Moruzzi, 1,
San Cataldo, 56124 Pisa, Italy
alberto.gotta@isti.cnr.it

Javier Herrero Sánchez
INFOGLOBAL, Spain

Du Hongfei
UniS - University of Surrey, CCSR,
Centre for Communication Systems
Research, Guildford,
Surrey GU2 7XH, UK
H.Du@surrey.ac.uk

Stylianos Karapantazis
AUTh - Aristotle University of
Thessaloniki, Thessaloniki,
Panepistimioupolis, 54124, Greece
skarap@auth.gr

Georgios Koltsidas
AUTh - Aristotle University of
Thessaloniki, Thessaloniki,
Panepistimioupolis, 54124, Greece
fractgkb@auth.gr

Victor Y. H. Kueh
UniS - University of Surrey, CCSR,
Centre for Communication Systems
Research, Guildford,
Surrey GU2 7XH, UK
victor_unis@yahoo.co.uk

Michele Luglio
UToV - University of Rome "Tor
Vergata",
Via del Politecnico, 1,
00133 - Roma, Italy
luglio@uniroma2.it

Vincenzo Mancuso
UToV - University of Rome "Tor
Vergata",
Via del Politecnico, 1,
00133 - Roma, Italy
vincenzo.mancuso@ieee.org

Mario Marchese
CNIT - University of Genoa
Research Unit, Via Opera Pia, 13,
16145, Genova, Italy
Mario.Marchese@unige.it

Giada Mennuti
CNIT - University of Florence
Research Unit, Via di S. Marta, 3,
50139, Firenze, Italy
giada@lenst.det.unifi.it

Maurizio Mongelli
CNIT - University of Genoa
Research Unit, Via Opera Pia, 13,
16145, Genova, Italy
Maurizio.Mongelli@unige.it

Antoni Morell
UAB - Universitat Autónoma de
Barcelona,
Dpt. Telecommunications and
Systems Engineering,
Engineering School,
Bellaterra 08193 - Barcelona, Spain
antoni.morell@uab.es

José Ignacio Moreno Novella
UC3M - Universidad Carlos III de
Madrid,
Avda. Universidad 30, 28911
Leganés, Spain
jmoreno@it.uc3m.es

Seounghoon Oh
RWTH - Rheinisch-Westfälische
Technische Hochschule
Aachen /
COMNETS, Kopernikusstr. 16,
D-52074 AACHEN, Germany
oh@comnets.rwth-aachen.de

Antonio Pantò
CNIT - University of Catania
Research Unit, Viale A. Doria, 6,
95125, Catania, Italy
antonio.panto@cnit.it

Cristina Párraga Niebla
DLR - German Aerospace Center,
Insitute of Comms. and
Navigation, Oberpfaffenhofen, 82234
Wessling, Germany
cristina.parraga@dlr.de

Veronica Pasqualetti
CNIT - University of Siena
Research Unit, Via Roma, 56,
53100, Siena, Italy
pasqualetti@unisi.it

Tommaso Pecorella
CNIT - University of Florence
Research Unit, Via di S. Marta, 3,
50139, Firenze, Italy
pecos@lart.det.unifi.it

Francesco Potortì
CNR-ISTI - National Research
Council (CNR), ISTI
Institute, Via G. Moruzzi, 1,
San Cataldo, 56124 Pisa, Italy
Potorti@isti.cnr.it

Cesare Roseti
UToV - University of Rome "Tor
Vergata",
Via del Politecnico, 1,
00133 - Roma, Italy
roseti@ing.uniroma2.it

Aduwati Sali
UniS - University of Surrey, CCSR,
Centre for Communication Systems
Research, Guildford,
Surrey GU2 7XH, UK
A.Sali@surrey.ac.uk

Gonzalo Seco Granados
UAB - Universitat Autónoma de
Barcelona,
Dpt. Telecommunications and
Systems Engineering,
Engineering School,
Bellaterra 08193 - Barcelona, Spain
gonzalo.seco@uab.es

Petia Todorova
FhI - Fraunhofer Institute for Open Communication Systems - FOKUS, Kaiserin - Augusta - Alee 31, 10589 Berlin, Germany
Petia.Todorova@fokus.fraunhofer.de

Orestis Tsigkas
AUTh - Aristotle University of Thessaloniki, Thessaloniki, Panepistimioupolis, 54124, Greece
torestis@auth.gr

Alessandro Vanelli-Coralli
UoB - University of Bologna DEIS/ARCES,
Viale Risorgimento, 2,
40136 - Bologna, Italy
avanelli@deis.unibo.it

María Ángeles Vázquez Castro
UAB - Universitat Autónoma de Barcelona,
Dpt. Telecommunications and Systems Engineering,
Engineering School,
Bellaterra 08193 - Barcelona, Spain
angeles.vazquez@uab.es

Fausto Vieira
UAB - Universitat Autónoma de Barcelona,
Dpt. Telecommunications and Systems Engineering,
Engineering School,
Bellaterra 08193 - Barcelona, Spain
fvieira@sunaut.uab.es

List of Acronyms and Abbreviations

3G	*3rd Generation*
3GPP	*3rd Generation Partnership Project*
4G	*4th Generation*
AAA	*Authentication, Authorization and Accounting*
ABC	*Always Best Connected*
ABR	*Available Bit Rate*
AC	*Adaptive Coding*
ACK	*Acknowledgement*
ACM	*Adaptive Coding and Modulation*
ADSL	*Asymmetric Digital Subscriber Line*
AF	*Assured Forwarding*
AICH	*Acquisition Indicator Channel*
AIMD	*Additive Increase Multiplicative Decrease*
AP	*Access Point*
API	*Application Programming Interface*
APP	*Application layer*
APSK	*Amplitude and Phase Shift Keying*
AQM	*Active Queue Management*
AR	*Access Router*
ARP	*Address Resolution Protocol*
ARQ	*Automatic Repeat reQuest*
ASC	*Access Service Class*
ASD	*Aggregated System Demand*
ATM	*Asynchronous Transfer Mode*
AVBDC	*Absolute Volume Based Dynamic Capacity*
AWGN	*Additive White Gaussian Noise*
BCH	*Bose-Chaudhuri-Hocquenghem* (in Chapter 1)
BCH	*Broadcast Channel* (in Chapter 5)
BDP	*Bandwidth-Delay Product*
BE	*Best Effort*
BER	*Bit Error Rate*
BGAN	*Broadband Global Area Network*
BGAN-X	*BGAN Extension project*
B-ISDN	*Broadband Integrated Services Digital Network*
BLER	*Block Error Rate*
BM-SC	*Broadcast-Multicast Service Center*
BO	*Bandwidth Occupation*
BoD	*Bandwidth on Demand*
BPM	*BSM Protocol Manager*
BPSK	*Binary Phase Shift Keying*
BS	*Base Station*
BSA	*Broadband Satellite Access*
BSM	*Broadband Satellite Multimedia*
BSM_ID	*BSM Identifier*
BSS	*Broadcasting Satellite Service*
BTP	*Burst Time Plan*
CA	*Congestion Avoidance*
CAC	*Call Admission Control*
CBP	*Call Blocking Probability*
CBQ	*Class-Based Queuing*
CBR	*Constant Bit Rate*
CCM	*Constant Coding Modulation*
CDM	*Code Division Multiplexing*
CDMA	*Code Division Multiple Access*
CDMA/HDR	*CDMA/High Data Rate*
CDP	*Call Dropping Probability*
CDVT	*Cell Delay Variation Tolerance*
CEN	*European Committee for Standardization*
CENELEC	*European Committee for Electro-technical Standardization*
CEPT	*European Conference of Postal and Telecommunications Administrations*
CF/DAMA	*Combined Free/Demand Assignment Multiple Access*
C/I	*Carrier-to-Interference ratio*
CIF-Q	*Channel Condition - Independent Fair Queuing*
C/I PS	*C/I Proportional Scheduler*
CIST	*Common Internal Spanning Tree*
CLR	*Cell Loss Ratio*
CM	*Control Module*

CMF	*Control and Monitoring Functions*
C/N	*Carrier power-to-Noise power ratio*
CN	*Core Network*
COPS	*Common Open Policy Service*
COST	*Co-operation in the field of Scientific and Technical Research*
CP	*Complete Partitioning*
CQI	*Channel Quality Indicator*
CR	*Capacity Request*
CRA	*Continuous Rate Assignment*
CRC	*Cyclic Redundancy Check*
CS	*Complete Sharing*
C-SAP	*Control-SAP*
CSI	*Channel State Information*
cwnd	*congestion window*
DAMA	*Demand Assignment Multiple Access*
DBA	*Dynamic Bandwidth Allocation*
DBAC	*Dynamic Bandwidth Allocation Capabilities*
DBRA	*Dynamic Bandwidth and Resource Allocation*
DBS	*Direct Broadcast Satellite*
DBS-RCS	*DBS with Return Channel System*
DCA	*Dynamic Channel (or Capacity) Allocation*
DCCH	*Dedicated Control Channel*
DCH	*Dedicated Channel*
DDP	*Delay Differentiation Parameter*
DDQ	*Delay Differentiation Queuing*
DiffServ	*Differentiated Service*
DLL	*Data Link Layer*
DMBS	*Double-Movable Boundary Strategy*
DOCSIS-S	*Data Over Cable Service Interface Specification for Satellite*
DP	*Differentiation Parameter*
DPSK	*Differential Phase Shift Keying*
DRA	*Dynamic Resource Allocation*
DRT	*Delayed Real-Time*
DS	*Direct Sequence*
DSCH	*Downlink Shared Channel*
DSCP	*DiffServ Code Point*
DSNG	*Digital Satellite News Gathering*
DTCH	*Dedicated Traffic Channel*
D-TDMA	*Dynamic TDMA*
DTH	*Direct-To-Home*
DULM	*Data Unit Labeling Method*
dupACKs	*duplicate ACKs*
DVB	*Digital Video Broadcasting*
DVB-C	*DVB-Cable*
DVB-CAS	*DVB-Conditional Access System*
DVB-GBS	*DVB-Global Broadcast Service*
DVB-H	*DVB-Handheld*
DVB-RCC	*DVB-Return Channel via Cable*
DVB-RCL	*DVB-Return Channel for LMDS*
DVB-RCS	*DVB-Return Channel via Satellite*
DVB-RCT	*DVB-Return Channel via Terrestrial*
DVB-S	*Digital Video Broadcasting via Satellite*
DVB-S2	*DVB-Satellite version 2*
DVB-T	*DVB-Terrestrial*
DVB-TM	*DVB-Technical Module*
EBU	*European Broadcasting Union*
ECC	*Electronic Communications Committee*
ECN	*Explicit Congestion Notification*
ECSS	*European Co-operation on Space Standardization*
EDF	*Earliest Deadline First*
EF	*Expedited Forwarding*
EHF	*Extremely High Frequency*
EIRP	*Effective Isotropic Radiated Power*
EMC	*ElectroMagnetic Compatibility*
EqB	*Equivalent Bandwidth*
ERA	*European Research Area*
ERM	*EMC and Radio spectrum Matters*
ESA	*European Space Agency*
ETSI	*European Telecommunications Standards Institute*
EU	*European Union*
FA	*Fixed Assignment*
FACH	*Forward Access Channel*
FC	*FIFO Maximum Capacity*
FCA	*Free Capacity Assignment* (in Chapters 1, 7, 8 and 9)
FCA	*Fixed Channel Allocation* (in Chapter 2)
FCFS	*First Come First Served*
FCT	*Frame Composition Table*
FDD	*Frequency Division Duplexing*
FDM	*Frequency Division Multiplexing*
FDMA	*Frequency Division Multiple Access*
FEC	*Forward Error Correction*
FER	*Frame Erasure Rates* (in Chapter 3)
FER	*Frame Error Rate* (in Chapter 5)
FH	*Frequency Hopping*
FHO	*Fast HandOver*
FI	*Fairness Index*
F_id	*frame_ID*
FIFO	*First In First Out*
FL1-HARQ	*Fast L1 hybrid ARQ*
FMT	*Fade Mitigation Techniques*
F_nb	*frame_number*
FP	*Framework Programme*
FSK	*Frequency Shift Keying*
FSS	*Fixed Satellite Service*
FTP	*File Transfer Protocol*
FZC	*Forward Erasure Correction*
GB	*Guaranteed Bandwidth*
GEO	*Geosynchronous (Geostationary) Earth Orbit*
GM	*Guaranteed Minimum*
GOPs	*Group of Pictures*
GoS	*Grade of Service*
GPRS	*General Packet Radio Service*

GPS	*Generalized Processor Sharing*
GSM	*Global System for Mobile Communications*
GW	*Gateway or Traffic Gateway*
HCA	*Hybrid Channel Allocation*
HDTV	*High Definition Television*
HLS	*Hierarchical Link Sharing*
HNS	*Hughes Network Systems*
HP	*High-Priority*
HPA	*High Power Amplifier*
HPD	*Hybrid Proportional Delay*
HSDPA	*High Speed Downlink Packet Access*
HS-DPCCH	*High Speed Dedicated Physical Control Channel*
HS-DSCH	*High Speed-DSCH*
HS-PDSCH	*High Speed Physical Downlink Shared Channel*
HTML	*HyperText Mark-Up Language*
IAB	*Internet Architecture Board*
IBR	*Information Bit Rate*
ICMP	*Internet Control Message Protocol*
IEEE	*Institute of Electrical and Electronics Engineers*
IETF	*Internet Engineering Task Force*
IFR	*Increasing Failure Rate*
IM	*Inter-Modulation*
IMT	*International Mobile Telecommunications*
IntServ	*Integrated Service*
IP	*Internet Protocol*
IPA	*Infinitesimal Perturbation Analysis*
IP-CAS	*IP-based Conditional Access System*
IPoS	*Internet Protocol over Satellite*
ISDN	*Integrated Services Digital Network*
ISI	*Integral Satcom Initiative*
ISLs	*Inter-Satellite Links*
ISN	*Interactive Satellite Network*
ISO/OSI	*International Standard Organization/Open System Interconnection*
ISP	*Internet Service Provider*
IST	*Information Society Technologies*
ITU	*International Telecommunication Union*
ITU-D	*ITU - Telecommunication Development sector*
ITU-R	*ITU - Radiocommunication sector*
ITU-T	*ITU - Telecommunication sector*
IWFQ	*Idealized Wireless Fair Queuing*
IWU	*Inter-Working Unit*
KKT	*Karush-Kuhn-Tucker*
L1	*Layer 1 (physical layer)*
L2	*Layer 2 (link/MAC layer)*
L3	*Layer 3 (network layer)*
LAN	*Local Area Network*
LC	*LUI Maximum Capacity*
LDP	*Label Distribution Protocol*
LDPC	*Low Density Parity Check*
LEO	*Low Earth Orbit*
LLC	*Logical Link Control*
LLC/SNAP	*LLC/Sub-Network Access Protocol*
LMDS	*Local Multipoint Distribution System*
LoS	*Line of Sight*
LP	*Low-Priority*
LRD	*Long Range Dependent*
LSP	*Label Switched Path*
LSR	*Label Switching Router*
LTFS	*Long-Term Fairness Server*
LUI	*Last Useful Instant*
MAC	*Medium Access Control*
MAC-hs	*MAC/HS-DSCH*
MAN	*Metropolitan Area Network*
MBMS	*Multimedia Broadcast Multicast Services*
MBU	*Minimum Bandwidth Unit*
MCS	*Master Control Station*
MEO	*Medium Earth Orbit*
MF	*Multi-Frequency*
MF-TDMA	*Multi Frequency - Time Division Multiple Access*
MLI	*Maximum Legal Increment*
MLPQ	*Multi-Level Priority Queuing*
MMPP	*Markov-Modulated Poisson Processes*
MMS	*Multimedia Messaging Service*
MN	*Mobile Node*
MODCOD	*Modulation and Coding*
MOS	*Mean Opinion Score*
MPE	*Multi Protocol Encapsulation*
MPEG	*Moving Picture Experts Group*
MPEG2-TS	*Moving Picture Experts Group 2 - Transport Stream*
MPLS	*Multiprotocol Label Switching*
M-SAP	*Management-SAP*
MSL	*Minimum Scheduling Latency*
MSS	*Maximum Segment Size*
MSTP	*Multiple STP*
MTs	*Multicast Terminals*
MTCH	*MBMS point-to-multipoint Traffic Channel*
MTU	*Maximum Transfer Unit*
NBS	*Nash Bargaining Solution*
NCC	*Network Control Center*
NCR	*Network Clock Reference*
ND	*Neighbor Discovery*
NGN	*Next-Generation Network*
NoE	*Network of Excellence*
nrt-VBR	*non-real-time-VBR*
OBP	*On-Board Processor*
OC	*Optimized Centralized*
OFDM	*Orthogonal Frequency Division Multiplex*
OP	*Optimized Proportional*
PAB	*Proportional Allocation of Bandwidth*
PCPCH	*Physical Common Packet Channel*
pdf	*probability density function*
PDS	*Proportional Differentiated Service*
PDU	*Protocol Data Unit*
P-EDF	*Prioritized-EDF*
PEP	*Performance Enhancing Proxy*

PER	*Packet Error Rate*
PF	*Proportional Fair*
PG	*Processing Gain*
PHB	*Per-Hop Behavior*
PHY	*Physical layer*
PLFRAME	*Physical Layer Frame*
PLP	*Packet Loss Probability*
PLR	*Packet Loss Rate*
PMPP	*Pareto-Modulated Poisson Processes*
PN	*Pseudo Noise*
POTS	*Plain Old Telephone Service*
PRACH	*Physical Random Access Channel*
PRC	*Power Ramping Control*
PRMA	*Packet Reservation Multiple Access*
PRMA-HS	*PRMA with Hindering States*
PSK	*Phase Shift Keying*
PSNR	*Peak Signal to Noise Ratio*
PSTN	*Public Switched Telephone Network*
QAM	*Quadrature Amplitude Modulation*
QID	*Queuing Identifier*
QoS	*Quality of Service*
QoSMO	*QoS Mapping Optimization*
QPSK	*Quadrature Phase Shift Keying*
RA	*Random Access*
RAB	*Radio Access Bearer*
RACH	*Random Access Channel*
RAN	*Radio Access Network*
RAT	*Robust Audio Tool*
RB	*Reserved Bandwidth*
RBDC	*Rate Based Dynamic Capacity*
RCBC	*Reference Chaser Bandwidth Controller*
RC-PSTN	*Return Channel - PSTN*
RCQI	*Relative Channel Quality Index*
RCS	*Return Channel via Satellite*
RCST	*Return Channel Satellite Terminal*
RED	*Random Early Detection*
RF	*Radio Frequency*
RHC	*Receding Horizon Controller*
RLC	*Radio Link Control*
RNC	*Radio Network Controller*
RRM	*Radio Resource Management*
RSP	*Recovery Service Provider*
RSTP	*Rapid STP*
RSVP	*Resource Reservation Protocol*
RT	*Real Time*
RTD	*Round Trip propagation Delay*
RTO	*Retransmission TimeOut*
RTP	*Real-time Transport Protocol*
RTT	*Round Trip Time*
rt-VBR	*real-time-VBR*
SAC	*Satellite Access Control*
S-ALOHA	*Slotted-ALOHA*
SBFA	*Server-Based Fairness Approach*
S-CCPCH	*Secondary Common Control Physical Channel*
SCED	*Service Curve-based Earliest Deadline first*
SCPC	*Single Carrier Per Channel*
SCPS-TP	*Space Communications Protocol Specification-Transport Protocol*
SCr	*Service Credit*
SD	*Satellite-Dependent*
SDMA	*Spatial Division Multiple Access*
S-DMB	*Satellite Digital Multimedia Broadcasting*
SDR	*Satellite Digital Radio*
SDTV	*Standard Definition Television*
SF	*Spreading Factor*
SFM	*Stochastic Fluid Models*
S-HSDPA	*HSDPA via Satellite*
SI	*Satellite-Independent*
SIR	*Signal-to-Interference Ratio*
SI-SAP	*Satellite-Independent - Service Access Point*
SL	*Super-frame Length*
SLA	*Service Level Agreement*
S-MBMS	*Satellite MBMS*
SMEs	*Small and Medium Enterprises*
SMG	*Special Mobile Group*
SMS	*Short Message Service*
SNIR	*Signal to Noise and Interference Ratio*
SOHO	*Small Office - Home Office*
SP	*Simple Proportional*
SPC	*Smith Predictor Controller*
SR	*Slot Request*
SRD	*Short Range Dependent*
SS	*Slow Start*
ssthresh	*slow start threshold*
ST	*Satellite (interactive) Terminal*
STB	*Set-Top-Box*
STFQ	*Stochastic Fairness Queuing*
STP	*Spanning Tree Protocol*
S-UMTS	*Satellite-UMTS*
SWTP	*Satellite Waiting Time Priority*
TB	*Transport Block*
TBTP	*Terminal Burst Time Plan*
TC	*Transported Capacity*
TCA	*Traffic Conditioning Agreement*
TCP	*Transmission Control Protocol*
TC-SES	*Technical Committee for Satellite Earth Stations and Systems*
TCT	*Time Composition Table*
TDM	*Time Division Multiplexing*
TDMA	*Time Division Multiple Access*
TE	*Terminal Equipment*
Telnet	*TELetype NETwork*
TF	*Transport Format*
TFC	*Transport Format Combination*
TFCI	*Transport Format Combination Indication*
TFCS	*Transport Format Combination Set*
TFRC	*Transport Format and Resource Combination*
TIST	*Telecommunications, Information Science and Technology*
TM	*Transmission & Multiplexing*
TOS	*Type Of Service*
TR	*Trunk Reservation*
TS	*Time Slot*
TS_nb	*timeslot_number*

TTI	*Transmission Time Interval*
T-UMTS	*Terrestrial UMTS*
TWTA	*Traveling-Wave-Tube Amplifier*
UBR	*Unspecified Bit Rate*
UDP	*User Datagram Protocol*
UE	*User Equipment*
UL	*Upper Limit*
UMTS	*Universal Mobile Telecommunications System*
UPC	*Usage Parameter Control*
URAN	*UMTS Radio Access Network*
U-SAP	*User-SAP*
UT	*User Terminal*
VBDC	*Volume Based Dynamic Capacity*
VBR	*Variable Bit Rate*
VC	*Virtual Channel*
VCM	*Variable Coding and Modulation*
VLAN	*Virtual Local Area Networks*
VLL	*Virtual Leased Line*
VoIP	*Voice over IP*
VP	*Virtual Partitioning*
VPI/VCI	*Virtual Path Identifier/ Virtual Channel Identifier*
VPN	*Virtual Private Network*
VQM_P	*Peak Video Quality Measurement*
VR-JT	*Variable Rate - Jitter Tolerant*
VR-RT	*Variable Rate - Real Time*
VSAT	*Very Small Aperture Terminal*
VSF	*Variable Spreading Factor*
WAN	*Wide Area Network*
W-CDMA	*Wideband Code Division Multiple Access*
WCI	*Wireless Channel Information*
WFBoD	*Weighted Fair Bandwidth-on-Demand*
WFQ	*Weighted Fair Queuing*
WiFi	*Wireless Fidelity*
WiMAX	*Worldwide Interoperability for Microwave Access*
WLAN	*Wireless LAN*
WP	*Work Package*
WRR	*Weighted Round Robin*
XTP	*eXpress Transfer Protocol*

Part I

Resource Management Framework for Satellite Communications

1

INTRODUCTION TO SATELLITE COMMUNICATIONS AND RESOURCE MANAGEMENT

Editor: Giovanni Giambene[1]

Contributors: Paolo Chini[1], Giovanni Giambene[1]

[1]CNIT - University of Siena, Italy

1.1 Satellite communications

Multimedia communications have been widely supported by terrestrial infrastructures that employ optical fibers in backbone links to achieve huge capacity. A technological alternative is represented by the use of satellites for providing multimedia broadband services to fixed and mobile users in several scenarios where terrestrial networks cannot be used or are congested.

Today, still a large number of persons living in remote areas or in underdeveloped regions do not have a realistic perspective of achieving access to high-speed Internet for many years. This problem constitutes a serious obstacle to making the benefits of the Information Society available to all. Such *digital divide* problem can be solved by satellite communications that can easily reach the different regions on the Earth by providing everywhere the same service types. Satellites are an important delivery platform of information society services, such as interactive TV and mobile, high-speed Internet access.

The most important reasons for the diffusion of satellite communications can be summarized as follows [1]:

- *Ubiquitous coverage*: a single satellite can reach every potential user across an entire continent. This is a very significant feature, especially in low population density areas or over the see, where the realization of terrestrial infrastructures would be not viable.

- *Support to mobile users*: a mobile user, which is situated in the satellite coverage area, can easily communicate with other fixed or mobile users.
- *Reduced cost*: with satellite communications, cost is independent of the distance. Moreover, satellite networks can easily cover a great part of the Earth, thus reaching a very big potential market of customers. This is an important opportunity in order to provide services at affordable costs.
- *Variety of connectivity*: it is possible to provide, in a simple and economic way, *point-to-multipoint* and *broadcast* communications, without complex *multicast routing* protocols (used in meshed terrestrial networks).
- *Rapid deployment and easy management of the network*: once a satellite is launched it can immediately reach a high number of users. With satellites, multimedia services can be provided to a wide multitude of users on broad areas in a quicker way than using a terrestrial infrastructure.
- *Bandwidth flexibility*: it is possible to provide simplex, duplex, narrow-band, symmetric and asymmetric bandwidth. Moreover, satellites can allow a broadband access to end-users, thus representing a possible solution to the "last mile" problem.

Very good books in the field of satellite communications, providing excellent basis on this field are detailed in references [2]-[7].

Satellites are situated on suitable orbits around the Earth; on the basis of their altitude, they can be classified into three main categories [1] (see Figure 1.1):

- *Low Earth Orbit* (LEO) satellites at a height between 500 and 2,000 km of altitude, i.e., below the Van Allen radiation belts. The Earth rotation period is about 100 minutes and the satellite visibility time is around 15 minutes. These orbits can be polar or inclined.
- *Medium Earth Orbit* (MEO) may be circular or elliptical in shape at a height between 8,000 and 12,000 km of altitude (between the two Van Allen radiation belts). The rotation period is 5-12 hours and the satellite visibility time is 2-4 hours.
- *Geosynchronous Earth Orbit* (GEO) is on the Earth's equatorial plane at a height of about 35,780 km with a rotation period of 24 hours and a satellite visibility time of 24 hours. Many GEO satellites are allocated on distinct slots on the equatorial plane orbit. The GEO satellite altitude and the equatorial orbit have been determined to allow that GEO satellites rotate at the same speed of the Earth. Hence, a GEO satellite remains in a stationary position in the sky with respect to a fixed point on the Earth; this is a desired feature for telecommunication purposes.

The balance between the gravity force versus the Earth and the centrifugal one determines the satellite orbital speed. The three Kepler's laws regulate the satellite orbital motion.

A satellite communication system is formed by a number of satellites, typically with the same orbit type (i.e., GEO, MEO or LEO) that cover a

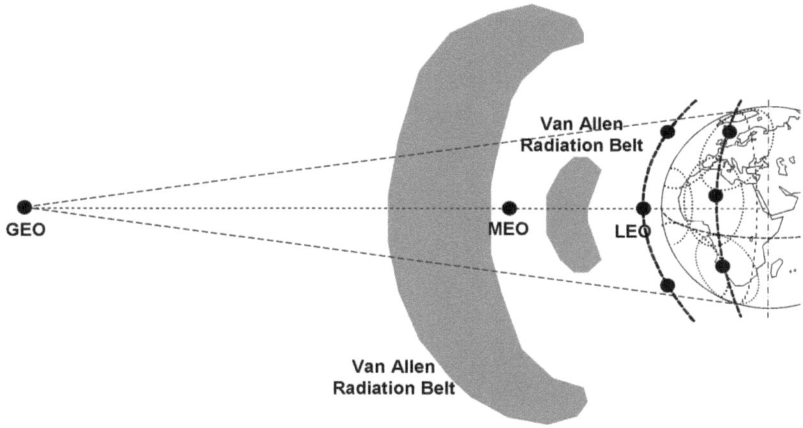

Fig. 1.1: Description of satellite orbit types.

region or the whole Earth, thus forming a *constellation*.

Three GEO satellites are sufficient to cover all the Earth, excluding Polar Regions. GEO satellites are well suited for global-coverage broadcast/multicast services and also for regional mobile and fixed communication services. MEO and LEO satellites are non-stationary with respect to a user on the Earth; hence, different satellites alternatively provide telecommunication service coverage to a given area on the Earth. A global MEO system needs a constellation of 10-12 satellites to assure a minimum elevation angle greater than 30°. LEO systems are characterized by constellations of more than 40 satellites with minimum elevation angle from 10° to 40°. A minimum elevation angle of about 40° (30°) is recommended in the MEO (LEO) case in order to have high link availability and acceptable delay variations. Moreover, LEO and MEO satellite systems allow lower propagation delays and hence, lower end-to-end latency in transferring data than GEO satellites.

GEO satellites are very big and can host a huge payload; high power and large antennas are needed to assure a reliable link with Earth stations. MEO satellites are smaller than GEO ones, so that launching operations are less expensive. Finally, LEO satellites are smaller and less expensive to build and to launch than GEO and MEO. Launchers allowing the transport of multiple satellites permit to reduce the cost to have an operational LEO satellite constellation.

The coverage area (footprint) of a satellite is divided into many cells (each irradiated by an antenna spot-beam) in order to concentrate the energy on a small area. Thus, it is also possible to shape the area served by a satellite on the Earth. Moreover, multi-spot-beam coverage permits remarkable advantages, like an efficient distribution of resources (e.g., reusing the same frequency) or a lower cost of the Earth terminal equipment (e.g., antennas with small size, since narrower surfaces are irradiated on the Earth, thus having a higher power per surface unit).

Frequency bands (of interest for satellite communications) and related designations are listed below [1],[3],[5]:

- L band from 1 to 2 GHz
- S band from 2 to 4 GHz
- C band from 4 to 8 GHz
- X band from 8 to 12 GHz
- Ku band from 12 to 18 GHz
- K band from 18 to 26 GHz
- Ka band from 26 to 40 GHz
- V band from 40 to 75 GHz.

These bands, composing the microwave spectrum, are actively used in commercial and military satellite communications. The typical frequency band allocations for satellite communications, adopted for different services, are detailed below considering uplink/downlink cases:

- *Fixed Satellite Service* (FSS): 6/4 GHz (C band), 8/7 GHz (X band), 14/12-11 GHz (Ku band), 30/20 GHz (Ka band), 50/40 GHz (V band). These services concern communications with fixed terrestrial terminals; moreover, they are often broadband (typically in the range of 1-200 Mbit/s) due to both the available *Radio Frequency* (RF) bandwidth and suitable link performance by using terrestrial fixed directional antennas. Even if these services have been originally allocated to GEO satellites, also non-GEO system allocations are possible.
- *Broadcasting Satellite Service* (BSS): 2/2.2 GHz (S band), 12 GHz (Ku band), 2.6/2.5 GHz (S band). These services deal with direct broadband broadcast transmissions through public operators. In particular, the Ku band segment of BSS has been reserved for orbit positioning and dedicated channels for individual nation's employment. This service has been mainly allocated to GEO satellites, but, like in the FSS case, also non-GEO satellites are possible.
- *Mobile Satellite Service*: 1.6/1.5 GHz (L band), 30/20 GHz (Ka band). These services are related to communications with mobile Earth stations (e.g., ships, vehicles, aircrafts, and also persons). An example of mobile satellite service is the Inmarsat system, operating in the L band with GEO satellites for land-mobile services. These bands have been assigned later also to non-GEO satellite networks.

Note that L, S and C bands are already congested; X band is typically reserved for government use (military fixed communications); Ku band is used by the majority of satellite digital broadcast systems as well as for current Internet access systems. Finally, Ka band allows higher bandwidths with smaller antennas (with respect to Ku band), but presents the problem of significant signal impairment in the presence of bad wheatear conditions (e.g., rain).

A transponder is a receiver-transmitter unit on a communication satellite. It receives a signal from the Earth (uplink), manages it and retransmits it back to Earth at a different frequency (downlink). A satellite has several transponders in its payload. Two different types of transponders can be distinguished as follows:

- *Bent-pipe transponder* (i.e., the transponder acts as a simple repeater). On board, the signal is simply amplified and retransmitted, but there is no improvement in the signal-to-noise ratio since also background noise is amplified.
- *Regenerating transponder*: a transponder demodulates and decodes the received signal, thus performing signal recovery before retransmitting it. Since at some point base-band signals are available, other activities are also possible, such as routing and beam-switching (in case of multi-beam satellite antenna). Satellites with regenerating transponders and on board processing capabilities can also employ *Inter-Satellite Links* (ISLs) with other satellites of the same constellation, thus permitting the routing of the signal in the sky.

It is important to provide here some interesting data for current state-of-the-art GEO satellites.

- The Astra 1H satellite has 32 transponders with 24/32 MHz bandwidth (total bandwidth of 1 GHz). Each transponder has a traffic capacity of 25-30 Mbit/s.
- The AmerHis satellite (51 transponders) has a hybrid payload with 4 channels, each with 36 MHz for a total capacity of 174 Mbit/s. Moreover, there is a DVB-RCS transponder that can manage up to 64 carriers, each with 0.5 Mbit/s and a DVB-S transponder with a capacity of 54 Mbit/s; see the following Section 1.4 for more details on DVB-RCS and DVB-S systems.

Tables 1.1 and 1.2 below provide a survey of some satellite communication systems that are currently operational or planned [8],[9]; for the definition of the different access techniques, please refer to the following Section 1.3.

A typical satellite network architecture is shown in Figure 1.2, where we can see the Earth station permitting the interconnection via a gateway to the terrestrial core network.

Satellite communications are broadcast in nature. Hence, satellites do not offer an adequate reliability from the security and privacy standpoint. Practically, it is possible that a malicious user can hear what the others are communicating. Therefore, it is necessary to adopt appropriate cryptography algorithms to control network accesses and to protect transmissions.

Recently, the *Broadband Global Area Network* (BGAN) system has acquired momentum to provide several services via Inmarsat-4 satellites (e.g.,

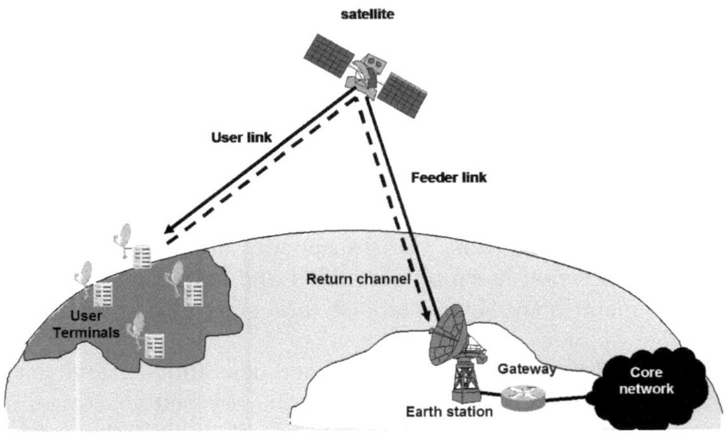

Fig. 1.2: Basic satellite network architecture.

System	Orbit type, altitude [km]	Services	Access scheme	Frequency bands
GlobalStar	48 LEO, 1414	Mobile satellite system voice and data services	Combined FDMA & CDMA (uplink and downlink)	Uplink: 1610.0-1626.5 MHz (L band) Downlink: 2483.5-2500 MHz (S band)
Iridium	66 LEO, 780	Mobile satellite system voice and data services	FDMA/ TDMA - TDD for both uplink and downlink	Uplink: 1616-1626.5 MHz (L Band) Downlink: 1610-1626.5 MHz (L Band)
ICO (new ICO)	12 MEO (10 active), 10355 (changed to 10390 km, late 1998)	ICO is planning a family of quality voice, wireless Internet and other packet-data services	FDMA/ TDMA - FDD	Uplink: 1980-2010 MHz Downlink: 2170-2200 MHz (C/S bands)

Table 1.1: Description of the characteristics of the main satellite communication systems (operational or planned) for non-GEO orbits.

telephony and ISDN calls; Internet/Intranet connection; SMS and MMS; UMTS location-based services like information on maps or local travel information), firstly to fixed terrestrial user terminals, and secondly to mobile terminals on planes, ships or land areas. BGAN satellites operate in the L band. It is possible to adapt the transmission power, bandwidth, coding rate and modulation scheme to terminal capabilities and to channel conditions, in order to achieve high transmission efficiency and flexibility. The baseline system allows communications from 4.5 to about 512 kbit/s to 3 classes

System	Orbit type, altitude [km]	Services	Access scheme	Frequency bands
Spaceway	16 GEO + 20 MEO, 36000 - 10352	With Spaceway, large businesses, telecommuters, *Small Office - Home Office* (SOHO) users and consumers will have access to two-way, high-data-rate applications such as desktop videoconferencing, interactive distance learning and Internet services	Uplink: FDMA/ TDMA Downlink: TDMA	Uplink: 27.5-30 GHz Downlink: 17.7-20.2 GHz Ka band
Thuraya	2 GEO	Voice telephony, fax, data, short messaging, location determination, emergency services, high power alerting	FDMA	Uplink: 1626.5-1660.5 MHz Downlink: 1525-1559 MHz L/C bands
Eutelsat (operator)	GEO satellites (e.g., Hotbird 4, Hotbird 6) equipped with the Skyplex regenerating transponder	Single digital TV programme broadcasting, digital radio broadcasting, interactive multimedia services and Internet connectivity	Uplink: DVB-RCS (TDMA) Downlink: DVB-S	Uplink: 13.75, 14-14.50, 29.50-30 GHz Downlink: 10.70, 10.86-12.75, 19.70-20.20 GHz Ku and Ka band
Wildblue	GEO (Anik F2)	High-speed broadband Internet access, satellite television, distance learning and telemedicine	Uplink: TDMA Downlink: MF-TDMA	Uplink: 5.9-6.4 GHz (C band), 14-14.5 GHz (Ku band), 28.35-28.6 and 29.25-30 GHz (Ka band) Downlink: 3.7-4.2 (C band), 11.7-12.2 (Ku band), 18.3-18.8 and 19.7-20.2 GHz (Ka band)
IPStar	GEO	Broadband access, Intranet and VPN, Broadcast/Multicast, Video on Demand, Voice, Leased Circuit/Trunking, Video Conferencing	Uplink: MF-TDMA Downlink: TDM/ OFDM	Uplink: 13.775-13.975, 14-14.5 GHz Downlink: 10.95-11.2, 11.5-11.7, 12.2-12.75 GHz
Inmarsat	11 GEO (10 active sats.): 4 Inmarsat-2, 5 Inmarsat-3, 2 Inmarsat-4	Simultaneous voice & data, Internet & Intranet content and solutions, Video-on-demand, videoconferencing, fax, e-mail, phone and LAN access	TDMA	Uplink: 1.626-1.66, 1.98-2.025 GHz Downlink: 1.525-1.559, 2.16-2.22 GHz

Table 1.2: Description of the characteristics of the main satellite communication systems (operational or planned) for GEO orbits.

of portable terminals. The enhanced system (BGAN-X, *BGAN Extension project*) has been developed to serve omni-directional and directional mobile terminals, extending the classes from 3 to 11.

1.2 Basic issues in the design of satellite communication systems

Satellite communications represent an attractive solution to provide broadband and multimedia services. To make the upcoming satellite network systems fully realizable, meeting new services and application *Quality of Service* (QoS) requirements, many technical challenges have to be addressed as described below [1]-[5].

Round Trip propagation Delay (RTD)

RTD is the propagation delay along a link (back and forth). In the satellite case, its value depends on the satellite orbit, the relative position of the user on the Earth, and the type of satellite [1],[3],[5]. In particular, if the satellite is regenerating, RTD involves a single hop from the Earth to the satellite and back to the Earth; whereas, if the satellite is bent-pipe, RTD typically involves a double hop (from Earth to satellite to Earth and back) since layer 2 control functions are in the Earth station. In case of GEO regenerating satellites, RTD varies in the range 239-280 ms. In particular, RTD is 239.6 ms for an Earth station placed on the Earth equator in the point below the satellite; whereas, RTD is about 280 ms for an Earth station placed at the edge of the satellite coverage area (i.e., seeing the satellite with the minimum allowed elevation angle). Note that RTD can be also referred to an end-to-end connection, involving many links (the satellite type is not relevant for such RTD). In the GEO case, this end-to-end RTD value (between a message transmission and the reception of the relative reply) varies from 480 to 558 ms; this value can increase due to processing, queuing and on-board switching operations.

The RTD values increase with the satellite orbit altitude and reduces with the elevation angle. LEO and MEO satellites are situated at low altitudes, so they allow lower RTD values than GEO. High RTD values cause several problems for both interactive and real-time applications (e.g., an evident and troublesome echo in phone calls); moreover, also reliable transport layer protocols can experience problems since the end-to-end delay loop is dominated by the propagation delay contribution due to the satellite segment. The maximum RTD value (RTD_{max}) for a given satellite constellation also depends on the minimum elevation angle (*mask angle*), i.e., the elevation angle at the edge of coverage. The RTD_{max} characteristics for LEO satellite systems are described in Figure 1.3.

Atmospheric effects

The effects of atmosphere (subdivided in troposphere and ionosphere) can be summarized as follows [2]:

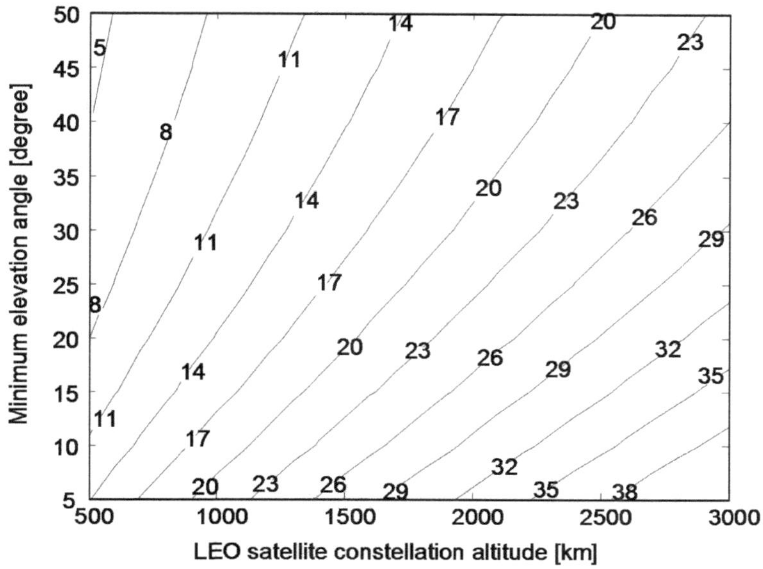

Fig. 1.3: RTD_{max} level curves in ms for LEO satellite constellations in the plane Minimum elevation angle [in degrees] versus LEO satellite constellation altitude [in km].

- *Atmospheric gasses.* Oxygen (dry air) and water vapor determine an attenuation of the electromagnetic signal that depends on the transmission frequency: below 10 GHz, it is possible to ignore the influence of the atmospheric gasses; between 10 and 150 GHz, molecular oxygen dominates the total attenuation (in this region the local attenuation peaks are at 22.3 GHz -Ka band- and at 60 GHz -V band-, respectively due to water vapor and molecular oxygen); whereas, above 150 GHz, the effect of water vapor is dominant.
- *Rain attenuation.* This type of attenuation is the most significant one among the atmospheric effects. There are several prediction models to establish the quantity of rain fall attenuation, depending on some parameters, such as the rain fall rate probability distributions, the slant path length, and the rain height. With these parameters it is possible to characterize the level of rain and the relative attenuation (e.g., rain, widespread rain, showery rain, rainstorm, etc.).
- *Fog and clouds.* The attenuation effects of fog and clouds are not so important for systems operating below 30 GHz; while, they are significant above 30 GHz. This type of attenuation is related to frequency, temperature and liquid water density (expressed in g/m^3). Empirical models (one of them is recommended by ITU) are used to predict fog and clouds attenuation.

- *Scintillation.* This is a phenomenon that affects satellite communication systems operating above 10° elevation angle and below 10 GHz (Ku band). This effect consists of small and quite rapid fluctuations due to some irregularities in the troposphere refractive index. As for the reception in a mobile environment, the signal can be faded and enhanced by these fluctuations.

Channel losses

In satellite networks, *Bit Error Rate* (BER) is very high, due to the above-mentioned atmospheric effects. The quality of the satellite link can be subject to rapid degradation that can cause long sequences of erroneous bits. These burst errors cause an on-off behavior for the channel. With the use of *Forward Error Correction* (FEC) codes (e.g., Reed-Solomon codes, convolutional codes, etc.), it is possible to reduce remarkably BER at the expenses of a lower information bit-rate (i.e., part of the available capacity is spent in sending redundancy bits).

Satellite lifetime

Satellites have an average life span due to the components' ageing process, the effect of radiations, the necessity of new components, etc. GEO satellites have a lifetime in the range of 10-15 years. MEO satellites have an operational period of 10-12 years. Finally, LEO satellites are efficient between 5 and 8 years, mainly due to radiation effects.

1.3 Multiple access techniques

Multiple access is the ability of a large number of Earth stations to simultaneously interconnect their respective multimedia traffic flows via satellite [1],[10]. These techniques permit to share the available capacity of a satellite transponder among several Earth stations. The most common techniques are:

- *Frequency Division Multiple Access* (FDMA),
- *Time Division Multiple Access* (TDMA),
- *Code Division Multiple Access* (CDMA),
- A mix of the above schemes (e.g., combining TDMA and CDMA or FDMA and TDMA).

These different multiple access techniques are surveyed below. Note that another form of multiple access is also allowed in the presence of a multi-spot-beam antenna on the satellite. This technique is called *Spatial Division Multiple Access* (SDMA) [11]. With a multi-spot-beam antenna, some beams may re-use the same frequencies, provided that the cross-interference (due to beam radiation pattern side-lobes) is negligible. Usually, beams separated by more than two or three half-power beam-widths can use the same frequencies; this frequency reuse technique permits increasing the utilization of air interface resources.

FDMA

In FDMA, the total bandwidth is divided into equal-sized parts; an Earth station is permanently assigned with a portion around a carrier or carriers. FDMA requires guard bands to keep the signals well separated. The traffic capacity of an Earth station is limited by its allocated bandwidth and the *Carrier power-to-Noise power ratio* (C/N). The carrier frequencies and the bandwidths assigned to all the Earth stations constitute the satellite's frequency plan. FDMA requires the simultaneous transmission of a multiplicity of carriers through a common *Traveling-Wave-Tube Amplifier* (TWTA) on the satellite. The TWTA is highly non-linear (it produces maximum output power at the saturation point, where the TWTA is operating in the non-linear region of its characteristics) and the *Inter-Modulation* (IM) products generated by the presence of multiple carriers produce interference. The only way to reduce IM distortion is to lower the input signal level, so that the TWTA can operate in a more linear region. For a given carrier, the dB difference between the single-carrier input power level at saturation and the input power level for that particular carrier in multi-carrier FDMA operations is called *input backoff*. The corresponding output transmission power reduction in dB is called *output backoff*.

TDMA

In TDMA, the total bandwidth is usually divided into time *slots*, organized according to a periodic structure, called *frame*. Each slot is used to convey one packet. Hence, TDMA is well suited for packet traffic. In TDMA uplink transmissions, Earth stations take turns sending bursts through a common satellite transponder. As for TDMA downlink transmissions from a satellite, only one carrier is used. Hence, TDMA provides a significant advantage, since it permits a transponder's TWTA to operate at or near saturation, thus maximizing downlink C/N. However, interference is not totally eliminated, since it is present in the form of inter-symbol interference that must be minimized by means of appropriate filtering. TDMA is easy to reconfigure for changing traffic demands, it is robust to noise and interference and allows mixing multimedia traffic flows.

While in TDM (*Time Division Multiplexing*) all data come from the same transmitter and the clock and time frequencies do not change, in TDMA each frame contains a number of independent transmissions. Each station has to know when to transmit and must be able to recover the carrier and the data synchronization for each received burst in time to sort out all desired base-band channels. This task is not easy at low C/N values. A long preamble is generally needed, which decreases system efficiency.

A group of Earth stations, each at a different distance from the satellite, must transmit individual bursts of data in such a way that bursts arrive at the satellite in correspondence with the beginning of the assigned slots. Stations must adjust their transmissions to compensate for variations in satellite movements, and they must be able to enter and leave the network without disrupting its operation. These goals are accomplished by exploiting the TDMA organization in frames, which contain reference bursts that permit establishing absolute time for the network.

Reference bursts are generated by a master station on the ground in a centralized-control satellite network. Each burst starts with a preamble, which provides synchronization and signaling information and identifies the transmitting station. Reference bursts and preambles constitute the frame overhead. The smaller the overhead, the more efficient the TDMA system, but the greater the difficulty in acquiring and maintaining synchronism.

Time access to the satellite link can be managed either in centralized or in distributed mode. Centralized control is generally more robust. On the other hand, the distributed control is more responsive to traffic variations, since it allows an update in one RTD.

CDMA

The signals are encoded, so that information from an individual transmitter can be detected and recovered only by a properly synchronized receiving station that knows the code used ("scrambling code") for transmissions. In a decentralized satellite network, only the pairs of stations that are communicating need to coordinate their transmissions (i.e., they need to use the same code). The concept at the basis of CDMA is spreading the transmitted signal over a much wider band (*Spread Spectrum*). This technique was developed as a jamming countermeasure for military applications in the 1950s. Accordingly, the signal is spread over a band *PG* times greater than the original one, by means of a suitable 'modulation' based on a *Pseudo Noise* (PN) code. PG is the so-called *Processing Gain*. The higher the PG, the higher the spreading bandwidth and the greater the system capacity. Suitable codes must be used to distinguish the different simultaneous transmissions in the same band. The receiver must use a synchronous code sequence with that of the received signal, in order to de-spread correctly the desired signal. There are two different techniques for obtaining spread spectrum transmissions:

- *Direct Sequence* (DS), where the user binary signal is multiplied by the PN code with bits (called *chips*) whose length is basically PG times smaller that that of the original bits. This spreading scheme is well suited for *Binary Phase Shift Keying* (BPSK) and *Quadrature Phase Shift Keying* (QPSK) modulations.

- *Frequency Hopping* (FH), where the PN code is used to change the frequency of the transmitted symbols. We have a fast hopping if frequency is changed at each new symbol, whereas a slow hopping pattern is obtained if frequency varies after a given number of symbols. *Frequency Shift Keying* (FSK) modulation is well suited for the FH scheme.

Comments and comparisons among the access techniques

The drawback of TDMA is the need to size Earth stations for the entire system capacity (transponder bandwidth), even though the single terminal uses a small portion of that. An interesting solution is given by the hybrid combination of *Multi-Frequency* (MF) with TDMA systems, which takes some advantages of both FDMA and TDMA [12]. In MF-TDMA the transponder spectrum is divided into several carriers, thus allowing the sizing of the station on a narrower bandwidth. Each carrier, in turn, is shared in TDMA mode. The transmission of the traffic occurs in time slots that may belong to different carriers. When a single modulator is used, slots of a transmission need not to overlap in time (i.e., simultaneous transmissions on different frequencies are not allowed). The MF-TDMA technique efficiently supports traffic streaming, while maintaining flexibility in capacity allocation.

1.4 Radio interfaces considered and scenarios

Different standardized air interfaces are available for satellite communication systems. In particular, this book is focused on both the satellite extension of the terrestrial *Universal Mobile Telecommunications System* (UMTS) [1] and the *Digital Video Broadcasting via Satellite* (i.e., DVB-S, DVB-S2 and DVB-RCS) [13]-[16]. In addition to this, scenarios have been considered that combine together different aspects, such as: satellite orbit type, mobile or fixed users, adopted air interface. In particular, the following scenarios have been identified:

- **Scenario 1**: *Satellite-UMTS* (S-UMTS) for mobile users through GEO bent-pipe satellite;
- **Scenario 2**: DVB-S/DVB-RCS for fixed broadband transmissions via GEO bent-pipe satellite;
- **Scenario 3**: LEO constellation with regenerating satellites for the provision of multimedia services to mobile users adopting handheld devices.

1.4.1 S-UMTS

Satellite communication systems should be able to provide to mobile users the same access characteristics of the terrestrial counterparts. We refer here to the provision of 3rd *Generation* (3G) mobile communication services through satellites. In particular, the interest is on the extension of the UMTS standard to the satellite context (S-UMTS). The ETSI S-UMTS Family G specification set aims at achieving the satellite air interface fully compatible with the terrestrial W-CDMA-based UMTS system [17]-[20]. S-UMTS will not only complement the coverage of the *Terrestrial UMTS* (T-UMTS), but it will also extend its services to areas where the T-UMTS coverage would be either technically or economically not viable.

The satellite radio access network of the S-UMTS type should be connected to the UMTS core network via the Iu interface [1],[21]. S-UMTS is expected to be able to support user bit-rates up to 144 kbit/s that appear to be sufficient to provide multimedia services to users on the move, employing typically small devices [22].

With the evolution of terrestrial 3G systems standardization, the *High Speed Downlink Packet Access* (HSDPA) has been defined to upgrade current terrestrial 3G (W-CDMA) systems to provide high bit-rate downlink transmission to users. HSDPA's improved spectrum efficiency enables users with downlink speeds typically from 1 to 3 Mbit/s. Hence, capacity-demanding applications are possible, such as video streaming. The mandatory codec for streaming applications is H.263, with settings depending on the streaming content type and the streaming application.

The novel HSDPA air interface is based on the application of *Adaptive Coding and Modulation* (ACM) and multi-code operation depending on the channel conditions (forward link) that are feed back by the *User Equipment* (UE) to the Node-B. The interest in this book is on the study for the possible extension of HSDPA via satellite, as an upgrade of S-UMTS specifications. In this case, all resource management functions for the S-HSDPA air interface are managed by the base station (i.e., Node-B) on the Earth that is directly linked to the *Radio Network Controller* (RNC) that operates as a gateway towards the core network. More details on this study will be provided in Chapter 5.

1.4.2 DVB-S standard

DVB-S has been designed for primary and secondary distribution in the bands of FSS and BSS [13]. Such systems should be able to provide direct-type services (*Direct-To-Home*, DTH) both to the single consumer having an integrated receiver-decoder, to systems with a collective antenna and to the terminal stations of cable-TV. The frequency bands for feeder and user links may occupy Ku/Ku, Ku/Ka and K/Ka bands.

Below the transport layer and the IP layer the *Multi Protocol Encapsulation* (MPE) provides segmentation & reassembly functions for the generation of *Moving Picture Experts Group 2 - Transport Stream* (MPEG2-TS) packets of 188 bytes (fixed length). A TCP header of 20 bytes, an IP header of 20 bytes and an MPE header + CRC trailer of 12 + 4 bytes are added to packets from the application layer; the resulting blocks are fragmented in payloads of MPEG2-TS packets. All the data flows transported in single

MPEG2-TS are of the TDM type. In the channel adaptation section, packets are processed in several steps, such as: channel encoding (outer Reed-Solomon coding, convolutional interleaver, inner convolutional encoding, puncturing), base-band shaping of impulses, and QPSK modulation. The resulting DVB-S transmissions via satellite are very robust, considering a minimum BER of about 10^{-11}. As an example, a typical data rate of about 38 Mbit/s is obtained with modern satellite transponders that have a bandwidth of about 33 MHz [13].

1.4.3 DVB-RCS standard

One of the reasons for the definition of a DVB standard with satellite return channel (DVB - *Return Channel via Satellite*, DVB-RCS) has been the increasing request of interactive applications and services with major informative volumes ([1]) that could not be achieved with a DVB-S-based system, where the return channel (realized through a terrestrial link via modem) cannot permit an adequate bit-rate capacity (maximum 64 kbit/s).

The specifications of DVB-RCS use and modify the DVB-S ones [14],[15]; moreover, they are independent of frequency, making easier to realize network and security mechanisms with an efficient transport layer. The DVB-S channel has been named *Forward Channel*, while the *Return Channel* is related to the link from the end-user back to the content network (see Figure 1.4). The return channel has a variable bit-rate up to a maximum of 2 Mbit/s and can dynamically assign its time-frequency resources (according to an MF-TDMA air interface) to the requesting terminals. The *Return Channel Satellite Terminal* (RCST) transmission capacity is constrained. According to the standardization, RCSTs can be single-user (144-384 kbit/s) or corporate (2 Mbit/s).

The standard [14],[15] defines a reference model for the *Interactive Satellite Network* (ISN) architecture, composed of a certain number of RCSTs, a GEO bent-pipe satellite, and the following elements:

- *Network Control Center* (NCC): it provides *Control and Monitoring Functions* (CMF); moreover, it produces timing & control signals that one or several *Feeder Stations* transmit for the ISN operations.
- *Traffic Gateway* (GW): it is a router that sends/receives data to/from the RCSTs, managing the exchange of data with public, proprietary and private providers.
- *Feeder*: it is the Earth station that transmits *Forward Link* (DVB-S) signal, where user data and ISN timing & control signals are multiplexed together.

---

[1] Recently, also other systems have been standardized for broadband satellite access such as DOCSIS-S and IPoS [23].

Figure 1.4 shows a simplified version of the DVB-S/DVB-RCS system architecture where NCC, GW and Feeder are 'collapsed' into the NCC, i.e., in a single Earth station.

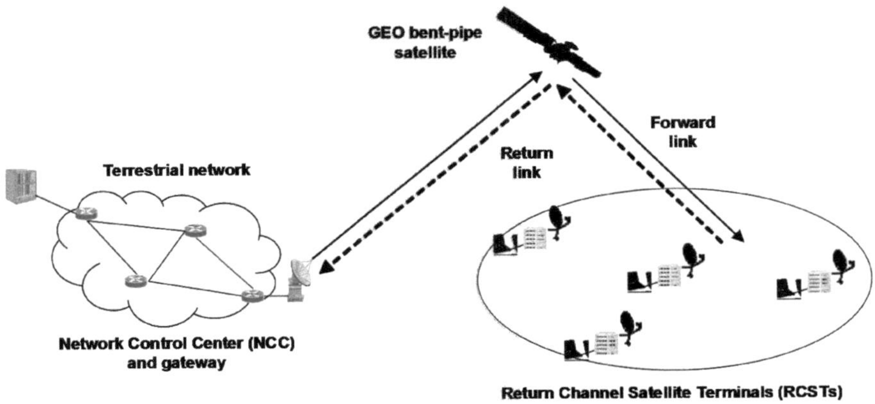

Fig. 1.4: Example of DVB-S/DVB-RCS system architecture.

Air interface characteristics of DVB-RCS

In order to operate successfully an ISN, it is important to use the satellite resources as efficiently as possible. Therefore, *Bandwidth on Demand* (BoD) schemes (also known with the name of *Demand Assignment Multiple Access*, DAMA, techniques) have been introduced in the DVB-RCS standardization in order to improve the utilization of satellite resources in the presence of distinct traffic classes.

The DVB-RCS standard specifies a MAC layer in which the NCC controls the allocation of the uplink capacity for RCST transmissions. BoD is defined as a set of MAC protocols and algorithms that allow an RCST to request resources to the NCC, when the RCST has traffic to pass to GW.

Return link transmissions are based on an MF-TDMA air interface, where RCSTs transmit their data using a range of carrier frequencies (with potentially different bandwidth size), each of them organized in super-frames, frames and time-slots. The NCC assigns to each active RCST a set of bursts, each of them is defined by frequency, bandwidth, starting time and duration. Different carriers can have the same or different timeslots characteristics, thus having a *fixed* or a *dynamic* timeslot structure. In the former case, timeslots have fixed characteristics, in terms of bandwidth and duration. Whereas, in the latter case, besides bandwidth and time-slot duration, both transmission rate and code rate can be changed in consecutive slots. Such flexibility allows a better RCST adaptivity to the variable requirements of multimedia transmissions.

The return link time and frequency organization of the air interface is depicted in Figure 1.5. Each super-frame is characterized by a *superframe_id*, and can be assigned to a group of RCSTs. In turn, each super-frame is divided in parts, characterized by a *superframe_counter* that can be divided in frames, identified by a *frame_number* (F_nb) or by a *frame_ID* (F_id). Frames can have different duration, bandwidth and composition of timeslots. Each frame is divided in timeslots characterized by a *timeslot_number* (TS_nb); also timeslots can be organized in slot groups with similar characteristics.

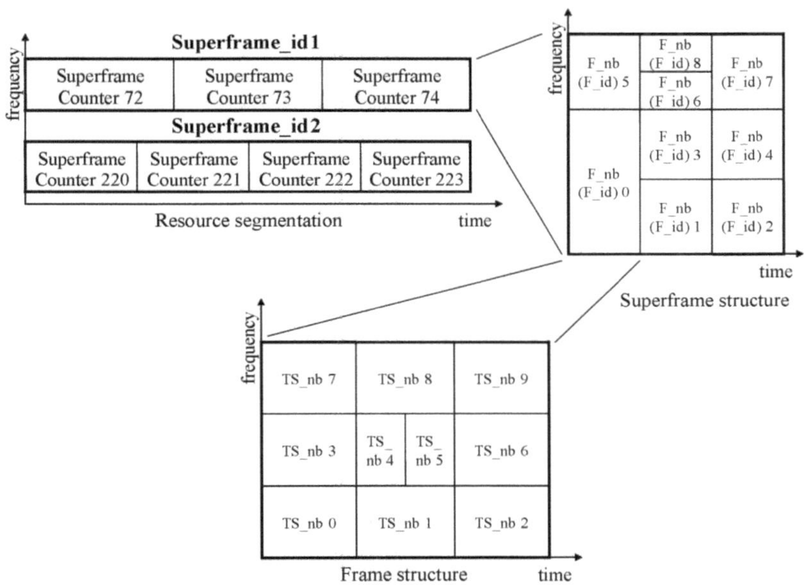

Fig. 1.5: Organization of the resources in the MF-TDMA air interface.

The RCST is responsible for analyzing, estimating and requesting the needed capacity for uplink transmissions (DAMA case), and for distributing the allocated capacity to the internal applications according to some rules. In particular, when an RCST has data to transmit, it first explicitly requests the needed capacity to the NCC (*Capacity Request*, CR, message). The NCC allocates return channel time slots based on each requests and informs all RCSTs of allowable transmission slots by using *Terminal Burst Time Plan* (TBTP) messages, sent regularly (e.g., once per super-frame) over the forward channel. Each RCST looks at the received TBTP and transmits data during the allocated time slots.

Allocation methods and traffic classes in DVB-RCS

Five capacity allocation methods (layer 2) are defined in the DVB-RCS standard [14],[15]:

- *Continuous Rate Assignment* (CRA),
- *Rate Based Dynamic Capacity* (RBDC),
- *Volume Based Dynamic Capacity* (VBDC),
- *Absolute Volume Based Dynamic Capacity* (AVBDC) and
- *Free Capacity Assignment* (FCA).

Note that CRA is a fixed capacity allocation, while RBDC, VBDC and AVBDC are DAMA schemes. Finally, with FCA the NCC assigns unutilized resources in a super-frame (after the fulfillment of the other request types), without any particular requests made by RCSTs. In allocating resources, the NCC adopts the following priority order:

$$CRA > RBDC > A(VBDC) > FCA.$$

Details on the capacity allocation methods are provided below.

***Continuous Rate Assignment* (CRA)**: CRA is a rate capacity that shall be provided in full for every super-frame while required. CRA is a fixed (and static) allocation of resources after an initial set-up phase with a negotiation between the RCST and the NCC. With CRA, a given number of time slots (i.e., packets) are continuously assigned to that RCST every super-frame until that RCST sends the assignment release message. CRA would typically be subscription-based: the user subscribed to a certain constant rate, and the RCST has automatically assigned this constant rate at log-on. CRA should be used for traffic, which requires a fixed guaranteed rate, with minimum delay and minimum delay jitter, such as the *Constant Bit Rate* (CBR) class of ATM networks. The CRA allocation method could also be used in conjunction with RBDC to manage a *Variable Bit Rate* (VBR) traffic that could not tolerate the request-allocation loop delay. In this case, CRA would guarantee a minimum bit-rate and RBDC should provide an additional dynamic capacity.

***Rate Based Dynamic Capacity* (RBDC)**: RBDC is a rate capacity that is dynamically requested by the RCST. RBDC capacity shall be provided in response to explicit CR messages from the RCST to the NCC, such requests being *absolute* (i.e., corresponding to the full rate currently being requested). Each request shall override all previous RBDC requests from the same RCST, and shall be subject to a maximum rate limit negotiated directly between the RCST and the NCC, $RBDC_{max}$. To prevent an RCST anomaly resulting in a hanging capacity assignment, the last RBDC request received by the NCC from a given RCST shall automatically expire after a time-out period, whose default value is 2 super-frames, such expiry resulting in the RBDC being reset to zero rate. CRA and RBDC could be used in combination, as previously explained. A typical application for RBDC over a GEO satellite could be to support the *Available Bit Rate* (ABR) traffic class of ATM networks.

***Volume Based Dynamic Capacity* (VBDC)**: VBDC is a volume capacity, dynamically requested by the RCST. VBDC capacity shall be provided in response to explicit CR messages from the RCSTs to the NCC, such requests being *cumulative* (i.e., each request shall add to all previous requests from the same RCST). The request indicates a total number of needed traffic slots (i.e., packets) that can be shared between several super-frames; successive VBDC requests add up. VBDC should be used only for traffic that can tolerate delay jitter, such as the *Unspecified Bit Rate* (UBR) traffic class of ATM or standard IP traffic. VBDC and RBDC can also be used in combination for ABR traffic, with the VBDC component providing a low priority capacity extension above the guaranteed limit of the RBDC category. MAC parameters are the minimum ($VBDC_{min}$) and the maximum ($VBDC_{max}$) volume request.

***Absolute Volume Based Dynamic Capacity* (AVBDC)**: AVBDC is a volume capacity that is dynamically requested by the RCST. This AVBDC capacity shall be provided in response to explicit CR messages from the RCST to the NCC, such requests being *absolute* (i.e., a request replaces the previous ones from the same RCST). The request indicates a total number of traffic slots that can be shared between several super-frames; a new AVDBC allocation cancels the previous ones. AVBDC is similar to VBDC and should be used instead of VBDC for the initial request or when the RCST senses that the VBDC request might be lost (re-initialization of a previous request); this might happen when requests are sent on contention bursts (see the next description on related signaling methods) or when channel conditions (e.g., packet error rate, E_b/N_0) are degraded. AVBDC is suitable to support the same traffic classes of VBDC.

***Free Capacity Assignment* (FCA)**: FCA is a volume capacity that shall be assigned to RCSTs from capacity, which would be otherwise unused. Such capacity assignment shall be automatic, not involving any requests from the RCSTs to the NCC. In particular, FCA should not be mapped to any traffic category since availability is highly variable. The assigned capacity is intended as a *bonus*, which can be used to reduce delays on any traffic type that can tolerate delay jitter. It should be noted that the term 'free' in FCA refers to 'spare' system capacity. CRA and FCA can also be viewed as two mechanisms to grant dynamically capacity to an RCST without explicit requests. FCA resources should be distributed to RCSTs according to the following criterions ranked by priority:

1. Performance optimization of TCP/IP in order to reduce the occurrence of TCP timeouts;
2. Equity (i.e., equal sharing of resources according to a round-robin scheme).

RBDC and VBDC methods are quite similar, but they differentiate on the basis of:

- The type of requested capacity (i.e., capacity expressed as a *bit-rate* in RBDC, or capacity expressed in terms of *packets* in VBDC);
- Request characteristics that are *absolute* in RBDC and *cumulative* in VBDC.

RBDC appears a more complex scheme since it involves a technique to estimate the requested bit-rate. With such a scheme, it is however, possible to follow better the bursty characteristics of the input traffic.

In order to send CR messages from the RCSTs to the NCC two signaling methods are available:

- *In-band signaling.* CRs are encapsulated in a *Satellite Access Control* (SAC) format and can be sent in SYNC bursts or normal MPEG2 data bursts using *Data Unit Labeling Method* (DULM), typically employed to send control and administrative information to the NCC.
- *Out-of-band signaling.* A minislot method is used (with or without contentions): minislots are periodically assigned to an RCST (or a group of RCSTs) for the transmission of shorter bursts than those used for traffic purposes.

To each transmission request made by an RCST, latency is associated mainly due to RTD. The *Minimum Scheduling Latency* (MSL) is the minimum delay between the computation of a CR and the time when it is possible to use the requested capacity by an RCST. In case of a bent-pipe satellite, MSL entails the following contributions (see Figure 1.6):

- CR evaluation and transmission;
- Round trip time between the RCST and the NCC (~500 ms for a GEO bent-pipe satellite);
- Processing delay on the NCC (~80 ms);
- TBTP transmission time from the NCC;
- TBTP processing delay on the RCST.

A typical choice for the super-frame length is 500 ms that also corresponds to the TBTP and CR transmission periodicity. A possible value for the frame length is 50 ms.

The DVB-RCS standard envisages 4 priorities (i.e., traffic classes) that are listed below in order of decreasing urgency level [15]:

- The *Real Time* (RT) class for the applications that require strong time constraints (e.g., VoIP and videoconference);

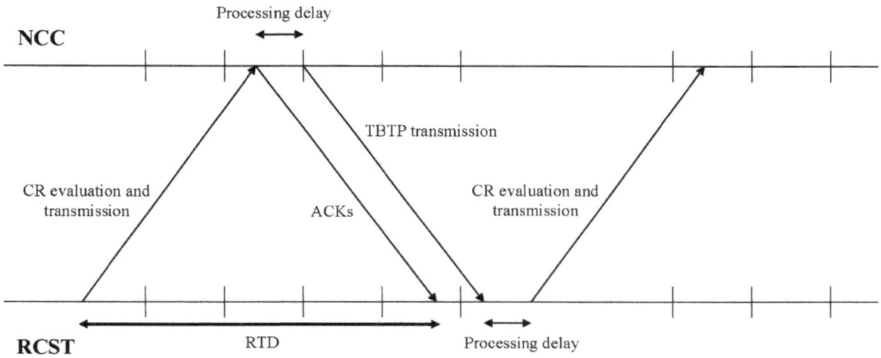

Fig. 1.6: Delay contributions in the process to allocate resources to RCSTs.

- The *Variable Rate - Real Time* (VR-RT) class is for variable bit-rate jitter-sensitive traffic;
- The *Variable Rate - Jitter Tolerant* (VR-JT) for variable bit-rate jitter-tolerant traffic (e.g., FTP application);
- The *Jitter Tolerant Priority* traffic class.

An RCST may queue all traffic arriving from the user interface, using separate queues for flows that are subject to different transmission priorities (i.e., service classes) [15]. As an example, one layer 2 queue shall be provided for each of the priorities (i.e., RT, VR-RT, VR-JT, JT); each queue should be served with a capacity allocation method (or a combination of them). For instance: CRA for RT, RBDC for VR, VBDC/AVBDC+FCA for JT.

Typically, at the IP level 4-16 queues can be managed according to specific IP QoS classes; while at layer 2, typically 4 queues are envisaged [24],[25]. Hence, the IP QoS service classes (i.e., layer 3 queues) need to be adequately mapped into equivalent MAC QoS classes (i.e., layer 2 queues).

Traffic generated at the RCST is first classified and packets are stored into one of several layer 3 queues. From layer 3 we have MPE encapsulation (see Figure 1.7) and the generation of layer 2 packets (e.g., MPEG2-TS) provided to suitable queues, waiting for transmission.

In a connectionless network, the prioritization of voice packets in both directions is crucial in order not to degrade the voice quality. Thus, the priority element plays an important role in the BoD architecture and must be present in all steps of the transmission.

1.4.4 DVB-S2 standard

After 10 years from the definition of DVB-S in 2003, the European DVB consortium has developed a second-generation standard for satellite broadcast transmissions, named DVB-S2 [16]. Such system employs the most recent

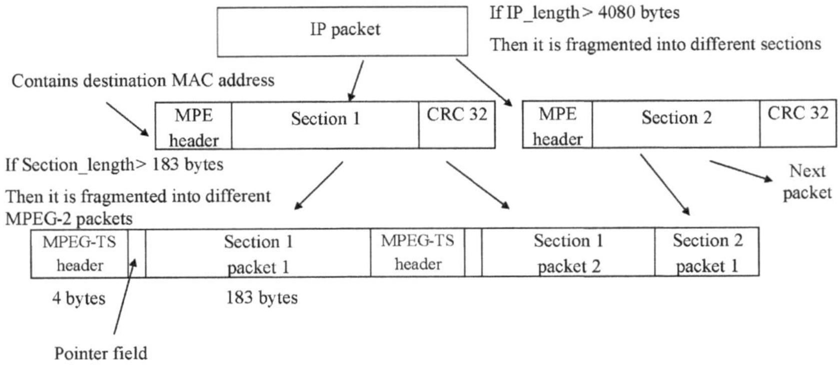

Fig. 1.7: MPE encapsulation for IP traffic.

advances in channel coding (e.g., *Low Density Parity Check*, LDPC, described below) combined with several modulation types (i.e., QPSK, 8PSK, 16APSK and 32APSK).

Besides broadcasting services, DVB-S2 can be employed for interactive point-to-point applications (e.g., Internet access) by using new modulation schemes and new operation modes that permit to optimize the modulation and coding schemes depending on channel conditions. In order to allow that DVB-S continues to operate during the transition period, the DVB-S2 standard also provides transmission means compatible with the satellite decoders of first-generation (*Set-Top-Box*, STB).

A DVB-S2 transmitter is composed by the following functional blocks that are described below [16],[26]: mode adaptation, stream adaptation, FEC encoding, modulation mapping, physical layer framing, base-band filtering and quadrature modulation.

Mode adaptation

There are three (application-dependent) operation modes for DVB-S2: *Constant Coding Modulation* (CCM), *Variable Coding and Modulation* (VCM) and *Adaptive Coding and Modulation* (ACM) [27].

- CCM is a constant protection system, which represents the simplest mode of DVB-S2; it is similar to the DVB-S one, since all data frames are modulated and coded with the same fixed parameters. Unlike DVB-S, DVB-S2 uses an LDPC inner error correction code.
- VCM can be applied to give distinct error protection levels to different services (e.g., robust protection for SDTV and less-robust protection for HDTV, audio, multimedia). In fact, the DVB-S2 standard supports the transmission of different services on the same carrier, each operating with its own modulation scheme and coding rate. VCM performs a kind of

multiplexing operation at the physical layer to provide distinct services with different characteristics.

- ACM is a functionality offered by DVB-S2, in case of interactive and point-to-point applications, when a return channel is available. ACM permits to change dynamically the coding rate and the modulation scheme on the basis of the measured channel conditions at the site that must receive the frame. The sender site dynamically acquires information on the receiving conditions by means of the return channel.

The following description considers the different service scenarios where DVB-S2 can be used. In particular, the following application areas are considered: *Broadcast Services*, *Interactive Services*, *Digital TV Contribution and Satellite News Gathering* and other *Professional Services/applications*. More details are provided below in relation to the operation modes.

- *Broadcast Services* are provided via DVB-S2 with the flexibility of VCM. There are also *Backwards Compatible-Broadcast Services* used for a joint interoperability with DVB-S decoders, and optimized *Non-Backwards Compatible-Broadcast Services*.
- *Interactive Services* are designed to operate with existing DVB return channel standards (e.g., RC-PSTN, RCS, etc.). DVB-S2 can use both CCM and ACM.
- *Digital TV Contribution and Satellite News Gathering* applications refer to point-to-point, or point-to-multipoint communications of multiple or single MPEG-TS, by means of CCM or ACM modes.
- *Professional Services/applications* mainly consists of professional point-to-point and point-to-multipoint applications (e.g., data content distribution); for these services, DVB-S2 uses CCM, VCM or ACM techniques.

Stream adaptation

This operation is applied to perform padding (to complete a base-band frame) and base-band scrambling.

FEC encoding

FEC permits to achieve excellent performance also in the presence of high levels of noise and interference. FEC is achieved with the concatenation of BCH (*Bose-Chaudhuri-Hocquenghem*) outer codes and LDPC inner codes. This technique permits to achieve a performance quite close to the Shannon limit. BCH outer codes are used to avoid error floors at low BER values. The selected LDPC codes [12] operate with code rates of 1/4, 1/3, 2/5, 1/2, 3/5, 2/3, 3/4, 4/5, 5/6, 8/9 and 9/10, depending on the adopted modulation and the system requirements. In particular, coding rates 1/4, 1/3 and 2/5 are used, combined with QPSK modulation in the presence of poor link conditions.

Depending on the application area, the FEC coded blocks have very large lengths (64800 bits for delay-tolerant applications, or 16200 bits). In the VCM and ACM cases, FEC and modulation mode can be varied in different frames, but they are constant in a frame.

Finally, bit interleaving shall be applied to 8PSK, 16APSK and 32APSK FEC coded bits.

Modulation mapping

Four constellations can be used for the transmitted payload, depending on the application area (see Figure 1.8) [28], as described below:

- QPSK and 8PSK are typically suggested for broadcast applications, since they have a quasi-constant envelope so that they can operate inside the non-linear region of satellite transponders (i.e., close to the saturation). Gray mapping can be used for these modulations.
- The 16APSK and 32APSK modes, mainly proposed for professional applications (these modulations could also be used for broadcasting), require a higher level of available C/N and the adoption of advanced pre-distortion methods to reduce the non-linearity effects in transponders.

DVB-S2 is expected to achieve spectral efficiencies ranging from 0.5 bit/s/Hz up to 4.5 bit/s/Hz.

Physical layer framing

This sub-system, synchronously with the FEC frames, generates the *Physical Layer Frame* (PLFRAME), supporting also some tasks, such as: dummy PLFRAME insertion, physical layer signaling, optional pilot symbols insertion and physical layer scrambling for energy dispersion.

A DVB-S2 system can be used with two configurations: single carrier per transponder and multi-carrier per transponder (the bandwidth of the transponder is divided with *Frequency Division Multiplexing*, FDM, among different carriers and related bands).

In case of ACM mode, the DVB-S2 air interface varies flexibly coding and modulation techniques to maximize performance and coverage. This is achieved through the TDM transmission of a sequence of PLFRAMEs, where the coding and modulation format can change for each new PLFRAME.

Base-band filtering and quadrature modulation

This function is used for a tighter bandwidth shaping (squared-root raised cosine) and to generate the radio frequency signal.

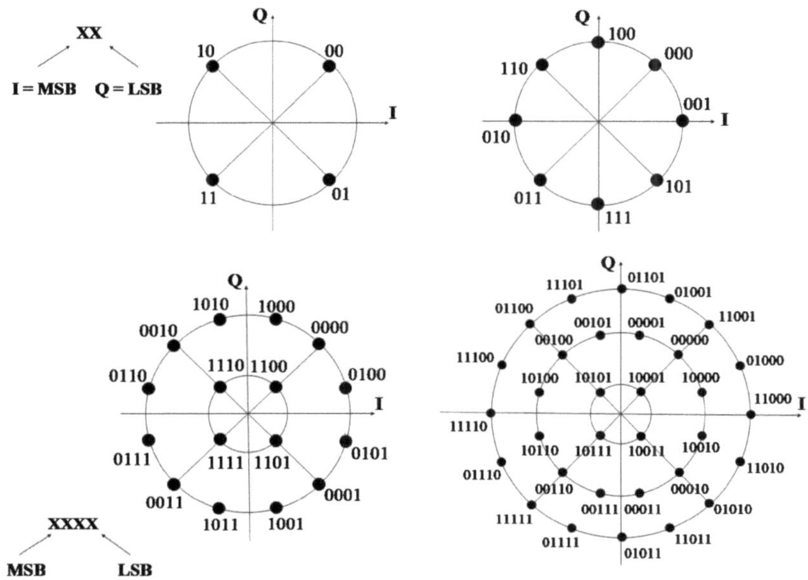

Fig. 1.8: The four possible DVB-S2 constellations before physical layer scrambling.

1.4.5 Numerical details on the selected scenarios for performance evaluations

This sub-Section provides some basic characteristics and numeric values for the parameters that have been used when evaluating the performance of the techniques proposed in the following Chapters of this book for the different scenarios. The details are provided below.

Scenario 1: S-UMTS as well as S-HSDPA

- GEO satellite
- Multi-spot-beam satellite antenna
- Bent-pipe satellite
- Terrestrial gateway containing the scheduler (MAC layer)
- Direct return link via satellite for channel quality measurements in case of point-to-point services
- Mobile users (mean speed equal to 60 km/h)
- GOOD-BAD Markov channel model (typically, 6 s mean GOOD duration and 2 s mean BAD duration) [29]
- IP-based traffic flows with UMTS transport layer encapsulation
- Traffic sources: video sources (sum of ON/OFF Markovian sources) [30] and Web sources (2-MMPP arrival process of Pareto-distributed datagrams) [31].

Scenario 2: DVB-S/DVB-RCS

- GEO satellite
- Single beam or multi-spot-beam satellite antenna
- Bent-pipe satellite
- Architecture involving an NCC and at least a GW
- Fixed users
- Direct return link for channel quality measurements; typically, Ka band is used (maximum capacity 2 Mbit/s)
- Forward link in K band
- Channel model: only troposphere effects (rain scintillation and gas) have to be considered. Basically an *Additive White Gaussian Noise* (AWGN) model has been adopted with a given packet error rate (uncorrelated losses)
- IP-based traffic flows with MPE encapsulation and generation of packets according to the MPEG2-TS format
- Traffic sources of the FTP type (elephant TCP connections).

Scenario 3: LEO constellation

- A Teledesic-like LEO system (the Boeing design with 288 satellites): altitude of 1375 km, and satellite capacity of 32 Mbit/s
- Multi-spot-beam satellite antenna
- End-users must switch from spot-beam to spot-beam and from satellite to satellite, resulting in frequent intra- and inter-satellite handovers
- We assume a two-dimensional mobility model: users move in straight lines and at constant speed (satellite ground track speed composed with the Earth rotation speed)
- All the spot-beam footprints are identical in shape and size (approximated by rectangles, 1790 km $\times$ 1790 km)
- Traffic assumptions (study made in Chapter 8, Section 8.6): non-real-time traffic for email or FTP and real-time multimedia traffic, e.g., interactive voice and video applications. For each class: (*i*) new calls arrive in the footprints according to independent Poisson processes; (*ii*) call holding times are exponentially distributed. Within each traffic class, three different user types are considered that are differentiated depending on the call holding time and bit-rates.

1.5 Satellite networks

A satellite network can play several roles [32]. In particular, it can be used as *Access Network* for final users or it can be part of the *Core Network*. Some examples are shown in Figure 1.9.

The ETSI TC-SES/BSM (*Satellite Earth Stations and Systems / Broadband Satellite Multimedia*) working group had the task to focus on IP layer

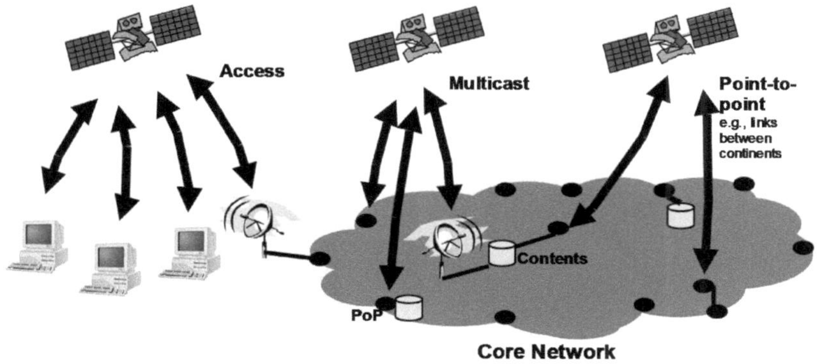

Fig. 1.9: Examples for the use of satellite links in telecommunication networks.

interworking, to define a new network architecture and to include alternative families of lower layer air interfaces [33]. A *Broadband Satellite Multimedia* (BSM) network is divided into 5 domains, as specified in ETSI TR 101 984 [32]:

- *User Domain*, representing the group of end-users;
- *Access Domain* that denotes the access network that is used to connect to the service provider (e.g., ADSL, UMTS, satellite);
- *Distribution Network*: this is an intermediate network that is interposed between the access network and the core network;
- *Core Network*, representing the backbone transport network that is used to connect the routers on a geographical area;
- *Content Domain* that represents the area where contents and information are stored to be made available to users.

The user requesting contents should access them feeling like as he/she was directly connected to the source of the information, the *Content Domain*; practically, many domains are traversed that are transparent to the user.

Let us now consider the BSM network functions from the protocol stack standpoint (see Figure 1.10) that can involve different layers, as specified in ETSI TR 101 985 [34]:

- The BSM network operates at layer 2, like a bridge.
- The BSM network operates at layer 3, so that the satellite Earth stations are routers.
- The BSM network operates at a layer above the 3rd one: the satellite Earth stations are gateways. In this case, these stations can perform a more accurate routing based not only on the IP datagram header, but also on information of higher layer headers. In such a case, the Earth station can implement special functions, like *Performance Enhancing Proxies* (PEPs)

that are important in order to improve the TCP performance in satellite networks.

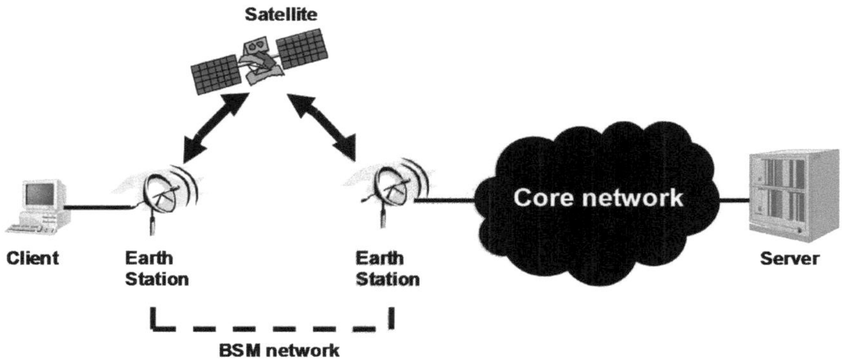

Fig. 1.10: BSM general network architecture.

Very Small Aperture Terminal (VSAT) networks are a special case of BSM networks where the user terminal employs a small antenna (i.e., VSAT) and simplified equipment so as to reduce costs. This small satellite terminal can be used for one way and/or interactive communications. VSATs can support several applications, such as: satellite news gathering, supervisory control and data acquisition, inquiry/response, TV and audio broadcasting, data distribution. VSAT networks are based on GEO satellites (typically of the bent-pipe type) according to a star topology: an Earth station acts as a hub (= gateway to the terrestrial network and master control station), receiving and transmitting all the data fluxes from/to VSATs. The forward link (from the hub to VSATs) is via GEO satellite. The return link (from VSAT to the hub) is typically via a terrestrial *Public Switched Telephone Network* (PSTN) link (to simplify the antenna design on the VSAT). Hence, forward and return links have an asymmetrical capacity; anyway recent advances in this field also allow the return link via satellite. Referring to the network architecture in Figure 1.10, the VSAT includes the client and the Earth station on the left; whereas, the hub coincides with the Earth station on the right. Different VSAT platforms use various technologies in order to access the satellite radio space segment and to share it among multiple users. One of the problems that VSAT networks have faced during their evolution has been the lack of compliance to any specific standards. In the last years, standardization bodies have established new standards to support satellite Internet [23]. The DVB standard has been the first one to be published, and ETSI adopted DVB-RCS for satellite return link transmissions. Another standard is IPoS (*Internet Protocol over Satellite*) developed by HNS (*Hughes Network Systems*) and standardized by ETSI. Finally, DOCSIS-S (*Data Over Cable Service Interface Specification for Satellite*), a modification to the DOCSIS cable-modem has been proposed for adapting it to the transmissions over satellite.

Let us focus on satellite IP networks. The ETSI TC-SES/BSM working group has defined the protocol stack architecture shown in Figure 1.11 where lower layers depend on satellite system implementation (*Satellite-Dependent*, SD, layers) and higher layers are those typical of the Internet protocol stack (*Satellite-Independent*, SI, layers). These two blocks of stacked protocols are interconnected through the SI-SAP (*Satellite-Independent - Service Access Point*) interface. Only a small number of generic functions need to cross the SI-SAP; in particular: address resolution, resource management, traffic classes QoS.

The SI-SAP interface is logically divided into three SAPs, each of them with a suitable function and security characteristics, as described in the ETSI TS 102 465 standard [35]. In particular, we have:

- SI-U-SAP (*User-SAP*): transfer of IP packets between the users;
- SI-C-SAP (*Control-SAP*): transfer of control data and of service signaling for SI-U-SAP;
- SI-M-SAP (*Management-SAP*): transfer of management information.

The protocol stack organization defined by TC-SES/BSM (see Figure 1.11) has been taken as the basis for the organization of the work in this book, where after a first part with introductory concepts, the second part deals with SD layers and the third part focuses on SI protocol layers. More details on the BSM protocol stack are provided in the following sub-Section.

1.5.1 SI-SAP interface overview

SI-SAP defines an interface between SI upper layers and SD lower layers, that applies to all air interface families for satellite communication systems [32],[34]. SI-SAPs correspond to the endpoints of BSM bearer services. SI-SAP is used to define standard SI bearer services that are built upon lower layer transmission services. Point-to-point, point-to-multipoint, multipoint-to-multipoint and broadcast bearer services are defined as the edge-to-edge services provided by the BSM sub-network. SI-SAP provides an abstract interface allowing BSM protocols (BSM address resolution, BSM resource management, etc.) to perform over any BSM family (i.e., layer 1 and 2 technology) [33]. For traffic handling purposes, SI-SAP uses a *BSM Identifier* (BSM_ID) and *Queuing Identifiers* (QIDs):

- The BSM_ID uniquely identifies a BSM network point of attachment and allows IP layer address resolution protocols (equivalent to *Address Resolution Protocol*, ARP for IPv4 and *Neighbor Discovery*, ND for IPv6) to be used over the BSM.

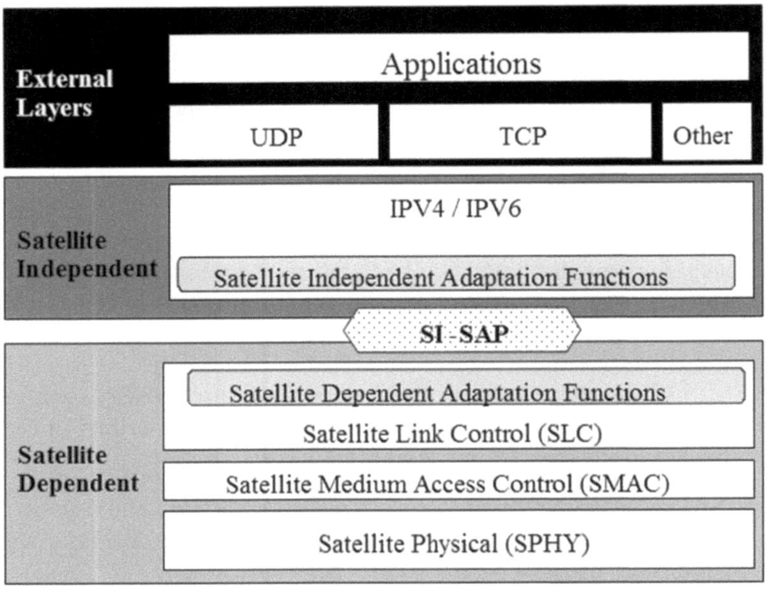

Fig. 1.11: Standardized protocol stack.

- QIDs are *abstract queues* (SI-SAP level) that represent the layer 2 queues in a general way to allow the mapping with layer 3 ones (note that using a QoS support mechanism at layer 3, different queues are needed). QIDs are a way to hide specific SD layer implementations (i.e., BSM technology) from the IP layer. Each QID queue is characterized by QoS-specific parameters (flowspecs, path label or *Differentiated Service*, DiffServ, marking) and is associated to lower layer transfer capabilities (i.e., capacity allocation methods) and buffer management policies [36]. The SD layers are responsible for assigning satellite capacity to these abstract queues (e.g., in DVB-RCS we can consider the allocation methods such as CRA, VBDC, etc., and combinations of them). The mapping of IP queues to QIDs is flexible: there is no strict constraint for a one-to-one mapping, but we may also consider that more IP queues correspond to the same QID (in this case, a *scheduler* should be used at layer 3 to determine the service order of the different queues to be mapped to the same QID). BSM networks use a suitable and general categorization of traffic flows in traffic classes that can be mapped to classical IP QoS classes [25]. In particular, 8 traffic classes, i.e., service priority levels, are defined from 0 for emergency services to 7 for low priority broadcast/multicast traffic.

Other functional blocks are involved in the management of the queues in BSM protocol architecture; the interested reader may refer to [36].

All the BSM services (data transfer, address management, group advertisement, etc.) use SI-SAP *primitives* [37]. These primitives are classified into functional groups within the *User plane* (U), *Control plane* (C) and *Management plane* (M). The primitives (exchanged between the upper layers and the lower layers) are of the following four types:

- The REQUEST primitive type is used when the SI layer is requesting a service from the SD layer.
- The INDICATION primitive type is used by the SD layer to notify the SI layer of activities. This type may either be related to a REQUEST type at the peer entity, or may be an indication of an unsolicited lower layer event.
- The RESPONSE primitive type is used by the SI layer to acknowledge the receipt of the INDICATION type from the SD layer.
- The CONFIRM primitive type is used by the SD layer to confirm that the activity requested by a previous REQUEST primitive has been completed.

The services provided at the SI-SAP level are divided into functional groups for U-, C-, and M-plane, as described below [37]. Each service uses one or more different types of the above-mentioned primitives.

- **U-plane services**
  - *Data Transfer*: These services are used to send and receive user data via the SI-SAP. Data transfer services can be used for both unicast and multicast data transfer.
- **C-plane services**
  - *Address Resolution*: A mechanism to associate a BSM_ID address to a given IPv4 unicast or multicast address. A successful address resolution service returns the associated BSM_ID. The BSM_ID can be either a *Unicast_ID* for unicast services or a *Group_ID* for multicast services.
  - *Resource Reservation*: These services are used to open, modify and close SD layer queues (for both unicast and multicast flows) to be used by SI layers. This function assigns the QID and defines or modifies the properties of the abstract queue that is associated with that QID. Resource reservation is required only for sending data (not for receiving data).
  - *Group Receive/Send*: They are mechanisms to activate and configure the SD layers to receive/send a needed multicast service. These services are used to associate a multicast group address (e.g., an IPv4 Class D address, or an IPv6 multicast address) with a series of SD parameters.
  - *Flow Control*: These primitives allow activating and adjusting the SD layers to provide SI-SAP flow control for a specific QID (i.e., on one or more of the SD layer queues).
- **M-plane services**
  - At present, no M-plane services are defined in the standard.

1.6 Novel approaches for satellite networks

The increasing demand for multimedia broadband services and high-speed Internet access via satellite requires the definition of an optimized satellite protocol stack as well as the full integration of the satellite network with terrestrial ones. Consequently, two innovative approaches are at present conceived [38], namely *horizontal approach* and *vertical approach*.

1.6.1 Horizontal approach

We expect that different wireless technologies (e.g., wireless local area networks, cellular systems, satellite networks) need to co-operate to allow the best radio coverage to the users, depending on their locations, mobility characteristics, applications, user profile, etc. This is in accordance with the *Always Best Connected* (ABC) paradigm. Therefore, it is necessary that the use of the resources in the different *Radio Access Networks* (RANs) be globally coordinated by means of a *resource brokerage function*. Such intelligence is centralized and allocates sessions to RANs or switches them from one to another, when some conditions are meet.

1.6.2 Vertical approach

The ISO/OSI reference model and the Internet protocol suite are based on a layering paradigm. Each protocol solves a specific problem by using the services provided by modules below it and gives a new service to upper layers. The disadvantages of such approach can be detailed as follows:

- The needs of a service provided by the communication system to its users are defined at the top-level, but the hierarchy and the overall system performance are built upon the bottom-level.
- The bottom level does not communicate directly, but through intermediate layers with the top-level. Information is lost during this layer-by-layer top-down conversion.
- Layers are independently optimized. However, in many cases, the close interaction among them should be considered.

A strict modularity and layer independence may lead to non-optimal performance in IP-based next-generation satellite communication systems. Finally, since both radio and power resources are strongly constrained on the satellite, a protocol optimization is mandatory. Such optimization requires a vertical design of the air interface protocol stack. Such cross-layer approach entails new interfaces across the layers, which exchange control information beyond the standard ISO/OSI structure. Cross-layer interfaces can be between or beyond adjacent protocol layers. Although interfaces between adjacent layers are in general preferable, there can be the need for efficient and direct interactions between non-adjacent layers [39]; in general, a layer should be aware of the internal state of the other layers of the protocol stack. For instance, OSI layer 3 (e.g., IP) and above often need direct interfaces to OSI layer 2, e.g., for handover support. Another example concerns transmission parameters (e.g., transmission mode, channel coding and persistency degree for link layer retransmissions) that must be related to application characteristics (e.g., type of information, source coding, etc.), network characteristics, user preferences and context of use. Finally, lower layers (i.e., 1 and 2) should be aware of higher layer (i.e., 3 and 4) behaviors in order to take appropriate decisions on traffic management.

Cross-layer methods can be classified into two broad groups as follows:

- *Implicit cross-layer design*: there is no exchange of signaling among different layers during operation, but in the design phase cross-layer interactions are taken into consideration for a joint optimization.
- *Explicit cross-layer design*: signaling interactions among (non-)adjacent protocol levels are employed so that the internal state of a protocol can be made available to the protocols at different layers for dynamic adaptations.

The above distinction of methods is at the basis of all the different cross-layer schemes presented in the following Chapters of this book.

As for explicit cross-layer methods, new interfaces are needed beyond adjacent layers. Different solutions have been proposed to support the cross-layer exchange of signaling information; an interesting method has emerged from the following papers [40]-[43], where a 'global coordinator' of the different layers is considered allowing to acquire status information from the different protocols to store it in a shared memory and to set the internal state of the protocols to be adaptable to different events. A possible implementation of the global coordinator could be to exploit the capabilities of the management plane of the protocol stack that can interact and coordinate the different layers. The management plane could exploit the control service access points between layers to send control broadcast signaling to all the layers for their respective actions. More details on these aspects are provided in Chapter 4.

Finally, referring to the ETSI TC-SES/BSM protocol stack architecture shown in Figure 1.11, it is important to note here that the cross-layer methods involving SD and SI layers will require the adoption of suitable primitives crossing the SI-SAP interface. In order to explain BSM cross-layer interactions and their relations to SI-SAP, some examples are proposed below for both implicit and explicit cross-layer design cases.

BSM Protocol Manager

The *BSM Protocol Manager* (BPM) has been conceived in the BSM protocol stack to maintain QoS and evaluate the BSM performance [44]. BPM resides above the SI-SAP and defines how IP protocols and packet markings are interpreted and transmitted through the BSM, which SI protocols are used and how they in turn trigger the SD functions (see Figure 1.12). BPM has interfaces at different levels of the BSM protocol stack. In particular, BPM interacts with a specific middleware to establish transport level and application level PEPs, communicates with bandwidth brokers and potentially with service discovery and security/authentication functions. BPM directly interacts with IP protocols, including *Multiprotocol Label Switching* (MPLS) for route discovery and *Integrated Service* (IntServ) or *Differentiated Service* (DiffServ) models (see Chapter 8, Section 8.2, for more details). For all these reasons the BPM could represent a viable solution to implement the so-called 'global coordinator' (explicit cross-layer approach design). In such a case suitable primitives should be designed to support cross-layer signaling through the C-plane.

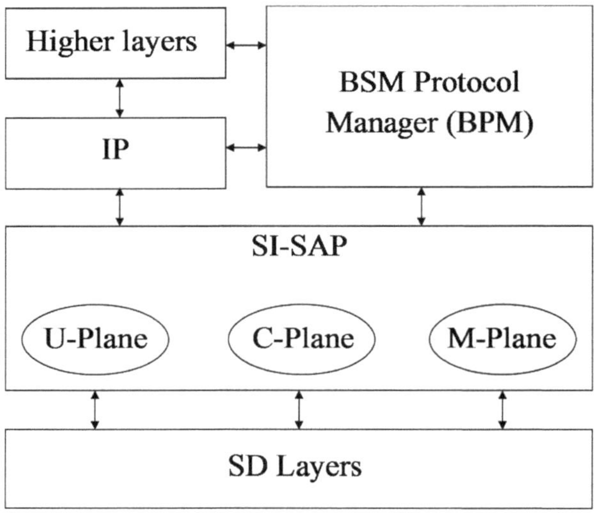

Fig. 1.12: Protocol manager and BSM protocol architecture [44].

Implicit cross-layer design examples

An example of implicit cross-layer (i.e., joint protocol design/optimization) could be PHY layer adaptation in the presence of ACM with thresholds between modes that have been selected to optimize the transport layer performance [39].

Explicit cross-layer design examples

It is possible to distinguish between explicit cross-layer involving layers across SI-SAP or involving layers in SD or SI.

An example of the first method is when cross-layer signaling is provided through the SI-SAP to connect the management of the layer 3 queues with that of layer 2 ones (e.g., information regarding the length of the layer 2 queue, used by layer 3). This mechanism is important to support QoS. In particular, the SI layer requests layer 2 queue status information through a C-plane primitive (namely, a REQUEST primitive) and the SD layer answers by means of another C-plane primitive (namely, a CONFIRM primitive). As already stated, it is possible that BPM manages this information exchange.

Finally, an example of cross-layer information exchange not involving the use of SI-SAP is that between layer 1 and layer 2. Such signalling can be used for the MODCOD switching of DVB-S2. In such case, a C-plane primitive is used to request, to notify or to update information (respectively REQUEST, INDICATION, and RESPONSE primitives).

1.7 Conclusions

Satellite systems are an attractive solution to provide multimedia communication services in wide areas of the Earth, also reaching those regions that lack of terrestrial telecommunication infrastructures. In this framework, this Chapter has provided an introduction to the features of satellites for communications, including: orbit types (GEO, MEO, LEO), atmospheric attenuation phenomena and related packet losses, multiple access schemes and the air interfaces of main interest for this book (i.e., S-UMTS and DVB-S/-S2/-RCS).

In this Chapter, a special attention has been also given to the basic aspects (characteristics, constraints, etc.) related to the management of satellite resources in S-UMTS and DVB-S/-S2/-RCS systems. Such information will be essential in the study of the resource management schemes that will be carried out in the next Chapters, each addressing these techniques from a different prospective. This Chapter has also provided three main scenarios with numerical details that will be adopted for numerical evaluations in this book.

Within the research community a range of issues are currently being investigated that are expected to improve the efficiency and the capacity of satellite communication systems. Towards this aim, this Chapter has introduced the novel cross-layer approach for the optimized design of the satellite air interface; many techniques based on this new paradigm will be described throughout this book.

References

[1] A. Andreadis, G. Giambene. *Protocols for High-Efficiency Wireless Networks.* Kluwer Academic Publishers, Norwell, MA, USA, 2003.

[2] R. E. Sheriff, Y. F. Hu. *Mobile Satellite Communication Network.* Wiley & Sons, Ltd, Baffins Lane, Chichester, England, 2001.

[3] L. Harte, S. Kellogg, R. Dreher, T. Schaffnit. *The Comprehensive Guide to Wireless Technologies: cellular, PCS, paging, SMR, and satellite.* Apdg Publishing, 2000.

[4] B. Elbert. *The Satellite Communication. Ground Segment and Earth Station Handbook.* Artech House, Norwood, MA, USA, 2001.

[5] A. Jamalipour. *Low Earth Orbital Satellites for Personal Communication Network.* Artech House, Norwood, MA, USA, 1998.

[6] G. Maral, M. Bousquet. *Satellite Communications Systems.* 3rd Edition, John Wiley & Sons, Chichester, England, 1998.

[7] S. L. Kota, K. Pahlavan, P. A. Leppänen. *Broadband satellite Communications for Internet Access.* Kluwer Academic Publishers. New York, 1994.

[8] Web site with URL:
http://www.ee.surrey.ac.uk/Personal/L.Wood/constellations/tables/.

[9] Web sites on planned or operational satellite communication systems with URLs:
http://www.spaceandtech.com/spacedata/constellations/
http://www.iridium.com/
http://www.boeing.com/
http://www.comlinks.com/satcom/spacew.htm
http://www.thuraya.com/content/technology.html
http://www.wildblue.cc/aboutwb.htm
http://www.ipstar.com/en/ipstar_space.asp.

[10] P. Barsocchi, N. Celandroni, E. Ferro, F. Davoli, G. Giambene, A. Gotta, F. J. González Castaño, J. I. Moreno, P. Todorova, "Radio Resource Management Across Multiple Protocol Layers in Satellite Networks: a Tutorial Overview", *International Journal of Satellite Communications and Networking*, Vol. 23, No. 5, pp. 265-305, September/October 2005.

[11] U. Vornefeld, C. Walke, B. Walke, "SDMA Techniques for Wireless ATM", *IEEE Communications Magazine*, Vol. 37, No. 11, pp. 52-57, November 1999.

[12] J. Gilderson, J. Cherkaoui, "Onboard Switching for ATM via Satellite", *IEEE Communications Magazine*, Vol. 35, No. 7, pp. 66-70, July 1997.
[13] ETSI, "Digital Video Broadcasting (DVB); Framing Structure, Channel Coding and Modulation for 11/12 GHz Satellite Services", EN 300 421, V1.1.2, (1997).
[14] ETSI, "Digital Video Broadcasting (DVB); Interaction Channel for Satellite Distribution Systems", EN 301 790, V1.3.1 (2002-11).
[15] ETSI, "Digital Video Broadcasting (DVB); Interaction Channel for Satellite Distribution Systems; Guidelines for the use of EN 301 790", TR 101 790, V1.2.1, (2003).
[16] ETSI, "Digital Video Broadcasting (DVB); Second Generation Framing Structure, Channel Coding and Modulation Systems for Broadcasting, Interactive Services, News Gathering and other Broadband Satellite Applications (DVB-S2)", EN 302 307.
[17] ETSI, "Satellite Earth Stations and Systems (SES); Satellite Component of UMTS/IMT2000; G-family; Part 1: Physical Channels and Mapping of Transport Channels into Physical Channels (S-UMTS-A 25.211)", TS 101 851-1.
[18] ETSI, "Satellite Earth Stations and Systems (SES); Satellite Component of UMTS/IMT2000; G-family; Part 2: Multiplexing and Channel Coding (S-UMTS-A 25.212)", TS 101 851-2.
[19] ETSI, "Satellite Earth Stations and Systems (SES); Satellite Component of UMTS/IMT2000; G-family; Part 3: Spreading and Modulation (S-UMTS-A 25.213)", TS 101 851-3.
[20] ETSI, "Satellite Earth Stations and Systems (SES); Satellite Component of UMTS/IMT2000; G-family; Part 4: Physical Layer Procedures (S-UMTS-A 25.214)", TS 101 851-4.
[21] 3GPP, "Technical Specification Group Services and System Aspects, Iu Principles", 3G TR 23.930.
[22] P. Taaghol, B. G. Evans, E. Buracchini, R. De Gaudenzi, G. Gallinaro, J. Ho Lee, C. Gu Kang, "Satellite UMTS/IMT2000 W-CDMA Air Interfaces", *IEEE Communications Magazine*, Vol. 37, No. 9, pp. 116-126, September 1999.
[23] H. Skinnemoen, A. Jahn, J. Kenyon, A. R. Noerpel, "A Comparative Study of DVB-RCS, IPOS and DOCSIS for Satellite", in *Proc. of the* 23rd *AIAA&Ka Band Joint Conference*, Rome, September 25-28, 2005.
[24] M. Marchese, M. Mongelli, "On-Line Bandwidth Control for Quality of Service Mapping over Satellite Independent Service Access Points", *Computer Networks*, Vol. 50, No. 12, pp. 1885-2126, August 2006.
[25] ETSI, "Satellite Earth Stations and Systems (SES); Broadband Satellite Multimedia; Services and Architectures; BSM Traffic Classes", TS 102 295, V1.1.1, February 2004.
[26] The special issue of the International Journal of Satellite Communications and Networking on the DVB-S2 standard for broadband satellite systems, 2004.
[27] D. Breynaert, M. d'Oreye de Lantremange, "Analysis of the Bandwidth Efficiency of DVB-S2 in a Typical Data Distribution Network", in *Proc. of CCBN2005*, Beijing, March 21-23, 2005.
[28] A. Morello, V. Mignone, "DVB-S2 - Ready for Lift off", *EBU Technical Review*, October 2004.
[29] E. Lutz, D. Cygan, M. Dippold, F. Dolainsky, W. Papke, "The Land Mobile Satellite Communication and Recording, Statistics and Channel Model", *IEEE Transactions on Vehicular Technology*, Vol. 40, No. 2, pp. 375-386, May 1991.

[30] C. Blondia, O. Casals, "Performance Analysis of Statistical Multiplexing of VBR Sources", in *Proc. of INFOCOM'92*, pp. 828-838, 1992.

[31] A. H. Aghvami, A. E. Brand, "Multidimensional PRMA with Priorized Bayesan Broadcast", *IEEE Transactions on Vehicular Technology*, Vol. 47, No. 4, pp. 1148-1161, November 1998.

[32] ETSI, "Satellite Earth Stations and Systems (SES); Broadband Satellite Multimedia; Services and Architectures", TR 101 984 V1.1.1 (2002-11).

[33] ETSI, "Satellite Earth Stations and Systems (SES); Broadband Satellite Multimedia (BSM) Services and Architectures; Functional Architecture for IP Interworking with BSM Networks", TS 102 292 V1.1.1 (2004-02).

[34] ETSI, "Satellite Earth Stations and Systems (SES); Broadband Satellite Multimedia; IP over Satellite", TR 101 985 V1.1.2 (2002-11).

[35] ETSI, "Satellite Earth Stations and Systems (SES); Broadband Satellite Multimedia (BSM) Services and Architectures: Security Functional Architecture", TS 102 465 V0.4.2 (2006-01).

[36] ETSI, "Satellite Earth Stations and Systems (SES); Broadband Satellite Multimedia (BSM) Services and Architectures: QoS Functional Architecture", TS 102 462 V1.1.1 (2006-12).

[37] ETSI, "Satellite Earth Stations and Systems (SES); Broadband Satellite Multimedia (BSM). Common air interface specification; Satellite Independent Service Access Point (SI-SAP)", TS 102 357 V1.1.1 (2005-05).

[38] 3GPP, "Technical Specification Group Radio Access Network, Improvement of RRM Across RNS and RNS/BSS", TR 25.881, 2001 (release 5).

[39] G. Giambene, S. L. Kota, "Cross-Layer Protocol Optimization for Satellite Communications Networks: a Survey", *International Journal of Satellite Communications and Networking*, Vol. 24, pp. 323-341, September/October 2006.

[40] Q. Wang, M.-A. Abu-Rgheff, "Cross-Layer Signaling for Next-Generation Wireless Systems", in *Proc. of the IEEE Wireless Communications and Networking Conference* (WCNC), New Orleans, USA, March 16-20, 2003.

[41] M. Conti, J. Crowcroft, G. Maselli, G. Turi, "A Modular Cross-Layer Architecture for Ad Hoc Networks", Chapter 1 in *Handbook on Theoretical and Algorithmic Aspects of Sensor, Ad Hoc Wireless, and Peer-to-Peer Networks*, Jie Wu (editor), CRC Press, New York, 2005.

[42] G. Carneiro, J. Ruela, M. Ricardo, "Cross-Layer Design in 4G Wireless Terminals", *IEEE Wireless Communications Magazine*, Vol. 11, No. 2, pp. 7-13, April 2004.

[43] V. Vardhan, D. G. Sachs, W. Yuan, A. F. Harris, S. V. Adve, D. L. Jones, R. H. Kravets, K. Nahrstedt, "GRACE: A Hierarchical Adaptation Framework for Saving Energy", *Computer Science, University of Illinois Technical Report UIUCDCS-R-2004-2409*, February 2004.

[44] ETSI, "Satellite Earth Stations and Systems (SES); Broadband Satellite Multimedia; IP Interworking over satellite; Performance, Availability and Quality of Service", TR 102 157 V1.1.1 (2003-07).

2

ACTIVITY IN SATELLITE RESOURCE MANAGEMENT

Editor: Erina Ferro[1]

Contributors: Erina Ferro[1], Franco Davoli[2], Petia Todorova[3]

[1]CNR-ISTI - Research Area of Pisa, Italy

[2]CNIT - University of Genoa, Italy

[3]FhI - Fraunhofer Institute - FOKUS, Berlin, Germany

2.1 Introduction

The efficient exploitation of common resources is an important aspect in networking, at all protocol layers. In satellite networking, in particular, there are a number of physical layer issues that have to be addressed in the design of the system:

- Fading
- Delay spread
- Doppler shift
- Limited spectrum
- Path loss and thermal noise.

Given these issues, the goal of *Radio Resource Management* (RRM) is to optimize bandwidth (capacity) utilization and *Quality of Service* (QoS), in the presence of traffic flows generated by services with different requirements. Whenever resources or their modifications are requested, the goal of RRM is to optimize request satisfaction and, at the same time, to try to maintain a certain degree of fairness among all users.

End-user QoS in satellite/terrestrial networks depends on the QoS achieved at each layer of the network, based on satellite-dependent and independent functions to be performed at the layer interfaces. The co-operation of all network layers from top to bottom, as well as of every network element, is fundamental. Each layer should use efficient technologies and counteract any performance degradation factors in order to fulfill the user performance requirements.

As an example of co-operative work, the following actions are considered in order to optimize system performance.

- Bandwidth-efficient modulation and encoding schemes have to be used at the *physical layer*, to improve the *Bit Error Rate* (BER) and the power level performance under poor weather conditions, such as heavy rain.
- Guaranteed bandwidth must be provided at the *data link layer* by using efficient bandwidth-on-demand multiple access schemes and by studying the interaction of mechanisms in the presence of congestion and fading. The provision of a specific bandwidth to be offered by the physical layer to the upper layers implies the existence of a bandwidth allocation scheme that shares the available bandwidth among the different user terminals with different traffic classes.
- The *network layer* is the lowest layer that deals with source-to-destination delivery of connection requests (in circuit-switched networks) or packets (in packet-switched networks); it must know about the topology of the communication subnet and choose the appropriate paths through it. Efficient routing policies must be implemented at this level in order to select paths with the lowest congestion probability. Regarding IP traffic management, user mobility has to be adequately taken into account. Hence, network layer protocols must provide a prioritized management for traffic coming from users that incur in handover phases (such as in the presence of non-GEO satellites). Additionally, mechanisms for IP-layer QoS provision have to be adequately mapped to MAC layer RRM protocols); indeed, besides considering the protocol layering overhead, the service capacity to network layer queues is provided by MAC queues that, generally, are not in one-to-one correspondence to the former ones. This point will be highlighted in some detail in Chapter 8, Section 3.
- At the *transport layer*, TCP connections, which currently constitute the bulk of the traffic transferred in the Internet, tend to occupy all the available bandwidth. The nature of most TCP traffic is asymmetric, with data flowing in one direction and acknowledgments in the opposite direction. This translates into different bandwidth requirements from the sender and the receiver, respectively. Bandwidth assignment and link quality have a strong impact on the TCP goodput.
- At the *application layer*, different traffic types (e.g., real-time traffic and non-real-time traffic) must have specific service level agreements and a

monitoring action has to be performed jointly with the network layer in order to adaptively modify the service priority.

Several strategies for the optimization of resource management have been investigated; resource management schemes are strongly related to the traffic. For example, supporting high bit-rate switched traffic over the radio interface and maintaining the QoS requested by single applications put new requirements on resource management. In addition to the variation in the demands due to the multimedia traffic nature, there are other system variations that have a strong impact on the adopted RRM technique. These include changes in the link quality experienced by each terminal due to the weather conditions, mobility, jamming, and other factors. As a matter of fact, RRM policies, along with network planning and air interface design, determine QoS performance at the network level and the individual user level. The RRM techniques encompass frequency and/or time channels, transmitted power, and access to base stations. The goal is to control the amount of resources assigned to each user to maximize some performance indicators, such as the total network throughput, the total resource utilization and the total network revenue, or to minimize other indicators, such as the end-to-end delay and the real-time transmission jitter, subject to some constrains such as the maximum call dropping rate and/or the minimum signal-to-noise ratio.

The better the RRM technique adopted, the better the performance of the overall system. It is however clear that the overall performance might be improved by considering the co-operation of several protocol layers together, which is commonly called "*cross-layer approach*". In this case, new functions need to be introduced in the protocol stack to enable interactions even between non-adjacent protocol layers. In designing a cross-layer architecture for satellite networks (as in other cross-layer designs), the architectural implications and the principle of *layer separation* [1] should be carefully considered. Relatively few studies have been published to-date on cross-layer optimization in a satellite context (a recent survey can be found in [2]). Cross-layer approaches for the satellite environment are deeply surveyed in Chapter 4 and some numerical results are provided in the following Chapters.

Comprehensive surveys on satellite RRM can be found in [3]-[7]. Reference [8] provides an account on *Call Admission Control* (CAC) in the more general wireless environment. A possible grouping of the RRM techniques in the literature can be attempted in the following three categories:

1. *Frequency/time/space resource allocation schemes* (such as channel allocation, scheduling, transmission and coding rate control, beam and bandwidth allocation);
2. *Power allocation and control schemes*, which control the transmitter power;
3. CAC and *handover algorithms*, which control the access port connection.

An overview of the most recent research activities in the RRM field follows. Of course, the overview cannot be exhaustive, as new material is continuously produced.

2.2 Frequency/time/space resource allocation schemes

Papers [9] and [10] treat the RRM subject from the scheduling perspective.

In [9], the authors propose a transmission scheduling method that deals with the problem of determining *Super-frame Length* (SL), when allocating the return channel resources to the capacity requests from satellite interactive terminals. A main purpose of this method is to minimize the SL in order to reduce scheduling-wait-time as well as to improve resource utilization. This method provides great flexibility in scheduling, by limiting the SL as much as possible, and also achieves high resource utilization, by smoothing the time-varying demands with an overload control.

In [10], the packet-scheduling function has been investigated within the access scheme of a unidirectional satellite system, providing point-to-multipoint services to mobile users. It is interesting how the authors here regard the satellite system as an overlay multicast/broadcast layer, which complements point-to-point 3rd *Generation* (3G) mobile terrestrial networks. The satellite access scheme features maximum commonalties with the *Frequency Division Duplexing* (FDD) air interface of the *Terrestrial Universal Mobile Telecommunications System* (T-UMTS), also known as *Wideband Code Division Multiple Access* (W-CDMA), thus enabling close integration with the terrestrial 3G mobile networks and cost-efficient handset implementations. Attention focuses on one of the radio resource management entities relevant to this interface: the packet scheduler. The lack of channel-state information and the point-to-multipoint service set the difference between the packet scheduler in the satellite radio interface from its counterpart in point-to-point terrestrial mobile networks. The authors formulate the scheduler tasks and describe adaptations of two well-known scheduling disciplines, multilevel priority queuing and weighted fair queuing schemes, as candidates for the time-scheduling function.

Papers from [11] to [15] address the RRM problem from the transmission and rate control point of view.

Reference [11] models the Ka band channel by using a Markov process to capture the impact of the time correlation in weather conditions. A rate adaptation algorithm is developed to optimize the data rate, based on real-time feedback on the measured channel conditions. The algorithm achieves both higher throughput and link availability as compared to a constant rate scheme. In [12], the authors consider a resource allocation problem for a satellite network, where variations of fading conditions are added to those of traffic load. Two novel optimization approaches are addressed. The first one, considered in more detail in [13], is based on the minimization over a discrete constraint set, by using an estimate of the gradient, obtained through a "relaxed continuous extension" of the performance measure. The computation of the gradient estimation relies on the infinitesimal perturbation analysis. The second approach adopts an open-loop feedback control strategy, aimed at providing optimal reallocation strategies as functions of the state of the network. A functional optimization problem is proposed, and a neural network-based technique is used in order to approximate its solution.

In [14] and [15], the authors propose an adaptive global strategy, which handles link congestion and channel conditions in multimedia satellite networks. The overall control system also includes CAC, an aspect mentioned later in this Chapter. However, we include these papers in the present group, in order to emphasize the presence of adaptive coding. In [15], in particular, a performance comparison is presented for a fixed admission control strategy versus the new adaptive CAC scheme for a *Direct Broadcast Satellite* (DBS) network with *Return Channel System* (DBS-RCS). The traffic considered includes both *Available Bit Rate* (ABR) traffic and *Variable Bit Rate* (VBR) traffic. The dynamic channel conditions in the satellite link consider time-varying error rates due to external effects, such as rain. In order to maximize the resource utilization, for both fixed and adaptive approaches, assignments of the VBR services are determined based on the estimated statistical multiplexing gain and other system attributes, namely, video source, data transmission and channel coding rates.

Papers from [16] to [37] deal with the RRM topic from the bandwidth allocation point of view.

In interactive satellite networks, the delay between a request and the reception of the reply is a key issue, due to the basic latency of the satellite link. The solution offered in [16],[17] for GEO satellites comprises a prediction-based resource-allocation policy and a scheduling time period as short as possible. A resource-allocation problem is mathematically formulated as a non-linear integer programming problem, considering uncertain future traffic conditions, and the author develops a real-time heuristic solution algorithm. Computational complexity analysis and extensive simulation results demonstrate the very good performance of the proposed method in terms of computational efficiency and heuristic solution quality.

In [18],[19] the authors propose a scheme for *Dynamic Bandwidth Allocation Capabilities* (DBAC) that is not based on classical circuit-switching, but allows changing the capacity of each connection dynamically without tearing down and setting up the connection. The analysis of the proposed DBAC scheme shows a significant increase in the overall utilization of the capacity, compared to a plain circuit-switching solution.

The work in reference [20] focuses on resource allocation and CAC issues in broadband satellite networks; the authors propose a resource allocation algorithm that integrates three classes of services at the MAC layer: *Constant Bit Rate* (CBR), bursty data, and best effort services. They propose a *Double-Movable Boundary Strategy* (DMBS) in order to establish a resource-sharing policy among these service classes over the satellite uplink channel. DMBS is a dynamically controlled boundary policy, which adapts the allocation decision to variable network loading conditions. CAC and bandwidth allocation decisions are taken at the beginning of each control period, after monitoring the filling level of the traffic request queues. The authors define a threshold level for the bursty data request queue in order to regulate the CAC process. The impact of the queue threshold value on the performance of the DMBS allocation policy is evaluated. A dynamic variation of this metric is also proposed to enhance the system response for interactive applications.

Reference [21] provides an overview of *Broadband Satellite Access* (BSA) systems, with an emphasis on resource management and interworking techniques to support IP-based multimedia services. Some key innovations are described, including *Combined Free/Demand Assignment Multiple Access* (CF/DAMA) for dynamic satellite bandwidth allocation, and an architecture for DiffServ provisioning over BSA systems. A CF/DAMA scheme for dynamic satellite bandwidth allocation is also the subject of the work proposed in [22]; this scheme allows the return channel capacity to be efficiently shared among many user terminals.

In [23], the resource allocation problem that arises in the context of a *Medium Earth Orbit* (MEO) satellite system with half-duplex communication capabilities is addressed. MEO satellite systems are characterized by relatively large propagation delays and intra-beam delay variations, which result in resource consumption. The authors propose a channel classification scheme, in which the available carriers are partitioned into classes and each class is associated with a range of satellite propagation delays.

References [24] and [25] deal with the problem of QoS provisioning for packet traffic. In [24], the authors address the problem by considering a resource allocation scheme that takes advantage of proper statistical traffic modeling to predict future bandwidth requests. This approach takes into consideration DiffServ-based traffic management to guarantee QoS priority among different users. Moreover, the satellite onboard switching problem is also addressed by considering a suitable implementation of the DiffServ policy based on a cellular neural network.

In [25], the problem of providing guaranteed QoS connections over a *Multi Frequency - Time Division Multiple Access* (MF-TDMA) system that employs *Differential Phase Shift Keying* (DPSK) is studied. The problem is divided into two aspects: resource calculation and resource allocation. The authors present algorithms for performing these two tasks and evaluate their performance in the case of a Milstar *Extremely High Frequency - Satellite Communication* (EHF-SATCOM) system.

References [26] and [27] present an algorithm for resource allocation in satellite networks to obtain time/frequency plans for a set of terminals with a known geometric configuration under interference constraints. The goal is to maximize the system throughput while guaranteeing that the different types of demands are satisfied, each type using a different amount of bandwidth. The proposed algorithm relies on two main techniques. The first generates admissible configurations for the interference constraints, whereas the second uses linear and integer programming with column generation.

In [28], the authors consider the problem of how a *Geostationary Earth Orbit* (GEO) satellite should assign bandwidth to several service providers (operators) so as to meet some minimum requirements on one hand, and to perform the allocation in a fair way on the other. They provide a computational method to optimize allocation fairness in polynomial time, taking practical issues into account.

References [29] and [30] consider the problem of allocating the uplink bandwidth of a satellite transponder among hierarchies of Earth stations, for guaranteed bandwidth and best-effort traffic types. CAC actions are taken locally at the Earth stations within the allocated bandwidth partition, which is recomputed either periodically or upon request, by considering dynamic variations in traffic and channel parameters (with a cross-layer interaction between physical and MAC layers).

The work in [31],[32] proposes a new DiffServ-based scheme of bandwidth allocation during congestion, termed *Proportional Allocation of Bandwidth* (PAB). This method can be used in satellite networks based on GEO, MEO, and LEO (*Low Earth Orbit*) constellations, in order to transport IP traffic and to provide QoS. In PAB, during congestion, all flows get a share of IP available bandwidth, proportional to their subscribed information rate.

Reference [33] considers an architecture to interconnect remotely located heterogeneous terrestrial distribution nodes in a mesh topology, by means of an onboard regenerative satellite. An emulated DVB-S (*Digital Video Broadcasting via Satellite*) regenerative environment is created, by using an actual transparent GEO satellite. Furthermore, a dynamic bandwidth mechanism is proposed, to be applied directly on the DVB-S stream of the uplink of each distribution node. This mechanism enables the provision of interactive IP-based multimedia services, at a guaranteed QoS.

The work in [34] focuses on dynamic resource allocation algorithms for sharing the limited uplink resources of a future satellite system among many bursty users with varying QoS requirements. The data rates provided to each terminal are selected to differentiate multiple QoS priority levels, to provide fairness and to maximize system capacity under time-varying channel conditions and traffic loads.

In [35], *Weighted Fair Bandwidth-on-Demand* (WFBoD) technique is defined and analyzed. It is a resource management process for broadband multimedia GEO satellite systems that provides fair and efficient resource allocation, coupled with a well-defined MAC-level QoS framework (compatible with ATM and IP QoS frameworks) and a multi-level service segregation into a large number of users with diverse characteristics. WFBoD is also integrated with the CAC process. Simulation results show that WFBoD can guarantee QoS for both non-real-time and real-time VBR flows.

A consolidated approach for *Voice over IP* (VoIP) over satellite networks based on the ETSI DVB-RCS standard is adopted in [36]. This paper addresses the role of *Bandwidth on Demand* (BoD) in the optimization of VoIP bandwidth allocation, and assesses the impact of BoD mechanisms on voice quality. The tradeoff between voice quality and bandwidth efficiency is investigated under different DVB-RCS-specific capacity request/allocation strategies; it is demonstrated that DVB-RCS provides an efficient platform for the integrated support of a variety of satellite VoIP applications.

Reference [37] compares BoD in an MF-TDMA environment and *Single Carrier Per Channel* (SCPC) from a practical perspective and evaluates the economical advantages of BoD.

2.3 Power allocation and control schemes

Normally, the literature considers three types of uplink power control techniques [5]:

- *Open loop.* One station receives its own transmission carrier (relayed by the satellite) and relies on its measurement of beacon fading in the downlink, in order to perform uplink power control.
- *Closed loop.* Two Earth stations lie within the same beam coverage and an Earth station can receive its own transmission carrier. Uplink power control based on this carrier is erroneous due to changes in input and output backoffs under uplink and downlink fading. It must be based on the reception of a distinct carrier transmitted by another station.
- *Feedback loop.* A central control station monitors the levels of all carriers it receives, and commands the affected Earth stations to adjust their uplink powers accordingly. This technique has inherent control delays, and requires more Earth segment and space segment resources.

Regarding downlink, power control allocates additional power to the transmission carrier(s) at the satellite in order to compensate for rain attenuation. As downlink fading occurs, downlink carrier power degrades and sky noise temperature seen by the Earth station increases. Power control correction is required to maintain carrier to noise ratio.

Papers from [38] to [41] treat RRM from the power allocation and control scheme perspective.

In [38], the authors consider the problem of using narrow transmission spot-beams on the satellite to support a broad spectrum of users with small terminals at high rates. Since satellite transmitter resources are expensive and there can be many spot-beam coverage cells within the satellite service area, it is attractive to look for some form of agile-scanning beam system and to time-share these precious resources. An optimized design of both the multi-beam antenna pattern and the scheduling can further improve the efficiency of transmission and power management. The advantage of parallel multi-beams in terms of spectral efficiency and power gain is shown, and the issue of multi-beam power allocation based on traffic demands and channel conditions over satellite downlinks with power and delay constraints is addressed. The study indicates that the use of a parallel multi-beam scheme with optimum power allocation can achieve a substantial power gain and a reasonable proportional fairness. By coupling power allocation with multi-beam scheduling when there are less active beams than cells, the authors show that a modest number of active parallel beams suffices to cover many cells efficiently.

In [39], the author analyzes a power-sharing multiple-beam mobile satellite system in the Ka band with high traffic variations from one beam to another. In order to cope with the multiple-beam varying traffic problem, the author proposes an offset reflector antenna, fed through an equal phase-shift active array. This active array consists of hundreds to thousands of equal phase-shift elements.

A power allocation policy is developed in [40] for a multi-beam satellite downlink, which transmits data to different ground locations over time-varying channels. The packets destined to each ground location are stored in separate queues and the server rate for each queue depends on the power allocated to that server and the channel state, according to a concave rate-power curve.

A method for satellite network configuration is proposed in [41]. It controls the transmitted power of multiple Earth stations, and establishes the received power-differences among them to generate the capture effect.

2.4 CAC and handover algorithms

This topic is widely treated in Chapter 6. This Chapter only aims at providing an overview.

Arriving calls are granted/denied access to the network by the CAC scheme based on predefined criteria, taking into consideration network loading conditions. The traffic of admitted calls is then controlled by other RRM techniques, such as scheduling, handover, power, and rate control schemes. CAC is extensively studied as an essential tool for congestion control and QoS provisioning. In terrestrial wireless networks, CAC is more sophisticate than in cabled networks, due to unique features of wireless networks such as multiple access channel interference, channel impairments, handover requirements and limited bandwidth. As in terrestrial wireless networks [8], in satellite networks there are several reasons for using CAC schemes, including:

- *To control the handover failure probability in LEO constellations.* Blocking a new call is surely better than dropping an in-progress call; regardless of the CAC procedure used, the criterion is maintaining active calls in progress and blocking new calls that might lead to an increase of the call dropping probability.
- *To limit the network traffic level to guarantee packet-level QoS parameters (packet delay, delay jitter and throughput).* Some CAC procedures can estimate packet delay and delay jitter from available resources in multiple-class networks (see [8] and references [130],[131] therein).
- *To ensure a minimum transmission rate.* This can be achieved either by limiting network load (see [8] and references [7],[67],[132] therein), by minimizing the transmission rate degradation (i.e., the condition where the transmission rate is below a minimum value) (see [8] and references [128],[133] therein) or by estimating the allocated transmission rate as an admission criterion (see [8] and reference [101] therein).

CAC schemes can be classified according to various design options [8] (centralization, information scale, service dimension, optimization, decision time, information type, information granularity, considered link).

A number of policies have been derived for resource sharing in CAC, first for cabled networks, and then for wireless networks in general. The simplest CAC rule is *Complete Sharing* (CS), i.e., connections are simply admitted if sufficient resources are available at the time of the request, without considering the importance of a connection when they are allocated. In the CS policy, the only system constraint is the overall capacity C. In the presence of multiple services, this policy may suffer from some problems such as unfairness, in the sense that it can monopolize the resources, it may lead to poor resource utilization and, finally, it may yield poor long-run average revenue. As an almost opposite situation, in the *Complete Partitioning* (CP) type of policies, every traffic class is allocated a set of resources that can be used only by that class. Other policies have been derived to provide optimized access to resources, and Ross [42] provides an extensive discussion about a number of different solutions. Actually, optimal approaches should be based on Markov decision processes, given a certain cost function to be minimized (or maximized) as a performance index; however, they must take into detailed account any allowable network state and state transition, which is impractical even for networks of modest complexity. The functional form of the optimal policies is usually unknown. Therefore, a set of generally sub-optimal policies with fixed structure (which can be often described by a set of parameters) has been developed. They are simpler to implement and, in some special cases, do correspond to the optimal one; among others, we can cite the above mentioned CP, *Trunk Reservation* (TR) [43], *Guaranteed Minimum* (GM) [44], and *Upper Limit* (UL) policies [44],[45]. Comparisons have been made between these policies and the optimal one. The results indicate that CP, TR, GM, and UL policies outperform the CS one when significant differences among classes exist in requirements for bandwidth and offered load [46]. Obviously, once one of such fixed-structure policies has been selected, parametric optimization can be adopted in order to choose the "best" values of parameters that minimize a given cost function (or maximize a performance index).

As already mentioned, reference [15], besides considering adaptive coding, also treats the RRM problem from the CAC point of view. This is also done in [20] and [29],[30], among others. Reference [8] provides an account on CAC in the more general wireless environment.

In [47], the authors combine CAC with the issue of optimal energy allocation for communication satellites. The objective is to choose the requests for transmission to serve so that the expected total reward is maximized. The special case of a single energy-constrained satellite is considered. Rewards and demands from users for transmission (energy) are random and known only at request time. Using a dynamic programming approach, an optimal policy is derived that is characterized in terms of thresholds. Furthermore, in the special case where energy demand is unlimited, an optimal policy is obtained in closed form.

In [48], a real-time traffic handling strategy, including distributed CAC and traffic resource management schemes, is harmonized with an in-band signaling technique for burst-based bandwidth requests and an effective policy for the allocation of radio resources.

2.4.1 Handover algorithms

In wireless mobile networks, many users share radio bandwidth. An important property of the network is that user devices change their access points several times. As their coverage area changes continuously, in order to maintain connectivity, end-users must switch between spot-beams and satellites, and, thus frequent intra- and inter-satellite handover attempts occur. This fact causes technical problems, requiring fair sharing of bandwidth between handover connections and new connections. One of the main problems to be solved in RRM is the handover management strategy in order to provide low call dropping probability and to keep high resource utilization.

Several approaches for handover prioritization proposed for terrestrial cellular systems have been studied in the recent literature for mobile satellite systems. The solutions include the guard channel scheme, a handover queuing where the highest priority is offered to handover calls, which are organized in a separate queue, and novel CAC algorithms, taking into account handover calls.

In [49], user location information is exploited for adaptive bandwidth reservation for handover calls. In a beam, bandwidth reservation for handover is allocated adaptively, by calculating the possible handovers from neighboring beams. A new call request is accepted if its originated beam has sufficient amount of available bandwidth for new calls.

The key idea of the algorithm in [50] is that bandwidth has to be reserved in a particular number of beams S the call may handover into, in order to prevent handover dropping during a call. The balance between new call blocking and handover call blocking depends on the selection of predetermined threshold parameters for new and handover calls.

In [51], a probabilistic resource reservation strategy for real-time services was proposed, based on the concept of sliding windows to predict the necessary amount of reserved bandwidth for a new call in future handover beams.

In [52], CAC and handover are based on user location. The system traces all user locations in each beam and updates user handover-blocking parameters.

Reference [53] proposes an intra-satellite handover management scheme for LEO satellites, called Q-WIN, specifically tailored to the QoS needs of multimedia applications. This scheme is based on priority queues, combined with the sliding virtual window concept for call admission. Simulation results confirm that, compared to the allocation schemes, Q-WIN offers low *Call Dropping Probability* (CDP), thus providing for reliable handover of calls in progress, acceptable *Call Blocking Probability* (CBP) for new calls and high resource utilization.

In [54], a guaranteed handover scheme is proposed. According to this method a new call is admitted in the network only if there is an available channel in the current cell and, simultaneously, in the first transit cell. When the first handover occurs, a channel-reservation request is issued to the next candidate transit cell, and so on. If all channels are busy, the request is queued in a FIFO list, until the next handover occurrence. The call is not forced to terminate provided that an available channel has been reserved in the meanwhile.

In [55], different queuing policies for handover requests are proposed. The handover requests, queued up to a maximum time interval (which is a function of the overlapping area of contiguous cells), are served according to a FIFO or a *Last Useful Instant* (LUI) scheme (that is, a handover request is queued ahead of any other requests already in the queue that have a longer residual lifetime).

In [56], a novel inter-satellite handover management scheme tailored for multimedia LEO satellite systems is proposed and evaluated. This scheme relies on queuing handover requests of different service classes in separate queues. The queue that stores handover requests of real-time services receives higher priority.

2.5 RRM modeling and simulation

There is a wealth of work on RRM modeling and simulation. References from [57] to [60] are just a few examples.

In [57], the authors describe the modeling and simulation of an FDMA (*Frequency Division Multiple Access*) satellite BoD service. This class of resource allocation processes, which includes BoD applications, is identified and compared with common resource allocation processes. Within this class, the bidirectional and possibly asymmetric nature of resource requests, the existence of both booked (advance notification) and immediate resource requests, the allowance of modifications to resource requests and the multiple resource constraints (e.g., bandwidth and power) present unique modeling challenges. In particular, we can consider three fundamental components: modeling the resource requests, modeling the fundamental resource allocation algorithm and modeling the processing of individual resource requests.

In [58], the authors focus on modeling and evaluating the bandwidth requirements of the next-generation of satellite communication technologies, which will support future aeronautical applications. The authors' interest is on the real-time delivery of high-resolution weather maps to the cockpit as a particularly demanding future application. In such scenario, the use of LEO and GEO satellite networks for efficient data delivery is investigated. The authors propose a joint uni-cast and broadcast communication technique that offers bandwidth reduction.

In [59], a new analytical model for equal allocation of divisible computation and communication load is developed. Equal load allocation is attractive in multiple processor systems when real-time information on processor and link capacity, which is necessary for optimal scheduling, is not available. This model includes a detailed accounting of solution reporting time.

Reference [60] presents a generalized notation as well as graph algorithms for resource management problems. Impairment graphs can be used for frequency planning, whereas flow graphs are suitable for channel access problems. To evaluate the performance of the resource management, service criteria (such as blocking or *Carrier-to-Interference ratio*, C/I) or efficiency criteria (bandwidth requirements) are derived from the graphs. The resource management techniques are applied to satellite networks with non-GEO orbits that entail time-varying network topologies. As a simple example, the channel assignment and capacity optimization of the EuroSkyWay system are shown. For a deeper inspection, a comparison of *Fixed, Dynamic and Hybrid Channel Allocation* schemes (FCA, DCA, HCA) for a typical MEO satellite scenario is provided. The author also investigates satellite diversity and its impact on bandwidth requirement and transmission quality.

2.6 Related projects in Europe

A number of satellite-related projects have been funded by the European Commission in both the *Fifth* and the *Sixth Framework Programmes* (FP5, FP6), as well as in COST Actions. In sub-Sections 2.6.1-2.6.4, we limit our overview to a few FP6 projects. Additional information can be found in *http://cordis.europe.eu.int/en/home.html*. Finally, sub-Section 2.6.5 mentions a recent COST Action and sub-Section 2.6.6 describes a new initiative in the satellite field for the FP7 EU programme.

2.6.1 TWISTER: Terrestrial Wireless Infrastructure integrated with Satellite Telecommunications for E-Rural applications

http://www.twister-project.net/

TWISTER is a project led by EADS Astrium and was selected for co-funding by the European Commission in the 1st call for proposals of the Aeronautics and Space priority of FP6.

This project started on February 1, 2004, and will operate validation sites throughout Europe for 3 years, through the deployment of up to 105 satellite access points in combination with radio networks. These validation sites support innovative applications to meet the specific needs of rural user communities in the domains of agriculture, education, community services, healthcare and e-business. This project emphasizes usages that benefit from broadband access. The objective of TWISTER is to support the development and widespread adoption of satellite communication services (like educational and health care services between islands, or e-business) to deliver broadband services to rural areas. User satisfaction is evaluated to propose improvements and to specify a roadmap for further services deployment. The integration of space-based infrastructure with terrestrial systems aims at achieving a seamless broadband coverage in rural areas. TWISTER investigates a number of hybrid satellite-wireless architectures and validates their on-site performance. The TWISTER consortium, involving many actors in the telecom value chain (user communities, service providers, satellite operators, equipment manufacturers) creates the necessary conditions to deploy successfully such satellite solutions over Europe as a complement to terrestrial networks for the benefit of the population and the economy.

2.6.2 MAESTRO: Mobile Applications & sErvices based on Satellite & Terrestrial inteRwOrking

http://ist-maestro.dyndns.org/MAESTRO

The MAESTRO project aims at studying technical implementations of innovative mobile satellite systems, targeting close integration and interworking with 3G and beyond-3G mobile terrestrial networks. MAESTRO seeks to specify and to validate the most critical services, features, and functions of satellite system architectures, achieving the highest possible degree of integration with terrestrial infrastructures. It aims not only at assessing the satellite system technical and economical feasibility, but also at highlighting their competitive assets on the way they complement terrestrial solutions.

In the frame of the MAESTRO project, innovative and convergent solutions pursue: (*i*) the successful and cost effective deployment of 3G multimedia services over mobile satellite networks; (*ii*) the reduction of the digital divide between urban and rural areas and regions by ensuring service continuity over heterogeneous GPRS/UMTS networks.

2.6.3 SatNEx: Satellite Network of Excellence

http://www.satnex.org

SatNEx is an FP6 research *Network of Excellence* (NoE), funded by the European Commission, which combines the research excellence of 22 major players in the field of satellite communications [61]-[63]. The primary goal of SatNEx is to achieve a long-lasting integration of European research in satellite communications, and to develop a common knowledge base. This collected expertise will support the European satellite industry through standardization, collaboration/consultancy and training. Through co-operation of outstanding universities and research organizations with excellent expertise in satellite communications, SatNEx is building a European virtual center of excellence in satellite communications and will contribute to the realization of the *European Research Area* (ERA). A dedicated satellite platform links partners in a broadcast, multicast or unicast configuration, providing training and video-conferencing capabilities, and promoting the simplicity and cost-effectiveness of using satellites for this purpose. SatNEx has established an advisory board incorporating key representatives of the European space industry, satellite service providers, and standardization and regulation organizations. SatNEx is steered by these players in providing a critical mass of resources and expertise, to make Europe a world force in the field of satellite communications. Part of the SatNEx mission is to disseminate internal research and expertise.

2.6.4 NEWCOM: Network of Excellence in Wireless COMmunications

https://newcom.ismb.it/public/index.jsp

NEWCOM is a European NoE that links in a cooperative way many leading research groups addressing the strategic objective "Mobile and wireless systems beyond 3G", a frontier research area of the priority thematic area of IST. This network involves 54 partners from 18 countries, comprising 40 universities and 14 companies. The major objective is a 'distributed European university' with common research projects and, in the longer term, a shared doctoral school. This NoE is devoted to the terrestrial wireless environment. However, some of the research topics, such as cross-layer optimization and reconfigurable radio, share common aspects with the satellite world.

2.6.5 VIRTUOUS: Virtual Home UMTS on Satellite

http://www.ebanet.it/virtuous.htm

The VIRTUOUS project [64], ended in 2002, aimed at identifying, designing and demonstrating a feasible, pragmatic, smooth migration path towards *Terrestrial and Satellite UMTS* (T-UMTS and S-UMTS). VIRTUOUS pursued the achievement of the following specific objectives:

- Design, development and implementation of both a URAN (*UMTS Radio Access Network*) *Radio Technology Independent* part and two URAN *Radio Technology Dependent* parts, able to handle a terrestrial and a satellite link, respectively;
- Development of two hardware test beds, representative of satellite and terrestrial UMTS physical layers, respectively;
- Definition of the S-UMTS components;
- Design, development and implementation of appropriate terminal and network *Inter-Working Units* (IWUs), aiming at integrating the GPRS and the UMTS segments;
- Implementation, integration and testing of a demonstrator including three segments: GPRS, terrestrial UMTS and satellite UMTS;
- Trials of meaningful UMTS services, with voice over IP as a candidate application.

2.6.6 COST Actions

European *Co-operation in the field of Scientific and Technical Research* (COST) is an intergovernmental framework for the co-ordination of nationally-funded research at European level, based on a flexible institutional structure. Established in 1971, COST has developed into one of the largest frameworks for research co-operation. The 34 member countries of COST include the 25 EU member states, Bulgaria, Croatia, Iceland, Norway, Romania, Serbia and Montenegro, FYR of Macedonia, Switzerland and Turkey. Moreover, Israel is a co-operating state. COST also welcomes Institutions from non-COST countries to join individual actions for mutual benefit. COST networks are called Actions. Co-operation takes the form of concerted activities, i.e., the co-ordination of nationally funded research activities. Some of the early COST actions have helped to pave the way for other European research programs, such as the EU Framework Programs (launched in 1983) and the EUREKA initiative (started in 1985; see *http://www.eureka.be*). COST plays an important role in scientific and technical co-operation in Europe, encouraging European synergy and networking and helping further European integration.

COST covers a wide range of scientific and technological areas: agriculture, biotechnology and food sciences, chemistry, environment, forests and forestry products, materials, medicine and health, meteorology, physics, social sciences and humanities, *Telecommunications, Information Science and Technology* (TIST), transport and urban civil engineering.

For more information, the reader may visit the Web site *http://www.cost.esf.org/index.php.*

COST Action 272: "Packet-Oriented Service Delivery via Satellite"

http://www.tesa.prd.fr/cost272/

This COST Action ended in the first half of 2005 and was entirely devoted to study aspects related to packet transmission via satellite. The main objectives of COST Action 272 were the identification of key requirements, analysis, performance comparison, architectural design and protocol specification of packet-oriented satellite communication systems, with a clear focus on Internet-type system concepts, applications and protocols/techniques across the various layers. This Action firstly assessed the interesting satellite-specific market segments and came up with a clearly focused set of reference scenarios (global/regional, GEO/non-GEO, broadcast/multicast/interactive, QoS/best-effort, all-IP/hybrid, etc.) as a basis for further research and development work, also providing some interesting technical solutions. COST Action 272 was the continuation of COST Action 253 ("Service Efficient Network Interconnection via Satellite") [65] and the starting point for the SatNEx Consortium, which elaborated the SatNEx NoE proposal.

2.6.7 The ISI Initiative

http://www.isi-initiative.eu.org/

The *Integral Satcom Initiative* (ISI) is an open platform, started in 2005, whose membership embraces all relevant and interested private and public stakeholders. ISI collaborates and cooperates with the European Commission, the *European Space Agency* (ESA), the EU and ESA Member States and Associated States, the National Space Agencies, International Organizations, user Fora, and other European technology platforms. ISI fosters international cooperation under a global perspective. The ISI technology platform brings together for the first time in a unified, industry-led forum all research and technology aspects related to satellite communications, including mobile, broadband, and broadcasting applications. The purpose is to foster and develop the entire industrial sector, to maximize the value of European research and technology development, and to contribute to EU and ESA policies. The document in [66] specifies the *Strategic Research Agenda* of the ISI technology platform. It addresses the overall development of satellite communications and satellite broadcasting in Europe till about year 2020. In doing so, it shows that satellite communications and broadcasting has strategic relevance for Europe, and identifies medium and long term strategic objectives. Key research themes of ISI are cited in [66]; among them, RRM research topics are addressed in various points of the ISI research vision. In particular: (*i*) cross-layer design of RRM techniques, with cross-layer information coming from adaptive physical layer and QoS requirements from upper layers, to achieve optimum performance of mobile broadband multimedia satellite services, is one of the key research items; (*ii*) advanced RRM techniques can provide optimum use of the scarce spectrum resource and contribute to lowering the level of electromagnetic radiation in the hybrid terrestrial/satellite network environment; (*iii*) novel RRM protocols are considered, which include *Medium Access Control* (MAC) and *Usage Parameter Control* (UPC) mechanisms for the QoS provision under fairness constraints.

2.7 Conclusions

The goal of RRM is to optimize capacity utilization and QoS in satellite links, in the presence of traffic flows generated by services with different requirements. The best results are obtained with the cooperation of the protocols operating at different architectural layers, i.e., through a cross-layer approach, while maintaining the principle of layer separation. A possible grouping of the RRM techniques in the literature can be: frequency/time/space resource allocation schemes, power allocation and control schemes, and call admission control and handover algorithms. For each of these groups, this Chapter reviews the current results in the literature, even if the survey is far from being exhaustive.

Some ongoing research projects in Europe that consider the RRM problem are cited, and the reader is encouraged to visit their Web sites for further information. Among these projects, the SatNEx Network of Excellence deserves special attention. It combines the research activities of 22 European institutions, with proved excellence in satellite communications. The realization of this book has been made possible due to the SatNEx support.

References

[1] V. Kawadia, P. R. Kumar, "A Cautionary Perspective on Cross-Layer Design", *IEEE Wireless Communications Magazine*, Vol. 12, No. 1, pp. 3-11, January 2005.

[2] N. Celandroni, F. Davoli, E. Ferro, A. Gotta, "Networking with Multi-Service GEO Satellites: Cross-Layer Approaches for Bandwidth Allocation", *International Journal of Satellite Communications and Networking*, Special Issue on *Cross Layer Protocols for Satellite Communication Networks: Part I* (S. Kota and G. Giambene, Eds.), Vol. 24, No. 5, pp. 387-403, September/October 2006.

[3] M. Ibnkahla, Q. M. Rahman, A. I. Sulyman, H. A. Al-Asady, Y. Jun, A. Safwat, "High-Speed Satellite Mobile Communications: Technologies and Challenges", in *Proc. of IEEE*, Vol. 92, No. 2, pp. 312-339, February 2004.

[4] E. Modiano, "Satellite Data Networks", *AIAA Journal on Aerospace Computing, Information and Communication*, Vol. 1, pp. 395-398, October 2004.

[5] S. Kota, M. Marchese, "Quality of Service for Satellite IP Networks: a Survey", *International Journal of Satellite Communications and Networking*, Vol. 21, No. 4-5, pp. 303-349, July-October 2003.

[6] P. Barsocchi, N. Celandroni, E. Ferro, A. Gotta, F. Davoli, G. Giambene, F. J. González-Castaño, J. I. Moreno, P. Todorova, "Radio Resource Management Across Multiple Protocol Layers in Satellite Networks: a Tutorial Overview", *International Journal of Satellite Communications and Networking*, Vol. 23, No. 5, pp. 265-305, September/October 2005.

[7] D. Boudreau, G. Caire, G. E. Corazza, R. De Gaudenzi, G. Gallinaro, M. Luglio, R. Lyons, J. Romero-Garcia, A. Vernucci, H. Widmer, "Wide-Band CDMA for the UMTS/IMT-2000 Satellite Component", *IEEE Transactions on Vehicular Technology*, Vol. 51, No. 2, pp. 306-331, March 2002.

[8] M. H. Ahmed, "Call Admission Control in Wireless Networks: a Comprehensive Survey", *IEEE Communications Surveys and Tutorials*, Vol. 7, No. 1, pp. 50-69, 1st Quarter 2005.

[9] D. Kim, D.-H. Park, K.-D. Lee, H.-J. Lee, "Minimum Length Transmission Scheduling of Return Channels for Multicode MF-TDMA Satellite Interactive Terminals", *IEEE Transactions on Vehicular Technology*, Vol. 54, No. 5, pp. 1854-1862, September 2005.

[10] M. Karaliopoulos, P. Henrio, K. Narenthiran, E. Angelou, B. G. Evans, "Packet Scheduling for the Delivery of Multicast and Broadcast Services over S-UMTS",

International Journal of Satellite Communications and Networking, Vol. 22, No. 5, pp. 503-532, September/October 2004.

[11] J. Sun, J. Gao, S. Shambayatti, E. Modiano, "Ka-Band Link Optimization with Rate Adaptation", *IEEE Aerospace Conf.*, Big Sky, MT, March 2006.

[12] M. Baglietto, F. Davoli, M. Marchese, M. Mongelli, "Neural Approximation of Open-Loop Feedback Rate Control in Satellite Networks", *IEEE Transactions on Neural Networks*, Vol. 16, No. 5, pp. 1195-1211, September 2005.

[13] F. Davoli, M. Marchese, M. Mongelli, "Optimal Resource Allocation in Satellite Networks: Certainty Equivalent Approach versus Sensitivity Estimation Algorithms", *International Journal of Communication Systems*, Vol. 18, No. 1, pp. 3-36, February 2005.

[14] F. Alagöz, D. Walters, A. AlRustamani, B. Vojcic, R. Pickholtz, "Adaptive Rate Control and QoS Provisioning in Direct Broadcast Satellite Networks", *Wireless Networks*, Vol. 7, No. 3, pp. 269-281, May 2001.

[15] F. Alagöz, B. R. Vojcic, D. Walters, A. AlRustamani, R. L. Pickholtz, "Fixed versus Adaptive Admission Control in Direct Broadcast Satellite Networks with Return Channel Systems", *IEEE Journal on Selected Areas in Communications*, Vol. 22, No. 2, pp. 238-249, February 2004.

[16] K.-D. Lee, "An Efficient Real-Time Method for Improving Intrinsic Delay of Capacity Allocation in Interactive GEO Satellite Networks", *IEEE Transactions on Vehicular Technology*, Vol. 53, No. 2, pp. 538-546, March 2004.

[17] K.-D. Lee, "Correction to 'An Efficient Real-Time Method for Improving Intrinsic Delay of Capacity Allocation in Interactive GEO Satellite Networks'", *IEEE Transactions on Vehicular Technology*, Vol. 53, No. 6, pp. 1948-1948, November 2004.

[18] N. Blefari-Melazzi, G. Reali, "A Resource Management Scheme for Satellite Networks", *IEEE Multimedia*, Vol. 6, No. 4, pp. 54-63, October-December 1999.

[19] N. Blefari-Melazzi, G. Reali, "Improving the Efficiency of Circuit-Switched Satellite Networks by Means of Dynamic Bandwidth Allocation Capabilities", *IEEE Journal on Selected Areas in Communications*, Vol. 18, No. 11, pp. 2373-2384, November 2000.

[20] H. Koraitim, S. Tohme, "Resource Allocation and Connection Admission Control in Satellite Networks", *IEEE Journal on Selected Areas in Communications*, Vol. 17, No. 2, pp. 360-372, February 1999.

[21] T. Le-Ngoc, V. Leung, P. Takats, P. Garland, "Interactive Multimedia Satellite Access Communications", *IEEE Communications Magazine*, Vol. 41, No. 7, pp. 78-85, July 2003.

[22] N. Iuoras, T. Le-Ngoc, "Dynamic Capacity Allocation for Quality-of-Service Support in IP-Based Satellite Networks", *IEEE Wireless Communications Magazine*, Vol. 12, No. 5, pp. 14-20, October 2005.

[23] I. Koutsopoulos, L. Tassiulas, "Efficient Resource Utilization Through Carrier Grouping for Half-Duplex Communication in GSM-Based MEO Mobile Satellite Networks", *IEEE Transactions on Wireless Communications*, Vol. 1, No. 2, pp. 342-352, April 2002.

[24] F. Chiti, R. Fantacci, D. Tarchi, S. Kota, T. Pecorella, "QoS Provisioning in GEO Satellite with Onboard Processing Using Predictor Algorithms", *IEEE Wireless Communications Magazine*, Vol. 12, No. 5, pp. 21-27, October 2005.

[25] J.-M. Park, U. Savagaonkar, E. K. P. Chong, H. J. Siegel, S. D. Jones, "Allocation of QoS Connections in MF-TDMA Satellite Systems: a Two-Phase

Approach", *IEEE Transactions on Vehicular Technology*, Vol. 54, No. 1, pp. 177-190, January 2005.

[26] S. Alouf, E. Altman, J. Galtier, J.-F. Lalande, C. Touati, "Quasi-Optimal Resource Allocation in Multi-Spot MFTDMA Satellite Networks", in M. Chen, Y. Li, D.Z. Du, Eds., *Combinatorial Optimization in Communication Networks*, pp. 1-41, Kluwer Academic Publishers, 2005.

[27] S. Alouf, E. Altman, J. Galtier, J.-F. Lalande, C. Touati, "Quasi-Optimal Bandwidth Allocation for Multi-Spot MFTDMA Satellites", in *Proc. of IEEE Infocom 2005*, Miami, FL, Vol. 1, pp. 560-571, March 2005.

[28] C. Touati, E. Altman, J. Galtier, "Fair Bandwidth Allocation between Service Providers in a Geostationary Satellite Network", *INRIA Res. Rep. No. 4421*, March 2001.

[29] N. Celandroni, F. Davoli, E. Ferro, "Static and Dynamic Resource Allocation in a Multiservice Satellite Network with Fading", *International Journal of Satellite Communications and Networking*, Vol. 21, No. 4-5, pp. 469-487, July-October 2003.

[30] N. Celandroni, F. Davoli, E. Ferro, A. Gotta, "Adaptive Cross-Layer Bandwidth Allocation in a Rain-Faded Satellite Environment", *International Journal of Communication Systems*, Vol. 19, No. 5, pp. 509-530, June 2006.

[31] A. Durresi, P. K. Jagannathan, R. Jain, "Scalable Proportional Allocation of Bandwidth in IP Satellite Networks", in *Proc. of IEEE Aerospace Conf. 2003*, Big Sky, MT, Vol. 3, pp. 1253-1264, March 2003.

[32] P. K. Jagannathan, A. Durresi, R. Jain, "Stateless Proportional Bandwidth Allocation", in R. D. van der Mei, F. Huebner, Eds., *Internet Performance and Control of Network Systems III, Proc. SPIE*, Vol. 4865, pp. 25-36, 2002.

[33] G. Xilouris, A. Kourtis, G. Stefanou, "Dynamic Bandwidth Allocation for LAN2LAN Interconnection Using DVB-S Satellite Transmission", in *Proc. of the* 2nd *Internat. Working Conf. on Performance Modelling and Evaluation of Heterogeneous Networks* (HET-NETs '04), Ilkley, UK, pp. P88/1-P88/12, July 2004.

[34] A. Narula-Tam, T. G. Macdonald, E. Modiano, L. Servi, "A Dynamic Resource Allocation Strategy for Satellite Communications", in *Proc. of IEEE MILCOM*, Monterey, CA, Vol. 3, pp. 1415-1421, October/November 2004.

[35] G. Açar, C. Rosenberg, "Simulation Analyses of Weighted Fair Bandwidth-on-Demand (WFBoD) Process for Broadband Multimedia Geostationary Satellite Systems", *International Journal of Satellite Communications and Networking*, Vol. 23, No. 4, pp. 229-245, July/August 2005.

[36] H. Skinnemoen, A. Vermesan, A. Iuoras, G. Adams, X. Lobao, "VoIP over DVB-RCS with QoS and Bandwidth on Demand", *IEEE Wireless Communications Magazine*, Vol. 12, No. 5, pp. 46-53, October 2005.

[37] W. G. Hwang, "Bandwidth on Demand for Deployed-IP Users", *IEEE IT Professional*, Vol. 7, No. 1, pp. 21-26, January/February 2005.

[38] J. P. Choi, V. W. S. Chan, "Optimum Power and Beam Allocation Based on Traffic Demands and Channel Conditions over Satellite Downlinks", *IEEE Transactions on Wireless Communications*, Vol. 4, No. 6, pp. 2983-2993, November 2005.

[39] S. Egami, "A Power-Sharing Multiple-Beam Mobile Satellite in Ka Band", *IEEE Journal on Selected Areas in Communications*, Vol. 17, No. 2, pp. 145-152, February 1999.

[40] M. J. Neely, E. Modiano, C. E. Rohrs, "Power Allocation and Routing in Multibeam Satellites with Time-Varying Channels", *IEEE/ACM Transactions on Networking*, Vol. 11, No. 1, pp. 138-152, February 2003.
[41] S. Shimamoto, H. Kubota, H. Kuwabara, Y. Onozato, "A Study on Satellite Network Configuration Control Employing Transmission Power Control", *Electronics and Communications in Japan (Part I: Communications)*, Vol. 83, No. 7, pp. 103-112, July 2000.
[42] K. W. Ross. *Multiservice Loss Models for Broadband Telecommunication Networks*, Springer, London, 1995.
[43] P. B. Key, "Optimal Control and Trunk Reservation in Loss Networks", *Probability in the Engineering and Informational Sciences*, Vol. 4, pp. 203-242, 1990.
[44] G. Choudhury, K. Leung, W. Whitt, "An Algorithm to Compute Blocking Probabilities in Multi-Rate, Multi-Class Multi-Resource Loss Models", *Advances in Applied Probability*, Vol. 27, pp. 1104-1143, 1995.
[45] C. C. Beard, V. S. Frost, "Prioritized Resource Allocation for Stressed Networks", *IEEE/ACM Transactions on Networking*, Vol. 9, No. 5, pp. 618-633, October 2001.
[46] S. B. Biswas, B. Sengupta, "Call Admissibility for Multirate Traffic in Wireless ATM Networks", in *Proc. of IEEE INFOCOM*, Kobe, Japan, Vol. 2, pp. 649-657, April 1997.
[47] A. C. Fu, E. Modiano, J. N. Tsitsiklis, "Optimal Energy Allocation and Admission Control for Communications Satellites", *IEEE/ACM Transactions on Networking*, Vol. 11, No. 3, pp. 488-500, June 2003.
[48] A. Iera, A. Molinaro, S. Marano, "Traffic Management Techniques to Face the Effects of Intrinsic Delays in Geostationary Satellite Networks", *IEEE Transactions on Wireless Communications*, Vol. 1, No. 1, pp. 145-155, January 2002.
[49] S. Cho, "Adaptive Dynamic Channel Allocation Scheme for Beam Handover in LEO Satellite Networks", in *Proc. of IEEE Vehicular Technology Conf., 2000* (Fall VTC 2000), Boston, MA, pp. 1925-1929, September 2000.
[50] I. Mertzanis, R. Tafazolli, B. G. Evans, "Connection Admission Control Strategy and Routing Considerations in Multimedia (Non-GEO) Satellite Networks", in *Proc. of the* 47th *IEEE Vehicular Technology Conf.* (VTC 1997), Phoenix, AZ, pp. 431-436, April 1997.
[51] M. El-Kadi, S. Olariu, P. Todorova, "Predictive Resource Allocation in Multimedia Satellite Networks", in *Proc. of IEEE GLOBECOM*, San Antonio, TX, pp. 2735-2739, November 2001.
[52] H. Uzunalioglu, "A Connection Admission Control Algorithm for LEO Satellite Networks", in *Proc. of IEEE International Conference on Communications* (ICC 1999), Vancouver, BC, Canada, pp. 1074-1078, June 1999.
[53] S. Olariu, S. R. A. Rizvi, R. Shirhatti, P. Todorova, "Q-WIN - A New Admission and Handoff Management Scheme for Multimedia LEO Satellite Networks", *Telecommunication Systems*, Vol. 22, No. 1-4, pp. 151-168, January 2003.
[54] E. Del Re, R. Fantacci, G. Giambene, "Different Queuing Policies for Handover Requests in Low Earth Orbit Mobile Satellite Systems", *IEEE Transactions on Vehicular Technology*, Vol. 48, No. 2, pp. 448-458, March 1999.
[55] G. Maral, J. Restrepo, E. Del Re, R. Fantacci, G. Giambene, "Performance Analysis for a Guaranteed Handover Service in a LEO Constellation with a

Satellite-Fixed Cell System", *IEEE Transactions on Vehicular Technology*, Vol. 47, No. 4, pp. 1200-1214, November 1998.

[56] S. Karapantazis, P. Todorova, F. N. Pavlidou, "On Bandwidth Management and Inter-Satellite Handover in Multimedia LEO Satellite Systems", in *Proc. of the* 23rd *AIAA International Communication Satellite System Conf.*, Rome, Italy, September 2005.

[57] D. W. Petr, K. M. S Murthy, V. S. Frost, L. A. Neir, "Modeling and Simulation of the Resource Allocation Process in a Bandwidth-on-Demand Satellite Communications Network", *IEEE Journal on Selected Areas in Communications*, Vol. 10, No. 2, pp. 465-477, February 1992.

[58] O. Ercetin, M. O. Ball, L. Tassiulas, "Modeling Study for Evaluation of Aeronautical Broadband Data Requirements over Satellite Networks", *IEEE Transactions on Aerospace and Electronic Systems*, Vol. 41, No. 1, pp. 361-370, January 2005.

[59] K. Kwangil, T. G. Robertazzi, "Equal Allocation Scheduling for Data Intensive Applications", *IEEE Transactions on Aerospace and Electronic Systems*, Vol. 40, No. 2, pp. 695-705, April 2004.

[60] A. Jahn, "Resource Management Model and Performance Evaluation for Satellite Communications", *International Journal of Satellite Communications*, Vol. 19, No. 2, pp. 169-203, March/April 2001.

[61] B. G. Evans, "SatNEx - A European Network of Excellence in Satellite Communications", *International Journal of Satellite Communications and Networking*, Vol. 23, No. 5, p. 263, September/October 2005.

[62] M. Werner, A. Donner, E. Lutz, R. Sheriff, F. Hu, R. Rumeau, H. Brandt, G. Maral, M. Bousquet, B. G. Evans, G. Corazza, "SatNEx - the European Satellite Communications Network of Excellence", in *Proc. of the* 59th *IEEE Vehicular Technology Conference (VTC 2004-Spring)*, Milan, Italy, Vol. 5, p. 2842, May 2004.

[63] R. E. Sheriff, Y. F. Hu, P. M. L. Chan, M. Bousquet, G. E. Corazza, A. Donner, A. Vanelli-Coralli, M. Werner, "SatNEx: A Network of Excellence Providing Training in Satellite Communications", in *Proc. of the* 61st *Vehicular Technology Conference* (VTC 2005-Spring), Stockholm, Sweden, Vol. 4, pp. 2668-2672, May 2005.

[64] F. Del Sorbo, F. Delli Priscoli, "Multimedia Services with QoS Support in an Integrated Terrestrial and Satellite UMTS Network Demonstrator: the IST VIRTUOUS Project", *International Journal of Satellite Communications and Networking*, Vol. 22, No. 1, pp. 139-156, January/February 2004.

[65] Y. Fun Hu, G. Maral, E. Ferro, "Service Efficient Network Interconnection via Satellite", *EU COST Action 253*, edited by John Wiley&Sons LTD, January 2002.

[66] "ISI Strategic Research Agenda", downloadable from the ISI Web site with URL: http://www.isi-initiative.eu.org/.

3

QoS REQUIREMENTS FOR MULTIMEDIA SERVICES

Editors: José Ignacio Moreno Novella[1], Francisco Javier González Castaño[2]

Contributors: Rafael Asorey Cacheda[2], Daniel Castro García[3], Antonio Cuevas[1], Francisco Javier González Castaño[2], Javier Herrero Sánchez[3], Georgios Koltsidas[4], Vincenzo Mancuso[5], José Ignacio Moreno Novella[1], Seounghoon Oh[6], Antonio Pantò[7]

[1]UC3M - Universidad Carlos III de Madrid, Spain

[2]UVI - Universidad de Vigo, Spain

[3]INFOGLOBAL, Spain

[4]AUTh - Aristotle University of Thessaloniki, Greece

[5]UToV - Università degli Studi di Roma "Tor Vergata", Italy

[6]RWTH - Rheinisch -Westfälische Technische Hochschule Aachen, Germany

[7]CNIT - University of Catania, Italy

3.1 Introduction

Internet development and an ever-increasing demand for bandwidth are boosting the market for satellite solutions. Technological progress leading to new satellite capabilities and the availability of bandwidth at lower cost is enabling this growing role of satellites in the Internet world. Satellite solutions are being used for both broadcast/multicast applications and point-to-point services. End-user access combines multicast and point-to-point services while content distribution to the "*edge*" of the Internet (i.e., to service providers' points-of-presence serving access local loops) is a true multicast application.

Geostationary Earth Orbit (GEO) satellites and *Low Earth Orbit* (LEO) constellations essentially play a complementary role, in order to provide this complete range of services. Due to the large amount of capacity they provide and their low-latency characteristics, LEO systems are very well suited for point-to-point high-quality services while GEO solutions are very efficient for both broadcast/multicast offerings and access services including a significant percentage of multicast data. To support the different services it is important to consider their *Quality of Service* (QoS) requirements.

This Chapter mainly describes QoS requirements for multimedia services based on international standards. Section 3.2 shows a classification of applications according to error and delay tolerance, as well as performance characterization of traditional and multimedia applications. This work is based on the ITU G.1010 [1] standard that has been adopted by other standardization bodies like 3GPP. Section 3.3 presents main QoS support models over IP networks, while Section 3.4 shows main concepts for the transmission of multimedia and broadcast services over satellite networks. Finally, Section 3.5 presents experimental results of application performance over a real platform; the main interest here is to present QoS results on classical and emerging applications.

3.2 Services QoS requirements

Nowadays it is very important to support QoS in telecommunication systems, considering the requirements that should be met when a service is provided. This task should take into consideration that a user is not interested in the way a particular service is provided, but in the service quality level he/she finally delectates.

QoS refers to the capability of a telecommunication system to provide better service to selected traffic over heterogeneous networks (technologies or domains). The primary goal of QoS is to provide priority, including dedicated bandwidth, controlled jitter and latency (required by some real-time and interactive traffic), and improved loss characteristics. Moreover, it is important to assure that providing priority for one or more flows does not cause the failure of other flows. On intuitive level, QoS represents a certain type of requirements to be guaranteed to the users (e.g., how fast data can be transferred, how much the receiver has to wait, how correct the received data is likely to be, how much data is likely to be lost, etc.).

QoS requirements for multimedia traffic have been covered by different standardization groups, like ITU, ETSI or 3GPP. The main work provided by

ITU is in Recommendations Y.1541 [2], F.700 [3], and G.1010 [1]. Applications have been classified in eight groups, according to the error tolerance and delay, as summarized in Figure 3.1 [1],[4].

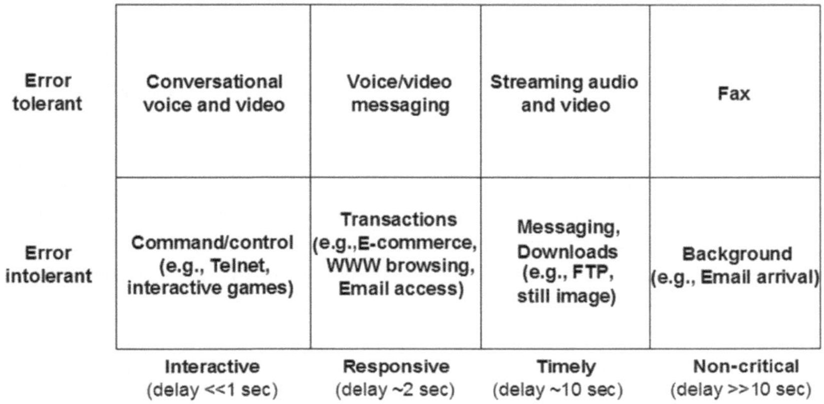

Fig. 3.1: End-user QoS categories mapping. This figure is reproduced with the kind permission of ITU.

Referring to the above Figure, it is possible to consider the following values on the ordinate axis for what concerns the error rates:

- Error tolerant applications
  - Conversational voice/video *Frame Erasure Rate* (FER) $< 3\%$
  - Voice/video messaging FER $< 3\%$
  - Streaming audio/video FER $< 1\%$
  - Fax *Bit Error Rate* (BER) $< 10^{-6}$
- Error intolerant applications
  - Information loss $= 0$.

The ETSI *Broadband Satellite Multimedia* (BSM) [5] working group provides technical reports and standards establishing a framework to specify QoS requirements for broadband satellite networks based on the Internet protocol suite. These standards (following those developed in ETSI and other bodies) identify how Internet quality-related standards can be adapted, translated or made transparent to satellite transmission protocols and equipment. Some of the results of this standardization work have been the definition of the protocol stack architecture shown in Chapter 1 (Section 1.5), where lower layers depend on satellite system implementation (*satellite-dependent layers*) and higher layers are those typical of the Internet protocol stack (*satellite-independent layers*).

The traffic classes established by BSM are based on ITU-T, Tiphon, 3GPP, and UMTS decisions, with adaptation to the satellite environment. In particular, the BSM standards deal with variable link layer conditions, high asymmetry and higher delay that are characteristics of satellite networks. The aim is to enable the satellite network and the *Internet Service Provider* (ISP) to ensure acceptable QoS levels and to relate these issues to the BSM architecture for broadband systems.

In UMTS and, by extension, in satellite networks, four basic *service classes* (layer 7) are defined [4]: *conversational*, *streaming*, *interactive* and *background*. It is interesting to note that there is no strict one-to-one mapping between these service classes and the namesake traffic classes (layer 2) [6]: an interactive application can very well use a bearer of the conversational traffic class, if the application/service or the user has tight requirements on delay. In the following sub-Sections the performance requirements for all four service classes are investigated from the user perspective.

Note that the delay values in the Tables of the following sub-Sections represent one-way delay (i.e., from originating entity to terminating entity).

3.2.1 Performance requirements for conversational services

The most common service in this category is real-time conversation, such as telephony speech. *Voice over IP* (VoIP) and video conferencing also belong to this category, with increasing relevance as the Internet is rapidly evolving. This is the only class whose characteristics are strictly determined by human perception (senses). Thus, this scheme has the most stringent QoS requirements: the transfer time should be low and, at the same time, the temporal relation of information entities of the stream should be preserved. The limit for acceptable transfer delay is very strict (failure to provide low transfer delays will result in unacceptable lack of quality). However, there are loose requirements on FER, due to the human perception. For real-time conversation, the fundamental QoS characteristics are:

- Preserving the temporal relation of information entities in the same stream;
- Conversational pattern (stringent and low delay).

Some application examples based on conversational services are: conversational voice, videophone, interactive games, two-way control telemetry and Telnet. Table 3.1 summarizes these applications providing the explicit requirements for each of them [1],[4].

Conversational voice

Audio transfer delay requirements [3] depend on the level of interactivity of end-users. To preclude difficulties related to the dynamics of voice communications, ITU-T Recommendation G.114 specifies the following general limits

Medium	**Application**	**Degree of symmetry**	**Data rate**	**Key performance parameters and target values**		
				End-to-end one-way delay	**Delay variation within a cell**	**Information loss**
Audio	Conversational voice	Two-way	4-25 kbit/s	< 150 ms preferred < 400 ms limit	< 1 ms	< 3% FER
Video	Videophone	Two-way	32-384 kbit/s	< 150 ms preferred < 400 ms limit Lip-synch: < 100 ms		< 1% FER
Data	Telemetry-two-way control	Two-way	< 28.8 kbit/s	< 250 ms	NA	Zero
Data	Interactive games	Two-way		< 250 ms	NA	Zero
Data	Telnet	Two-way (asymmetric)		< 250 ms	NA	Zero

Table 3.1: End-user performance expectations - conversational services.

for one-way transmission delay (assuming that echo control has been applied) [7]:

- 0 to 150 ms: preferred range (below 30 ms the user does not notice any delay at all, whereas above 100 ms the user does not notice delay if echo cancellation is provided and there are no distortions in the link)
- 150 to 400 ms: acceptable range (but with increasing degradation)
- Above 400 ms: unacceptable range

We should remember here that there are three types of satellite systems: LEO, MEO and GEO. Due to their different distance to Earth's surface, the propagation delay for the transmitted signal (from Earth to the satellite and back to Earth) varies from 10 ms to 250 ms (see Section 1.2). This means that for LEO and MEO satellite systems the preferred range described above is achievable. However, a GEO system cannot achieve an end-to-end delay below 250 ms. This means that, according to the satellite system used, the network designer should be very careful when selecting operational modes. Other classes have looser requirements and they may be supported by GEO satellites.

The human ear is highly intolerant to short-term delay variation (*jitter*), so it should be kept really low. It has been suggested that 1 ms is an adequate limit. However, the human ear is tolerant to moderate distortion of the speech signal. An acceptable performance is typically obtained with FER up to 3%. Finally, a connection for a conversation normally requires the allocation of symmetrical communication resources.

Videophone

Videophone requires a full-duplex system, carrying both video and audio, and it is intended for a conversational environment. Therefore, the same delay requirements of conversational voice will apply, i.e., no echo and minimal effect on conversational dynamics, with the added requirement that audio and video must be synchronized within certain limits to provide "*lip-synch*" (i.e., synchronization of the speaker's lips with the words the end-user hears). In fact, it will be difficult to meet these requirements, due to the long delays incurred in video codecs. Human eye is tolerant to some information loss, so that some degree of packet loss is acceptable. It is expected that high performance video codecs will provide acceptable video quality with FER up to about 1%. In satellite networks, the same considerations for conversational voice hold in this case.

Interactive games

Interactive games are games that use the network to interact with other users or systems. Requirements for interactive games are very dependent on the specific game considered in terms of bandwidth and delay. Many interactive games try to exchange high volumes of data, but demand very short delays, and a delay of 250 ms is reasonable.

Two-way control telemetry

Telemetry is a technology that allows the remote measurement, operation and reporting of information of interest. Two-way control telemetry is included here as an example of a data service that does require real-time conversational performance. Two-way control implies very tight limits on allowable delay and a value of 250 ms is proposed, but a key difference with voice and video services is that information loss cannot be tolerated. It is well known that the satellite channel is error-prone and in order to achieve zero information loss we need sophisticate error control techniques to ensure it. Delay is a relative issue for this class of traffic. As far as a satellite network can meet the deadlines that a particular telemetry service imposes, it can support that service.

Telnet

Telnet (*TELetype NETwork*) is a network protocol used on the Internet or local area network connections. In this context, Telnet refers to the program that provides the client part of the protocol. It allows a remote server access. Due to the interactivity of the program, Telnet needs a low delay to allow a user perception of interactivity. This application is included here with a requirement for a low delay in order to provide back instantaneous character echoes. By extension we could consider in the same service/application group any remote access applications like *rlogin* (*remote login*) or *ssh* (*secure shell*).

3.2.2 Performance requirements for interactive services

This second class comprises interactive services (i.e., a human or a machine request on-line data from a remote server). It is characterized by the request-response pattern of the end-user. An entity at the destination is usually expecting a response message within a certain period of time. The *Round Trip propagation Delay* (RTD) time is therefore one of the key attributes. Another characteristic is that the content of the packets must be transparently transferred (with a low BER). The resulting overall requirement for this communication scheme is to support interactive non-real-time services with low RTD.

For interactive traffic, the fundamental QoS characteristics are:

- The request-response pattern;
- Preserving payload content.

Some examples of this service type are: voice messaging and dictation, data, Web-browsing, high-priority transaction services (e-commerce) and e-mail (server access). The corresponding requirements are summarized in Table 3.2 [4].

Voice messaging and dictation

The requirements for information loss are essentially the same as for conversational voice, but, on the contrary, there is more tolerance to delay since there is no direct conversation involved. Therefore, the main task becomes to determine the delay that can be tolerated between the user, issuing a command to replay a voice message, and the actual start of the audio. There is no precise data on this, but a delay in the order of a few seconds is considered to be reasonable for this application.

Web-browsing

The main performance factor is the visualization response time, after a Web page has been requested. A value of 2-4 s per page is proposed. However, a decrease up to a target of 0.5 s would be desirable.

Medium	Application	Degree of symmetry	Data rate	Key performance parameters and target values		
				One-way delay	Delay variation	Information loss
Audio	Voice messaging	Primarily one-way	4-13 kbit/s	< 1 s (playback) < 2 s (record)	< 1 ms	< 3% FER
Data	Web-browsing - HTML	Primarily one-way		< 4 s/page	NA	Zero
Data	Transaction services - high priority e.g., e-commerce, ATM	Two-way		< 4 s	NA	Zero
Data	E-mail (server access)	Primarily one-way		< 4 s	NA	Zero

Table 3.2: End-user performance expectatives - interactive services.

3.2.3 Performance requirements for streaming services

This service class is mainly unidirectional with high continuous utilization (few idle/silent periods) and low time variation between information entities within a flow. However, there is no strict limit for delay and delay variation, since the stream is normally aligned at the destination. Additionally, there is no strict upper limit for the packet loss rate.

For real-time streams, the fundamental QoS characteristics are:

- Unidirectional continuous stream;
- Preserving time relation (variation) between information entities of the stream.

The resulting overall requirement for this communication scheme is to support real-time streaming services with continuous unidirectional data flows. Table 3.3 details some application examples and the corresponding limitations [4].

Note that Figure 3.1, Table 3.1, Table 3.2 and Table 3.3 derive from 3GPP TS 22.105 [4]. 3GPP^{TM} TSs and TRs are the property of ARIB, ATIS, ETSI, CCSA, TTA and TTC who jointly own the copyright in them. They are subject to further modifications and are therefore provided "as is" for information purposes only. Further use is strictly prohibited.

Medium	Application	Degree of symmetry	Data rate	Key performance parameters and target values		
				Start-up delay	Transport delay variation	Packet loss at session layer
Audio	Speech, mixed speech and music, medium and high quality music	Primarily one-way	5-128 kbit/s	< 10 s	< 2 s	$< 1\%$ Packet loss ratio
Video	Movie clips, surveillance, real-time video	Primarily one-way	20-384 kbit/s	< 10 s	< 2 s	$< 2\%$ Packet loss ratio
Data	Bulk data transfer/ retrieval, layout and synchronization information	Primarily one-way	< 384 kbit/s	< 10 s	NA	Zero
Data	Still image	Primarily one-way		< 10 s	NA	Zero

Table 3.3: End-user performance expectations - streaming services.

Audio streaming

Audio streaming is expected to provide better quality than conventional telephony, thus the packet loss requirements will be correspondingly tighter. However, there are no conversational elements involved and the delay requirements can be relaxed.

One-way video

The main distinguishing feature of one-way video is the absence of conversational elements. Therefore, the delay requirements will be not so stringent.

Still image

Regarding still images, the human eye is tolerant to information loss. However, single bit errors can cause large disturbances in still image formats. Therefore, it is generally expected that there will be zero errors in the transmission of still images. Delay requirements are low.

3.2.4 Performance requirements for background services-applications

This service class applies when the end-user, typically a computer, sends and receives data files in background. It is a classical data communication scheme where the destination is not expecting data within a certain deadline. Hence, the propagation delay (like that of satellite systems) is not that important in this case. However, error control is very important, since errors should be kept to very low levels (in the satellite scenario such feature calls for adequate coding protection and retransmission schemes).

For background traffic, the fundamental QoS characteristics are:

- The destination is not expecting data before a certain deadline;
- Preserving payload content.

The resulting overall requirement for this communication scheme is to support non-real time services without any special requirement on delay. A background application has no delay constraint. In principle, an essentially error-free delivered information is the only requirement for applications in this category. However, there is still a delay constraint, since data is effectively useless if it is received too late. Examples of these applications are: fax, low priority transaction services, e-mail (server to server), *Short Message Service* (SMS), download of databases and measurement records.

Fax

Fax is not normally considered a real-time communication. Nevertheless, there is an expectation that a fax transmission will take less than 30 s.

Low priority transaction services

An example in this category is SMS. An acceptable delivery delay is 30 s. Table 3.4 compares the applications on the basis of the service class and the associated delay requirement [8].

3.3 IP QoS frameworks/models

Many factors influence the user-perceived quality of a telecommunication service, from *codecs* employed to the performance of the network. The constraints and requirements have been presented in the previous Section 3.2. In this Chapter we will analyze the mechanisms designed for IP networks to achieve QoS. This Section addresses the IP layer and as such we keep it very general, so that the satellite network can be one of the possible scenarios.

It is well known that IP networks were not designed to provide any

Service class	**Conversational (delay $\ll$ 1 s)**	**Interactive (delay $\sim$ 1 s)**	**Streaming (delay $<$ 10 s)**	**Background (delay $>$ 10 s)**
Error tolerant	Conversational voice and video	Voice messaging	Streaming audio and video	Fax
Error intolerant	Telnet interactive games	e-commerce Web browsing	FTP, still image, paging	e-mail arrival notification

Table 3.4: Application examples in terms of QoS.

QoS guarantees. However, the applications traditionally using IP as a communication technology could perfectly cope with that lack; telephony or iterative applications over IP (that are nowadays beginning to be used) need transport networks with very strict QoS support. Such mechanisms vary from 100% guarantee solutions (employing techniques that can be assimilated to virtual circuit creation/provisioning) to other solutions not providing 100% guarantees. The over-provisioning approach is also considered but, of course, it cannot be applied in scarce-bandwidth radio access networks. Besides, offering different qualities for the data transport service will create new opportunities for providing several quality levels at different prices. We can conclude that, in the future, the IP data transport will be QoS-enabled.

The way to provide QoS in IP networks has been discussed for a long time. The most accepted solutions are IETF's IntServ [9] and DiffServ [10]: both IntServ and DiffServ endow the routers with QoS mechanisms, such as queuing, scheduling and shaping, as illustrated in Figure 3.2. These mechanisms are implemented in the router interfaces. The main difference between IntServ and DiffServ lies in the level of detail used by the classifiers and in the need to keep state information.

The IntServ model provides end-to-end QoS guarantees by reserving per-flow resources (normally using the RSVP protocol [11]) in all the nodes along the path. While this architecture provides excellent QoS guarantees, it has scalability problems in the network core because of per-flow state maintenance and per-flow operation in routers. It is worth noting that RSVP is not the only IP reservation protocol, but RSVP is by far the most accepted one and has become an "integral" part of IntServ networks. There exist even some commercial RSVP-enabled routers.

RSVP identifies a communication session by the combination of destination address, transport-layer protocol type, and destination port number. In IPv6 those two last parameters may be replaced by the flow label. RSVP is used to reserve resources in the routers along the path between the sender and the receiver(s). RSVP also allows freeing these resources when they are no

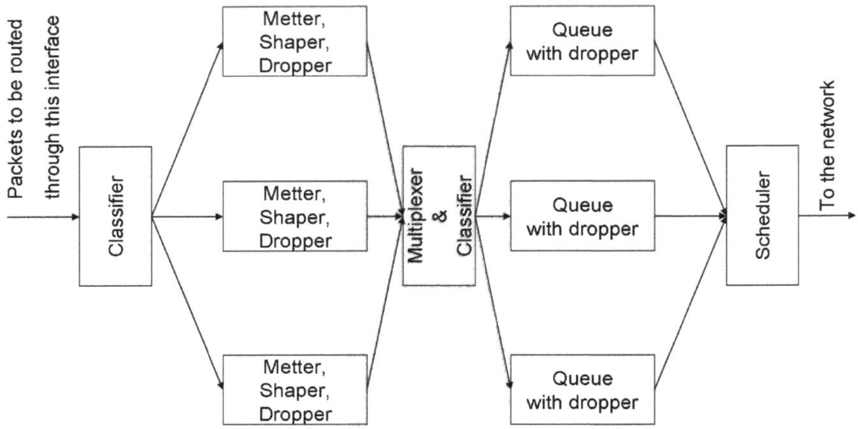

Fig. 3.2: QoS mechanisms in a router interface.

longer needed. Normally these reservations are to be policed and it is common to have an entity termed bandwidth broker (or, also, QoS broker) that takes the policy decision and communicates it to the routers. This entity will be studied later in this Section.

The primary messages used by RSVP are the "Path" message, which originates from the traffic sender, and the "Reservation" message, which originates from the traffic receiver(s). They are used in the resource reservation process. RSVP can also explicitly shut down the QoS sessions using RSVP teardown messages. Teardown messages can be initiated by an application in an end-system (sender or receiver) or a router as the result of state time-out. RSVP supports two types of teardown messages: "path-teardown" and "reservation-request teardown". Path-teardown messages delete the path state (deleting the reservation state), travel toward all receivers downstream from the point of initiation, and are routed like path messages. Reservation-request teardown messages delete the reservation state, travel towards all matching senders upstream from the point of teardown initiation, and are routed like corresponding reservation-request messages.

On the other hand, DiffServ requires no per-flow control in the core network and, consequently, routers do not maintain any per-flow state and operation; no reservation protocol exists. As a result, DiffServ is relatively scalable in the forwarding/data plane, but offers no strict QoS guarantees. The criterion to classify the packets in core routers relies on the *DiffServ Code Point* (DSCP) field in the packet header [12]. DSCP defines three classes of priority:

- *Best Effort* (BE): to provide the service in the same way as in the current Internet, where there are no QoS guarantees, IETF recommends that the DSCP value should be 000000 (bin).
- *Assured Forwarding* (AF): The AF group contains four independent classes, each with three different *drop precedence* values in the queues. There is no specified algorithm for each value, but the dropping probabilities must be increasing and the packets must be marked with AF DSCP value and must arrive to the destination in the proper order. In case of congestion, the dropping probability depends on the *drop precedence* value.
- *Expedited Forwarding* (EF): EF is designed as the best group. It should provide very small drop probability, latency and jitter. That is the reason why this service is sometimes regarded as a *Virtual Leased Line* (VLL). This *Per-Hop Behavior* (PHB) is predestined to handle real-time applications like video streams. When EF packets enter a DiffServ router, they should be handled in very short queues and quickly serviced to maintain lower latency, packet loss, and jitter. Throughput of the EF flow should be limited to the value that can be handled by each node. It is necessary to avoid the situation where the queue could overflow and cause flow degradation. IETF recommends that the EF class should be marked with the DSCP value 101110 (bin).

Routers should allocate enough resources for the high priority DSCPs, while the lower ones or the "*classical*" BE traffic (DSCP 0) may use spare resources. DiffServ networks require access control in the edge routers, so that only authorized users can inject packets with high priority DSCPs. Access control is enforced by the shapers. Depending on the type of edge routers, this access control can take place in different levels of detail. For instance, in edge routers connecting the core network to the users (*Access Routers*, ARs) this control follows a per-user and per-flow basis, since ARs will handle a small traffic load. However, for edge routers connecting the core network to the Internet or other domains, this access control can only proceed at a very rough level of detail.

Besides the QoS-enabled routers, another entity called QoS broker [13] is used to control and to manage the network. This entity, for scalability reasons, can be replicated in the network; moreover, the network can be hierarchically divided into several areas, as proposed in [14]. In a simplified way, the QoS broker manages and monitors the network resources in one particular domain of operation. It also monitors the edges for incoming and outgoing resource reservations/utilization. The information thereby acquired is used in conjunction with the policy system information to take admission control decisions and reconfigurations and to convey them to the routers. A QoS broker is then an entity that takes *Service Admission Control* decisions and performs network device configuration, according to a set of conditions imposed by the network administration entities (e.g., *Authentication, Authorization and Accounting*, AAA, System) with the goal of achieving end-to-end

QoS, also over different networks. The QoS broker may also be responsible for managing inter-domain communications with neighbor QoS brokers, so that QoS-enabled transport services are implemented in a coordinated way across various domains.

Since IntServ requires resource reservation, it is the most evident scenario to integrate a QoS broker. In IntServ a RSVP enabled router may consult the QoS broker (using the *Common Open Policy Service*, COPS, protocol) about the decision to take when receiving RSVP path or reservation messages. The decision taken by the QoS broker is normally conveyed in a COPS message and then enforced by the router. In DiffServ, the edge routers need to perform admission control and may also outsource the decision to the QoS broker. This process can take place when the DiffServ access router detects a new traffic; the level of detail to define new traffic may vary, as we just explained. QoS brokers functionally can go beyond taking policing decisions; generally they are also in charge of managing the network. The actual role of the QoS broker may adapt to the different scenarios and business models. For instance in the scenario described in Section 3.4.1, the "recovery provider" may consult a QoS broker before gathering data from the content providers and sending it to the satellite so that this broadcasts it.

Many existing approaches combine IntServ and DiffServ: IntServ in the access part of the network and DiffServ in the core network. Of course, solutions based on other paradigms also exist and are even complementary to these ones. For example, [15] proposes new routing schemes over high availability networks.

3.4 Broadcast and multicast services

In addition to DVB-S broadcast, satellite IP multicast for content distribution to the "*edge*" of the Internet and to corporate sites has numerous advantages over terrestrial technology. Satellites offer highly "*regular*" broadband data streams and a single transmission from a central operation center can be delivered to a high number of receiving sites. In addition to reducing costs, the single long hop of the satellite link replaces all the small hops of terrestrial content distribution and bypasses bottlenecks, thus improving QoS in many applications. Thus, satellite multicast for content distribution and satellite content delivery to mobile terminals (either broadcast or multicast) are interesting working areas. Clearly, reception is mainly possible when the satellite is in direct line-of-sight or attenuation is low. Hence, complementary terrestrial repeaters enhance the architecture by retransmitting the satellite signal.

When only a satellite signal is present (i.e., no terrestrial repeaters), satellite broadcasting systems may use time diversity to enhance availability. This technique broadcasts the same content twice, so that the two transmissions are uncorrelated with respect to mobile reception blockages. The receiver is able to combine them to provide seamless reception.

In the case of satellite broadcasting to mobile terminals using mobile communication modulations, the *client* could switch between two content sources with different QoS levels: satellite (or terrestrial-repeated satellite) and terrestrial wireless networks (when neither satellite nor terrestrial repeaters are available). This handover between physically different access interfaces is problematic for example in the case of UMTS and WiFi (again, the latter would provide a higher QoS level, at least in terms of regularity, if a satellite gateway is present).

When terminals support dual network access, e.g., satellite and terrestrial (WiFi, UMTS, etc.) links, it is quite critical to select the appropriate network for each application depending on both the available resources and the kind of application involved. In general, interface selection can be network-initiated or terminal-initiated. In the first case, the network operator decides the appropriate access network for each application, whereas in the second case the terminal will decide the best path. All these procedures must be performed during application initialization as well as during handovers, and must consider available access technologies, user profile (SLA, user requirements, etc.), and QoS capabilities depending on the available resources. In the case of multicast and broadcast services, terminal-initiated interface selection seems the natural approach, since it would be too difficult for a network operator to select *individually* optimum interfaces for the large user populations involved.

Satellites have traditionally served point-to-point communications (such as intercontinental telephony circuits) and unidirectional TV broadcast. *Very Small Aperture Terminals*, VSATs (i.e., narrowband data terminals in transactional mode), appeared in the 90's. With some exceptions, the medium access technology at that time neither allowed broadband service provision nor massive terminal deployment, but 10-to-100 units at most. On the other hand, equipment manufacturers developed proprietary platforms that could not interoperate. A high terminal/service cost kept related services within corporate markets, beyond the possibilities of SMEs. This situation has radically changed in the last six years, due to technological advances such as multiple access protocols. On one hand, VSAT terminal manufacturers (Hughes, Gilat [16], etc.) have developed fully bidirectional equipment (still proprietary) to provide broadband services to large user communities and, in some cases (Starband [17], DirectWay), at an acceptable cost even for residential users. On the other hand, a bit of new manufactures offer interoperable equipment based on the DVB standard, i.e., specifically, MPE (*Multi Protocol Encapsulation*) and RCS (*Return Channel via Satellite*).

The advent in 1997 of the MPE standard for DVB IP data encapsulation implied that equipment manufacturers should no longer supply the whole communication chain thanks to interoperability. Traditional head-end manufacturers began to include IP data insertion equipment in their catalogues (Thomcast [18], Divicom, Rohde & Schwarz, etc.), and some new ones completely centered their efforts in this direction (Logic Innovations,

Tandberg [19], etc.). In general, they did not provide user terminals, due to the deep differences between professional and user markets in terms of quality goals, sales support, etc. For this reason, many PC peripheral manufacturers entered the competition with DVB-S boards and boxes (Adaptec, Terratec [20], Technotrend, etc.).

MPE stimulated the entrance of satellite IP services into the mass market. For applications requiring interactivity (bidirectionality), the services initially relied on auxiliary terrestrial technologies for the return channel, wired (POTS, ISDN or Frame Relay) or wireless ones (GSM, GPRS or similar). There was a clear lack of a satellite technology to eliminate this terrestrial dependence. In 1999, the DVB-RCS standard covered this gap. Despite of some initial interoperation problems (usually leading to the election of the same supplier for the whole communications chain), the standard has matured in the last years. Several operators have selected it (Satlynx, Hispasat [21], etc.).

In the last two years the new protocol *DOCSIS for Satellite* (or DOCSIS-S) is emerging as an alternative to DVB-RCS, based on the well-known DOCSIS standard for cable networks and mostly promoted by American vendors and providers (Viasat [22] and WildBlue [23]). Compared with DVB-RCS, DOCSIS-S exploits the economies of scale of silicon designs for cable infrastructure, and takes advantage of a huge selection of *Operations and Business Support Systems* platforms from the cable market. However, DOCSIS-S is still a "vendor-promoted protocol", not a real standard; thus interoperability and availability of suppliers are an issue.

These new protocols enable new multimedia application scenarios based on multicast and broadcast distribution. One of these applications is distance learning with or without interactivity. In it, a teacher provides a lesson to a number of remote students using multicast video and audio streaming and additional aids such as a digital blackboard, slides, etc. When interactivity (return channel) is available, students may send questions to the teacher either by chat or by their own microphone and webcam, so that the other students may follow the question and the response. In this case, because of the delay induced by the satellite itself (500 ms for a GEO system), the media access protocol for the return channel (100 - 300 ms) and the video codecs (100 - 1000 ms), a voice handshake similar to a "walkie-talkie" must be implemented in order for the teacher and the student not to interfere. Also, when there is a large audience, the application must provide specific controls so that the teacher may act as moderator, granting or denying participation to the students. At present, distance learning systems (Centra [24]) and services (Hughes [25], Gilat [16]) are commercially available and widely deployed.

Another common multicast application not requiring real-time operation that largely benefits from a return channel when available is massive content distribution, where a central station delivers common multimedia contents to a large population of remote clients (with a reception acknowledge mechanism when interactivity is provided). The typical data losses and corruptions are avoided by *a*) adding redundant information to the data to be transmitted at the application level by means of convolutional coding, polynomial protection and interleaving, and *b*) implementing a return channel so that each remote client may inform the central station about the missing parts of the media content after receiving them and correcting the errors. Then, the central station re-sends those pieces of data grouped in overlapped parts, to avoid repeating the same datagrams. Massive content distribution software solutions are available from Kencast [26] and Tandberg [19], among others.

The DVB-RCS standard enables other innovative application scenarios for satellite content delivery to mobile terminals, such as *Delayed Real-Time* (DRT) services with QoS support for GEO satellite distribution systems. We describe them in the next sub-Section.

3.4.1 Delayed real-time service over GEO satellite distribution systems

The distribution of multimedia contents via satellite, even though it is one of the very first services envisioned by the satellite communication community, still represents a hot topic for satellite networks. There are many types of multimedia communication services; in this sub-Section, we address the class of DRT services, whose importance arises in the field of QoS-aware real-time communications.

DRT services fall in the category of streaming services whose requirements are discussed in sub-Section 3.2.3. DRT services have been conceived as an extension of unidirectional real-time broadcast and multicast services. So far, there are no standard architectures to support DRT, but diverse applications have been proposed in order to cope with given QoS requirements by means of specific application layer mechanisms. Instead of limiting DRT support to a mere application layer implementation, this Section presents an architecture that exploits both application and transport layer features. The proposed architecture assumes that DVB-RCS is deployed over a GEO satellite system. Nonetheless, it can be easily extended and adapted to any other layer 2 protocol stack suitable for broadcast and multicast applications, allowing customers to interact in real-time with the multimedia distribution farm (e.g., WiMAX or UMTS technologies).

A DRT service recovers from data losses and corruptions by using a buffer and, in turn, by introducing an artificial delay at the beginning of the play-out phase. A real-time return channel is fundamental, since the receiver must initiate a data recovery procedure after a data loss has been detected. In that case, additional resources can be invocated over the distribution channel, if available. The maximum possible duration for each recovery phase is determined by the length of the adopted buffer, and can be modulated by the choice of the codec (or codecs) for the multimedia streaming.

It is worth noting that multiple retransmissions could be requested at the same time by different users (e.g., by users belonging to the same multicast group), and, therefore, different retransmissions could partially overlap. Accordingly, retransmissions are executed in multicast and are requested through dynamically joining and pruning the multicast retransmission group. As a consequence, it is possible to design an architecture where a legacy satellite broadcast service is endowed with a specific multicast recovery algorithm able to mitigate the impact of network/satellite disruptions. This is the case of link failures due to user mobility and related shadowing effects. The reference scenario (Figure 3.3) is composed of three main elements:

- The *Content Provider* (we assume to have N content providers in the network);
- The *Recovery Service Provider* (just one in the network);
- The *users* (specifically, N groups of users, one group for each active content provider).

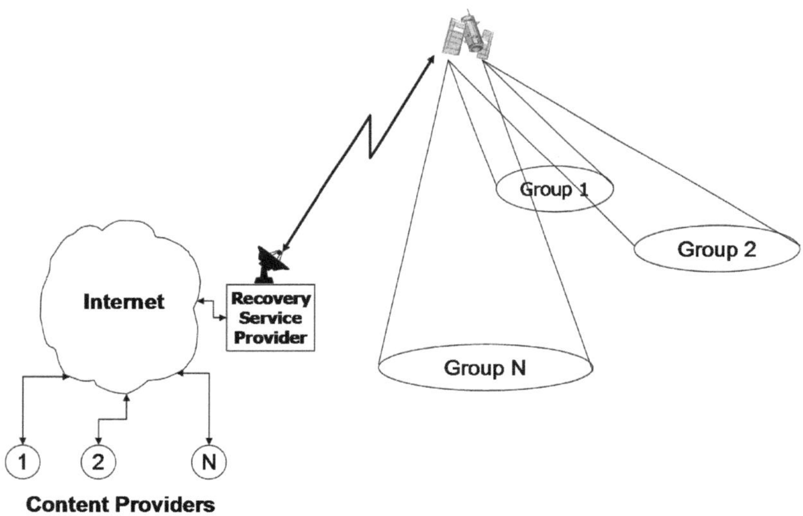

Fig. 3.3: DRT service architecture.

The *Content Providers* are the primary sources for video applications, i.e., they generate the real-time data. We can suppose that a content provider is located just before the satellite hop or, more generally, that the Internet spreads between them.

The *Recovery Service Provider* consists of a streaming proxy that has access to satellite resources and manages the retransmission priority. In fact, retransmission requests can be listed according to a priority that is related to the time constraints of the recovery phase, but also to the type of service and the customer class the service pertains to. It is worth noting that retransmission requests can be rearranged in time by the proxy, based on a metric that quantifies the importance of a data segment for a requesting customer, so that a simple FIFO scheduling of retransmissions is far to be optimal in terms of fairness, throughput and user satisfaction degree.

The user is actually to be considered as a set of customers (*Group 1*, *Group 2*, ... , *Group N*) located behind the satellite link, whose applications share some common bandwidth resources. Optimizing the usage of those resources is one of the goals of the envisaged architecture.

3.4.2 Scenario characterization and results

Every content provider sends a multimedia stream over the satellite link using a guaranteed bandwidth. According to Figure 3.3, there are N content providers and, therefore, N statically allocated channels. Data are sent to the streaming application after a playout delay (e.g., D seconds). Each receiver uses a local proxy buffer to store at most D seconds of streaming data, i.e., data to be played within D seconds. This "*elastic buffer*", that empties at constant rate and fills at variable rate, permits to continue the playout during the satellite channel outage, as long as sufficient information has been previously stored in the buffer. When a channel outage happens, the receiver (i.e., the proxy located at the receiver group) leaves a blank space in the application buffer and, when the channel is again available, sends a retransmission request to a *Recovery Service Provider* (RSP), in order to fill the hole in the elastic buffer. All the retransmissions use a shared channel, e.g., the $(N+1)$-th channel. We propose that, in this "*recovery*" channel, content providers retransmit the packets using a transport protocol with the *Additive Increase Multiplicative Decrease* (AIMD) scheme [27],[28]. In particular, the number of packets a sender can put on the network is limited by a *congestion window* (*cwnd*) that is managed as follows:

- Slowly (additively) increase the *cwnd* size as long as there is no congestion. Typically, the *cwnd* is increased by one packet for each window sent without a packet drop (in practice, $cwnd = cwnd + \alpha/cwnd$ as each ACK returns, with $\alpha = 1$).
- Quickly (multiplicatively) decrease the *cwnd* size as soon as congestion is detected. Typically, *cwnd* is halved for each window containing a packet loss ($cwnd = \beta/cwnd$, with $\beta = 0.5$).

In this way, the available bandwidth is fairly shared. After receiving a retransmission request, the RSP (which acts like a proxy for on-demand services) classifies the request according to the run-time estimated urgency. The urgency is calculated from the information requested and the time available for recovery purposes. Correspondingly, every user communicates a time interval and two timestamps conveyed by the retransmission request:

$$\Delta\mathrm{T}, [t_0, t_1] \tag{3.1}$$

where t_0 is the time when the broadcast connection became unavailable for the requesting receiver, t_1 is the time when the data to be retransmitted should be used by the multimedia player, and ΔT is the data window that is requested, i.e., the "*room*" to be filled in the receiver buffer, in playout seconds.

The RSP assigns a proper bandwidth to each retransmission, which is calculated from the corresponding urgency. The policy that determines the urgency of a request is based on both the difference (t_1-$t_{current}$) and ΔT, i.e., the intervals available to start and to complete the recovery procedure. This means that the urgency of a retransmission may change during the retransmission itself, so that bandwidth assignments have to be dynamically adjusted. Possibly, a policy function might run on the retransmissions codec, trying to accommodate multiple requests in the same channel.

Once the codec has been selected for a retransmission, the amount B of data to be sent is determined, and the following formula is used to compute the AIMD transmission parameters α and β:

$$B = r(\alpha, \beta) * (t_1 - t_{\mathrm{current}}) \tag{3.2}$$

where B is the amount of data to send at time t_1 and r is the rate to be achieved by means of an opportune choice of α and β.

A formula is shown in [29] that correlates the AIMD mean sending rate r with the control parameters, α and β, the loss rate p, the mean *Round Trip Time*, RTT, the mean timeout value, T_0, and the number b of packets each ACK acknowledges:

$$r(\alpha, \beta) = \frac{1}{TD_{\alpha,\beta} + TO_{\alpha,\beta}} \tag{3.3}$$

where:

$$TD_{\alpha,\beta} = RTT\sqrt{\frac{2b(1-\beta)}{\alpha(1+\beta)}p} \tag{3.4}$$

$$TO_{\alpha,\beta} = T_0 \min\left(1, 3\sqrt{\frac{(1-\beta^2)b}{2\alpha}p}\right) p(1+32p^2) \quad . \tag{3.5}$$

Thus, from the bandwidth value, the proxy calculates α and β parameters of the AIMD transport scheme, which will be communicated to every content provider that has to retransmit data.

Here we modeled the link with a good-bad process with exponentially distributed permanence times for both good and bad states. Real-time broadcast applications are always on, with a fixed bandwidth usage. Also the bandwidth available for retransmission is fixed and guaranteed by the distribution systems, and the playout delay of each receiving application is the same for all users. Furthermore, we represent each multicast group with a single user that acts as the worst-case user, so that the good-bad process actually refers to the time distribution of periods in which link failure occurs or not, for an entire multicast group. This assumption simplifies the simulative analysis while preserving the correctness of results; in fact, in our system, overlapping retransmission requests sum and turn into a single multicast retransmission. Finally, no codec adaptation has been considered.

As for the transport protocol, we have tested UDP-like retransmissions (the evaluation of TCP and AIMD-like protocols will be considered in a future study). However, preliminary results obtained with UDP, justify the study of connected transport protocols to enhance system performance.

As a reference, let us consider a scenario with $N = 10$ *Content Providers* generating an aggregate of 20 Mbit/s (each *Content Provider* generates at a fixed, but different rate of about 2 Mbit/s, to avoid synchronization effects), and a 6 Mbit/s bandwidth is guaranteed for recovery. The playout delay of users is 20 seconds, and the transport protocol is UDP. The average duration of the bad state of each link has been set to 5 s; we have obtained the results by changing the average duration of the good state and by collecting simulation results over 200000 seconds.

Figure 3.4 shows the aggregate amount of retransmitted data when the adopted retransmission priority is proportional to the bandwidth of the real-time stream. Curves are normalized to the aggregate number of bytes requested by users. The lower curve in the Figure represents data retransmitted for retransmissions that the system was able to complete. It is clear that a great number of retransmissions is stopped due to lack of resources as soon as the link error probability exceeds 0.2. Furthermore, for error probability greater that 0.1, the number of unrecoverable bytes increases (due to outage periods longer that the playout delay, which are now more frequent).

For the same scenario, Figure 3.5 depicts the aggregate delivered data and the amount of data lost due to link failures during the retransmission procedure. Lost data are normalized to retransmitted data and not to requested data, to give a correct measure of the needs of a connected transport protocol during the recovery procedure. Note that system performance is not satisfactory even with values of the link failure probability as small as 0.1, which is not so much for users.

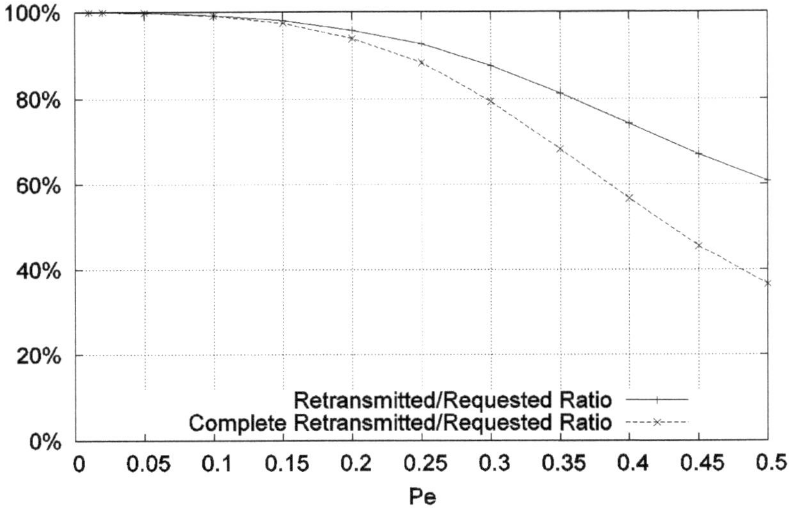

Fig. 3.4: Retransmitted data using a retransmission priority proportional to the required bandwidth.

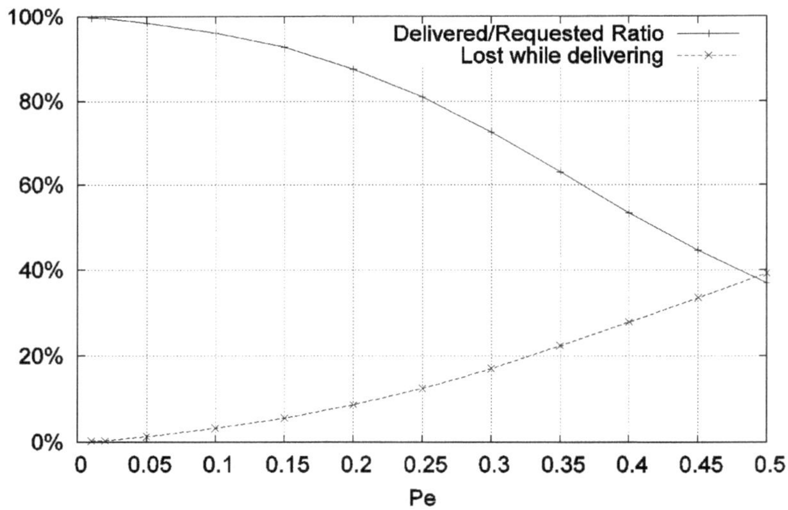

Fig. 3.5: Delivered and lost retransmitted data using a retransmission priority proportional to the required bandwidth.

3.5 Experimental results on QoS

Many of the application QoS requirement studies have been done in current-day Internet networks, for instance many of the considerations shown in Section 3.2. The aim of this Section is to describe the work carried out in a *Next-Generation Network* (NGN) prototype to characterize the application QoS requirements in such a kind of network. Results refer to real experiments on application behavior.

The test bed was an "*all IPv6*" network; Figure 3.6 depicts the network architecture. Two access technologies, one wired (Ethernet) and one wireless (IEEE 802.11), where employed. This can represent a subset of all the access technologies a future network operator may offer to its customers. Users, employing the appropriate devices could connect to any of the two networks. In the test bed, wireless connectivity is assured using commercial "SMC WLAN" cards with prism driver. Satellite links were not available in our test bed due to the complexity and high costs in using these links for experiments. We however believe that the obtained results may provide good insights also for general networks (including satellite links) in particular for what concerns the characterization of application behavior in NGNs with features such as mobility or QoS, using IP as convergence layer.

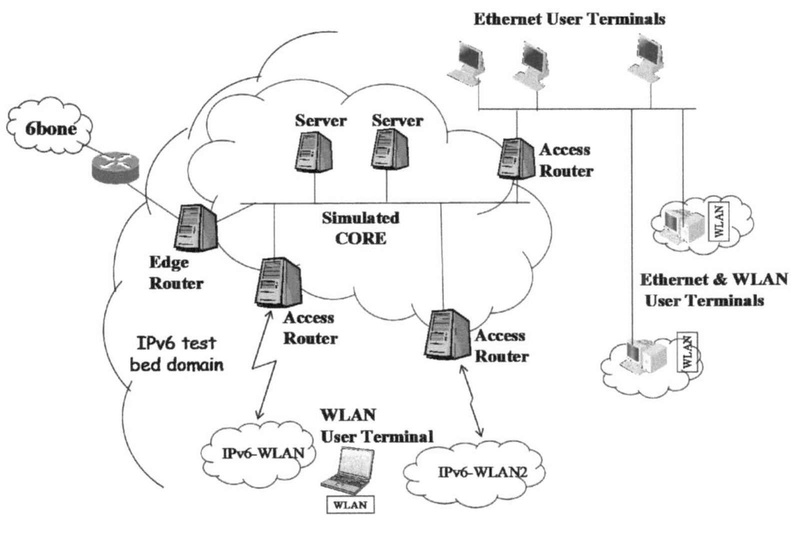

Fig. 3.6: NGN prototype test bed.

Our network was divided in 2 parts: (*i*) an "*access part*" where the users connect to via either Ethernet or WLAN (i.e., WiFi); (*ii*) the core network. The latter is connected to the "*6 bone*" (IPv6 Internet) via an *Edge Router* and to the "*access part*" and the users' terminals via the *Access Routers*. The core network hosts two servers supporting different functionalities of NGNs. These functionalities include aspects that should be present in next-generation commercial mobile networks, such as user authentication and accounting; mobility and QoS management were also controlled by these servers. All the nodes (including the routers) are general-purpose machines (Pentium III and IV PCs). All run Red Hat 7.2 with Linux-2.4.16 kernels. More details about the test bed can be found in [30].

QoS is based on DiffServ with access control. This access control is performed on the *Access Routers* on a per flow and per user basis. The *Access Router* outsources the admission decision to the QoS broker, an entity located in the core network able to take this decision and configure the routers with appropriate parameters.

The test bed here described is composed of general-purposes machines and it is just a mere representation of what a next network infrastructure may be, but we believe that the results obtained in it can provide us early and valuable hints about the applications specific QoS requirements when using NGNs.

We performed on-site real measurements of end-user performance perception and application characterization under different situations that can be present in NGNs, as detailed in [30] and [31].

The tested applications correspond to conversational services and interactive services. All of them were IPv6 applications. Conversational services were provided by *Robust Audio Tool* (RAT) for conversational voice and Quake 2 and Tetrisnet for games. Again, for interactive services we employed RAT (for audio streaming) and VideoLan for video streaming. Conversational and interactive services characterization was already described in Section 3.2; the added value of this Section is to show experimental studies obtained in an NGN prototype and check the differences.

Two kinds of tests were performed: the first was intended to characterize application behavior in terms of bandwidth needs (including burstiness); the second one experimented with user tolerance to delay, jitter and packet loss. We will show and analyze the results; the tests methodology is further detailed in [30].

For the first type of tests, *ethereal* [32], a network analyzer software, was used to capture the packets and *tcpstat* was adopted to analyze the application traffic. Two parameters were evaluated: packet size and packets per second. Mean, min, max, deviation and deviation/mean values were calculated for those two parameters. First, the results are presented and then some conclusions drawn. Audio stream has constant packet size and very small variation in packet rate. For video stream we have a nearly constant packet size and a small variation in packet rate. For conversational applications the results are as follows:

- Conversational voice presents a constant packet size, but also a high variation in packet rate.

- The Tetrisnet game generated a very low traffic, but with great variation in packet size and rate.
- Quake 2 generated more traffic and also had remarkable variations in packet size and a small variation in packet rate.

As a general conclusion, interactive applications have a higher bandwidth variation since they depend on user behavior: there is silence suppression, thus when the user does not talk no packets are sent. Moreover, Quake 2 bandwidth consumption depends on user activity: the more it interacts the larger the packets are, because more information needs to be sent (packets are sent at a rather constant rate). The bandwidth of the streaming application does not depend on user behavior, but only on the nature of scenes and audio. Obviously, the employed codecs play a fundamental role in determining application bandwidth consumption.

The results are as expected and similar to the ones obtained in the current Internet. However, there are some remarkable aspects worth to mention. For instance, mobility and overhead. Mobility in NGNs will be based on *Mobile IP* (MIPv6). This means adding, to the basic IP header the IP home address header and, also, generally the IPv6 routing header. For conversational applications with only audio, the payload is small and, as such, the ratio payload/overhead becomes very small. We also found NGNs specific results when dealing with applications adaptability. In NGNs, the users will roam between several access technologies with different performance characteristics. Applications should be able to cope with this heterogeneity adapting themselves, for instance in "layered" video, sending only detailed layers when the available bandwidth is high, for instance in downlink satellite links.

As aforementioned, the second type of tests evaluated user-perceived quality. NIST Net [33] is the software that can alter network conditions. It was employed to generate packet loss, delay and jitter in the test-bed network. Since NIST Net works only on IPv4 networks and the test-bed infrastructure was pure IPv6, a tunnel was set up. Table 3.5 presents the results. These results were as expected: conversational applications (Tetrisnet, Quake 2, and VoIP) have more strict requisites for delay and jitter. Tetrisnet is an exception, since it is an interactive application, but interaction speed is rather small (in the order of a second) so that delay requirements are very loose.

Application	Packet loss (%)	Delay/Direction (ms)	Jitter/Direction (ms)
Audio Stream	2	> 500	100
Quake 2	15	100	150
VoIP	10	150	50
Tetrisnet	20	> 500	> 500

Table 3.5: QoS requirements as measured in the NGN prototype.

The obtained requirements are similar to those presented in Section 3.2 for nowadays networks. The specific aspects of NGNs can be found mainly in the fact that network QoS is priced and tailored for the users. As such, we found that low profile users, "paying" less for the transport service where much more tolerant with their requirements. Besides, for some users, more than having better QoS, the important aspect was the unique NGN ability of supporting all kinds of applications and having seamless inter-technology handovers with the capability of taking the best profit from the available access technologies.

3.6 Conclusions

This Chapter stressed on the importance of providing QoS for data transport. Some applications have stringent QoS requirements, mainly related to delay and jitter. Satellite networks may suffer from too high delays so QoS aspects should be considered very carefully. On the other side, satellite networks are very well suited for multicast and broadcast transmissions as well as for DRT services. For about 6 years now, satellite networks are also a commercial solution for completely different scenarios: unicast bidirectional services like broadband Internet access. These scenarios, requiring strong QoS requirements, need a careful analysis and the implementation of mechanisms to support QoS as discussed in the next Chapters of this book.

References

[1] ITU-T Recommendation G.1010: "End-user multimedia QoS categories", URL: http://www.itu-t.org.

[2] ITU-T Recommendation Y.1541: "Network performance objectives for IP-based services", URL: http://www.itu-t.org.

[3] ITU-T Recommendation F.700: "Framework Recommendation for audio-visual/multimedia services", URL: http://www.itu-t.org.

[4] 3GPP, "Technical Specification Group Services and System Aspects Service aspects; Services and Service Capabilities", TS 22.105 V6.0.0 (2002-09) (Release 6), URL: http://www.3gpp.org.

[5] ETSI, "Satellite Earth Stations and Systems (SES); Broadband Satellite Multimedia (BSM) services and architectures; Functional architecture for IP interworking with BSM networks", TS 102 292, V1.1.1 (2004-02).

[6] 3GPP, "QoS Concept and Architecture", TS 23.107, URL: http://www.3gpp.org.

[7] ITU-T Recommendation G.114: "One-way transmission time", URL: http://www.itu-t.org.

[8] P. Barsocchi, N. Celandroni, F. Davoli, E. Ferro, G. Giambene, F. Castaño, A. Gotta, J. I. Moreno, P. Todorova, "Radio Resource Management across Multiple Protocol Layers in Satellite Networks: A Tutorial Overview", *International Journal of Satellite Communications and Networking*, Vol. 23, No. 5, pp. 265-305, September/October 2005. ISSN: 15442-0973.

[9] R. Braden *et al.*, "Integrated Services in the Internet Architecture: an Overview", IETF RFC 1633, June 1994.

[10] S. Blake *et al.*, "An Architecture for Differentiated Services", IETF RFC 2475, December 1998.

[11] R. Braden *et al.*, "Resource ReSerVation Protocol (RSVP) - Version 1 Functional Specification", IETF RFC 2205, September 1997.

[12] K. Nichols *et al.*, "A Two-Bit Differentiated Services Architecture for the Internet", IETF RFC 2638, July 1999.

[13] K. Nahrstedt *et al.*, "The QoS Broker", *IEEE Multimedia*, Vol. 2, No. 1, pp. 53-67, Spring 1995.

[14] G. Cortese *et al.*, "CADENUS: Creation and Deployment of End-User Services in Premium IP Networks", *IEEE Communication Magazine*, Vol. 41, No. 1, pp. 54-60, January 2003.

[15] G. Schollmeier *et al.*, "Providing Sustainable QoS in Next-Generation Networks", *IEEE Communications Magazine*, Vol. 42, No. 6, pp. 102-107, June 2004.
[16] Gilat Satellite Networks, URL: www.gilat.com.
[17] StarBand, URL: http://www.starband.com/.
[18] Thales Broadcast and Multimedia, URL: http://www.thomcast.com/.
[19] Tandberg, URL: http://www.tandbergtv.com.
[20] TerraTec Electronic GmbH, URL: http://www.terratec.net/.
[21] Hispasat, URL: http://www.hispasat.com.
[22] Viasat, URL: http://www.viasat.com.
[23] Wildblue, URL: http://www.wildblue.com.
[24] Centra, URL: http://www.saba.com/centra-saba/.
[25] Hughes, URL: http://www.hughes.com.
[26] Kencast, URL: http://www.kencast.com.
[27] E. Altman, C. Barakat, V. Manuel Ramos, "Analysis of AIMD Protocols over Paths with Variable Delay", *INFOCOM 2004*.
[28] L. Cai, X. Shen, J. W. Mark, J. Pan, "A QoS-Aware AIMD Protocol for Time-Sensitive Applications in Wireless/Wired Networks", in *Proc. of IEEE Infocom'05*, Miami, Florida, March 13-17, 2005.
[29] Y. R. Yang, S. S. Lam, "General AIMD Congestion Control", University of Texas, *Tech. Rep.* TR-2000-09, May 2000.
[30] P. Serrano *et al.*, "Medida y análisis del tráfico multimedia en redes móviles de cuarta generación", *Telecom*, I+D 2004, Madrid.
[31] A. Cuevas *et al.*, "Usability and Evaluation of a Deployed 4G Network Prototype", *Journal of Communications and Networks* (ISSN: 1229-2370), Vol. 7, No. 2, pp. 222-230, June 2005.
[32] Ethereal: A Network Protocol Analyzer, URL: http://www.ethereal.com/.
[33] NIST Net Home Page, URL: http://snad.ncsl.nist.gov/itg/nistnet/.

4

CROSS-LAYER APPROACHES FOR RESOURCE MANAGEMENT

Editor: María Ángeles Vázquez Castro[1]

Contributors: Franco Davoli[2], Erina Ferro[3], Giovanni Giambene[4], Petia Todorova[5], María Ángeles Vázquez Castro[1], Fausto Vieira[1]

[1]UAB - Universitat Autónoma de Barcelona, Spain

[2]CNIT - University of Genoa, Italy

[3]CNR-ISTI - Research Area of Pisa, Italy

[4]CNIT - University of Siena, Italy

[5]FhI - Fraunhofer Institute - FOKUS, Berlin, Germany

4.1 Introduction

The enormous advantages of physical layer adaptivity for adequate operation of wireless systems over widely-varying channel conditions have been widely proved. However, an optimal adaptation strategy for a given set of resource constraints requires a joint optimization across layers. Such a cross-layer optimization is becoming a new paradigm for wireless system design, which can be extraordinarily complex as the number of optimization parameters and layers grows.

In this Chapter, we present a comprehensive literature survey of existing cross-layer design approaches for resource management optimization in order to draw some preliminary conclusions on adaptive satellite systems.

4.2 Literature survey on cross-layer methods

Fade Mitigation Techniques (FMT) allow for adaptation to the dynamics of the physical system, thus introducing a new concept in system design, no longer based on worst-case behavior. Three different FMT types can be distinguished (see for instance [1]), each of them introducing a diverse degree and nature of adaptivity: power control techniques, diversity techniques and adaptive waveform techniques.

A conventional protocol stack employs independent design of protocol layers, thus precluding adaptation of the system to changing conditions. Cross-layer optimization offers a new paradigm for the design of next-generation wireless networks. As satellite-based systems evolve towards Internet-centric networks, system adaptivity poses new challenges; for example, dynamic resource management to provide the different QoS requirements and *Service Level Agreements* (SLAs), suitable for multimedia.

Cross-layer methods provide a natural solution to the challenges of adaptation to both system dynamics and the demands of highly dynamic applications. In order to optimize the overall performance, the joint adaptation of several layers must be coordinated, requiring a new cross-layer framework to be designed and standardized. It is important to realize that different communities have somewhat diverse perspectives on cross-layer optimization. For instance, the networking community has proposed developing protocols and mechanisms to adapt the network to the applications. Conversely, the video community has suggested adaptation of the source coding to the network, since Shannon's separation theorem does not apply to general time-varying channels, or to systems with a complexity or delay constraint. At the satellite-dependent layers (i.e., physical and MAC layers), there are proposals to adapt the radio resource management to pre-defined traffic profiles and to changing propagation conditions. In general, cross-layer design involves interactions among five key protocol layers: application layer (including presentation and session), transport layer, network layer, link (MAC) and physical layer.

A cross-layer approach requires the introduction of new control functions in the protocol stack in order to enable interactions between non-adjacent protocol layers. This is in itself an important topic of research and one that is currently not well understood in the general case. Initial solutions are therefore likely to be oriented for *ad hoc* optimizations for specific protocol stacks and may be suited to only a small number of system scenarios. Once the approaches are well understood, future work may seek to generalize the primitives and control exchanges.

In designing a cross-layer architecture for satellite networks, care must be taken to consider the implications and the principle of layer separation. In particular, it is important to define the extent to which parameters at a lower (e.g., physical) layer should influence control strategies at higher layers (e.g., network QoS, transport reliability, application data format) [2]. This may be dependent on the specific environment and on the type of control exerted on the system. Separation principles (which are also related to time scales) may be adopted in adaptive hierarchical control systems, whereby tighter (regulatory control) actions are taken at lower layers, and their effect is perceived through aggregate parameters. However, especially in satellite systems, the presence of protocol enhancing proxies with specific protocol stacks may mitigate the potential negative effects of cross-layer interactions on the network as a whole.

The cross-layer protocol design entails a protocol stack optimization on the basis of novel interactions even between non-adjacent protocol layers. Due to the specificity of the optimization process, the cross-layer design should be suitably tailored for each examined protocol stack and systems scenario. In particular, among these scenarios, we may consider two most significant cases: (*i*) DVB-S/-RCS (or DVB-S2) -based systems for GEO-based broadband communications; (*ii*) S-UMTS systems for GEO or non-GEO-based communications to mobile users.

In the following paragraphs, a preliminary literature survey is provided in order to illustrate the available cross-layer methods. The different proposed cross-layer approaches have been categorized according to the layers or layered functionalities that are jointly optimized.

Joint PHY/MAC optimization

In [3], the authors provide a cross-layer optimized design of the MAC layer under Rayleigh fading, based on a Markov chain formulation. System information and physical layer measurements are jointly considered with the intention of maximizing the overall throughput. In [4], a discussion on protocol harmonization for MAC and physical layer for IEEE 802.11 is addressed. The authors investigate the effects of packet length, transmit power and bit-error rate. Their results show that minimum energy is consumed for an optimal transmission power, which is proportional to the packet length. In [5], the joint effects of finite length queuing at MAC layer and adaptive coding and modulation are analyzed. The performance gain is quantified when applying cross-layer design to maximize throughput. In [6], the authors describe the flow of information between PHY and MAC layers in order to save power and to improve overall performance via an adaptive distributed MAC (uplink) protocol. Several authors propose link layer adaptation to reduce the transmission errors based on current channel conditions. In [7], around 50% improvement in goodput and 20% improvement in transmission range is shown to be obtained by using the optimal *Maximum Transfer Unit* (MTU) for a particular BER. In [8], it is shown that an 18-25% throughput gain may be obtained by increasing the frame length, depending on radio conditions. In [9], the authors focus on the cross-layer optimization of the scheduling policies to assure queuing stability. In [10], the issue of jointly optimal energy allocation and admission control for communication satellites in Earth orbit (LEO, MEO and GEO) is addressed. Using a dynamic programming approach, an optimal policy is derived.

In general, information about channel conditions can be used to adapt the coding or schedule transmission [11]-[13]. In [14], several levels of adaptation are proposed within each layer, fast and slow ones. The adaptation also covers the "hardware" layer. In [15], the authors propose a cross-layer design approach using perfect prediction-based wireless channel conditions to improve the performance of a multicast packet scheduler over satellite network environments in the downlink transmission. In [16], cross-layer methods are used to improve the efficiency of reliable multicast services supported by GEO satellites. The reliability issue has to be carefully taken into account, since satellite resources are expensive and link quality degrades significantly during adverse weather conditions. This paper proposes to remove at low layers, most of packet discarding, but introduces an additional protection for protocol headers. Moreover, at transport level erasure coding is used in combination with a hybrid-ARQ protocol. Such approach allows that applications (like massive file transfers) requiring full reliability are less demanding in terms of network resources.

Joint PHY/MAC/APP optimization

A coordinated cross-layer adaptation can be considered to meet QoS demands from the application layer. In [17], a mechanism is proposed to map QoS levels of scalable video to the QoS levels of the transmission, both being time-varying. Scheduling policies are derived allowing QoS mapping interaction between the video coder and the transmission module. In [18], a cross-layer framework for WLAN QoS support is proposed. The authors show that QoS at MAC layer can be optimized by taking advantage from layers 4-7 information. In [19], a joint cross-layer design for QoS content delivery is proposed. The authors derive a QoS-aware scheduler and power adaptation scheme at both uplink and downlink MAC layer to coordinate the behavior of the lower layers for an efficient utilization of resources. They show that the cross-layer design provides a good scheme for wireless QoS content delivery. In [20], power saving is proposed by using feedback from the application about delay sensitivity. Moreover, information about the type of coding used by a video-application could be used by the frame scheduler at the network interface to save power [21].

In a similar context, the problem of QoS mapping between adjacent layers has been recently treated in [22],[23]. Rather than considering specifically the network and the MAC layers, the problem is posed in a more general setting, as defined by the ETSI *Broadband Satellite Multimedia* (BSM) protocol architecture [24],[25], at the *Satellite Independent - Service Access Point* (SI-SAP). Specifically, the interworking between the *Satellite-Independent* (SI) and *Satellite-Dependent* (SD) architectural components is considered by taking into account both the change in encapsulation format and the traffic aggregation (in the passage from SI to SD queues). In the presence of IP DiffServ queues at layer 3, the problem consists in dynamically assigning the bandwidth (service rate) to each SD queue, so that the performance required in the SI IP-based SLA is guaranteed. By considering a fluid model and the loss volume as the performance indicator of interest, the *Infinitesimal Perturbation Analysis* (IPA) technique of Cassandras *et al.* [26] is applied. Assuming that the SI layer is properly configured, in order to satisfy the requirements (i.e., the IP buffers do not constitute a bottleneck for QoS) the MAC resource allocation is performed to maintain on-line the equalization between the loss volumes at the network layer and at the MAC layer. In doing so, the allocation is dynamically adapted, to follow both traffic and fading variations. More details on this scheme are provided in Section 8.4.

Joint optimization of layers involving transport layer

The transport layer is in charge of establishing end-to-end network connections. Transport protocols like TCP interpret large delays and packet losses, typical of a wireless channel, as a congestion event, thus affecting the TCP performance.

In [27], it is shown that increasing MAC level retransmissions, in order to avoid TCP retransmissions, decreases the power consumption. In [28] and [29], TCP windows are optimized according to the application priority. The bandwidth assignment problem for long-lived TCP connections in a faded satellite environment is addressed in [30], where cross-layer optimization approaches between physical and transport layers are presented. Another example of physical-transport cross-layer approach can be found in [31], where the authors demonstrate that it is possible to obtain a better performance for TCP connections by jointly choosing the bit error rate and the information bit-rate of satellite links that maximize the goodput of a single TCP connection, without touching the TCP stack.

In [32], an innovative resource allocation algorithm, based on a cross-layer interaction between TCP and MAC layers is proposed for a DVB-RCS scenario. Such an algorithm aims to synchronize the requests of resources with the TCP transmission window trend. The obtained results show that the scheme permits to reduce the delay, to increase the utilization of air interface resources, and to achieve a fair sharing of resources among competing flows. This approach calls for a TCP-driven *Dynamic Bandwidth and Resource Allocation* (DBRA) to be operated at layer 2 so as to reduce the queuing delay (layer 2) and congestion phenomena (with timeout expirations) [33]. More details on these techniques are shown in Section 9.4.

In split scenarios [34], the end-to-end TCP semantics is broken. The satellite link is isolated by the terrestrial segment and interconnecting routers (*Performance Enhancing Proxies*, PEPs) are used that close the TCP flow. PEPs are typically implemented at transport or application layer. Examples of transport layer PEPs are TCP spoofing and TCP connection-split proxies.

In both PEP types, the goal is to shield high-latency or lossy satellite network segments from the rest of the network, in a transparent way to applications. A critical issue in PEP is the design of buffers and related management rules and sizes. Interesting proposals envisage the adoption of *Active Queue Management* (AQM) at the MAC layer for improving the TCP performance. In AQM, when the router determines that the bandwidth is fully utilized, packets are dropped even when the queue is not full in order to reduce the data injection rate of the TCP sender [35].

In [36], experimental quantitative performance metrics can be found; they are obtained by using H.264 and UDP-Lite for the next-generation transport of IP multimedia. A cross-layer technique is proposed that features partial checksum coverage for the packet header allowing the application to signal implicitly the link CRC coverage. The sending end-host implicitly signals (i.e., without explicit control messages) by using a modified transport header, such as UDP-Lite. This work discusses the architectural implications for enhancing performance of a wireless and/or satellite environment.

Joint optimization of layers involving call admission control

Reference [37] presents an overview of high-speed mobile satellite communication systems, the technologies adopted or planned for deployments, and the challenges. Various physical channel models for characterizing the mobile satellite systems are presented. The most prominent technologies used in the physical layer, such as coding and modulation schemes, multiple-access techniques, diversity combining, etc., are discussed in the scenario of satellite systems. What is interesting in our context is the overview of cross-layer design methods employed in satellite systems, in particular those that involve joint network and physical layer optimizations, or joint MAC and physical layer optimizations. Specifically in the GEO satellite environment, different forms of parametric *Call Admission Control* (CAC) strategies have been proposed, among others, in [38],[39], and [40], which are all based on a cross-layer optimization. In [38], where the presence of both *Variable Bit Rate* (VBR) MPEG connections and *Available Bit Rate* (ABR) data has been considered, CAC is exerted with the goal of keeping the probability that the bandwidth dedicated to VBR exceed a given value below a predetermined threshold. A bandwidth expansion factor, whose value is adaptively adjusted on the basis of measurements, is used to account for statistical multiplexing effects in VBR traffic. FEC and MPEG coding rate adjustments are other corrective actions taken to cope with traffic and channel variations. The approach taken in [39] and [40] considers real-time *Reserved Bandwidth* (RB) and *Best Effort* (BE) traffic; however, no rate adjustment derived from application-level coding is assumed to be available for RB flows. Adaptive cross-layer bandwidth partitions are derived per station, based on stationary performance indexes, such as the call blocking probability for RB connections and the loss probability for data packets, which are recomputed at each significant change in fading or traffic intensities. The control architecture has a hierarchical structure, where CAC tasks are delegated to local controllers at the stations, and uplink capacity partitions for the Earth stations are adaptively determined by a *Master Control Station* (MCS). Owing to the dynamic fade changes, the bandwidth assigned to an Earth station may be temporarily insufficient to carry on the currently ongoing number of RB connections; since inelastic traffic is considered, in such cases one or more ongoing calls would be dropped. However, reallocations of the bandwidth partitions upon detection of significant changes in traffic intensities and fading classes do help in reducing the probability of this event. As regards the MCS, the bandwidth allocation is formulated as an optimization problem in a discrete setting (with the assignment granularity determined by the *Minimum Bandwidth Unit*, MBU); if the performance index is a separable function of the station parameters (e.g., a sum of terms, each depending only on the bandwidth to be assigned to a station), the problem can be numerically solved by applying dynamic programming over the stations [39],[40], possibly in a form that may greatly reduce the search space, by exploiting the presence of constraints.

It is worth noting that these model-based approaches can be by-passed by using a fluid approximation and by treating the bandwidth partitions as continuous variables. A gradient descent technique can be adopted, in conjunction with IPA for gradient estimation [27],[28]. The advantage of these methodologies is that they are measurement-based and they require neither the knowledge of any functional form of the performance index nor any characterization of the traffic sources.

A cross-layer radio resource management problem involving network and MAC layers has been extensively considered in [29],[41], and [42]. In particular, *Dynamic Capacity Allocation* (DCA) is applied, by computing bandwidth requests for each Earth station's DiffServ queue, which are passed to a centralized scheduler, typically residing in an MCS. The latter assigns the bandwidth proportionally to the requests received. The requests are computed on the basis of queuing models, capturing both *Short Range Dependent* (SRD) and *Long Range Dependent* (LRD) behaviors, and by using as QoS metric the probability of the length of each service queue to exceed a given threshold, depending on the service; this probability must be kept below a specified value, beyond which the station is considered in outage. The scheduling of the MAC queues must be such that this constraint is maintained for the IP-level queues [i.e., those corresponding to *Expedited Forwarding* (EF), *Assured Forwarding* (AF) and *Best Effort* (BE) services within a given Earth station]. The remaining capacity is assigned on a free basis, according to *Combined Free/Demand Assignment Multiple Access* (CF/DAMA). Only traffic is taken into account (fading variations are not considered), but, as noted in [29], the effect of fade countermeasures might be included as a reduction in the available uplink bandwidth.

Concluding comments

From this literature review, some general conclusions can be drawn as follows:

- Little work has been published to date on cross-layer optimization in the satellite context.
- Most of the cross-layer optimizations proposed in the terrestrial wireless realm involve physical layer and MAC layer. After these two layers, the application layer is also widely considered. TCP is a particular case in the sense that very different alternatives have been explored in order to optimize the TCP protocol itself, especially over satellite channels.
- Two main system performance parameters are optimized: QoS or service differentiation, in particular harmonization of QoS across layers, and throughput. A special attention is also paid to energy saving, which may not be directly applicable to a satellite scenario.
- A wide variety of methodologies are presented and therefore no mature general methodology seems to be available. Moreover, every published work seems to follow an *ad-hoc* cross-layer methodology for the particular case to be optimized.

4.3 The need of a cross-layer air interface design

The ISO/OSI reference model and the Internet protocol suite are based on a layering paradigm. The target of the ISO/OSI reference model was to define an 'open system' so that different network elements can interwork independently of manufacturers. The OSI protocol stack entails 7 different abstraction levels, addressing *separately* communication tasks. Each protocol solves a specific problem by using the services provided by modules below it and giving a new service to upper layers. The main interest here is on IP-based scenarios. The Internet protocol stack is slightly modified with respect to the ISO/OSI one and entails 4 layers, as depicted in Figure 4.1.

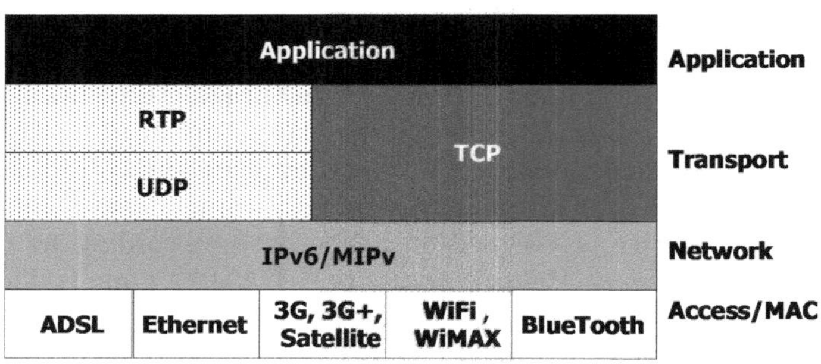

Fig. 4.1: Current view of the Internet protocol stack.

Standardization bodies define the different protocols that a system can use to exchange information. The implementation of interfaces is left free to manufactures, provided that they support the primitives that determine the service.

The disadvantages of the strict layered approach can be detailed as follows:

- The needs of a service provided by the communication system to its users are defined at the top-level. The hierarchy and the overall performance of the system is however build upon the bottom-level.
- The bottom level does not communicate directly, but through all higher layers with the top-level. Information is lost during this layer-by-layer top-down conversion.
- Layers are independently optimized.

The challenging characteristics of satellite communications are:

- Dynamically-varying channel characteristics; both slow and fast variability are present in a satellite scenario depending on whether mobile or fixed users are considered;
- Similar to terrestrial mobile channel, the satellite mobile channel lacks of reliability (need of countermeasures: coding, retransmissions, modulation techniques, diversity, etc.);
- Strong influence of intra-system interference levels;
- Bandwidth shortage and need of supporting broadband applications; necessity of managing the bandwidth in an efficient way;
- QoS support for multimedia traffic classes;
- Interoperability among different wireless networks (2.5G, 3G, 4G, WiFi, WiMAX, satellite, etc.).

A strict modularity and layer independence may lead to non-optimal performance in IP-based next-generation satellite communication systems. Furthermore, the growth of heterogeneous networks entails the need of adaptive actions. Finally, since both radio resources and power are strongly constrained, a system optimization is needed. In this framework, a better adaptation to system dynamics and traffic demands can be attained by employing a cross-layer approach with interactions even between non-adjacent protocol layers.

Without a cross-layer design in the air interface we can expect a loss of system efficiency according to some typical problems outlined below.

- IP packets lost due to errors induced by the wireless channel are interpreted as signals of congestion at the TCP level, thus lowering the bit-rate (congestion window). A long time is needed to recover (in terms of TCP goodput) after a loss event especially when multiple losses occur that cause a TCP timeout.
- Radio resources can be also allocated to mobile users that have bad channel conditions.

- Intra-system and inter-system handoff procedures can take a too long time that leads to connection interruption or higher layer protocol timeouts.

System efficiency is an important task in satellite communications where radio resources are costly and scarcely available. System efficiency is needed for allowing a mass-market diffusion of satellite services. Whereas, QoS support is the mandatory aspect requested by end-users who do not care about resource utilization, but expect a good service. Resource utilization and QoS support are typically conflicting needs; for instance, the best QoS condition for delay-intolerant traffic is to have a high amount of available resources, thus contrasting with system efficiency. These conflicting needs can be solved by means of a suitable cross-layer system design and by exploiting the multiplexing effect. In particular, the different layers of the OSI protocol stack should be jointly optimized or dynamically jointly adapted to find the best trade-off between resource utilization efficiency and QoS provision.

The idea behind cross-layer design is that we can obtain substantial gains in performance and efficiency by jointly optimizing the behavior of different layers. For example, source compression at the application layer can improve with knowledge of the transmission rate being used at the link layer. Moreover, the network layer can gain by looking both up and down the stack in order to obtain route diversity and multilink routing, where the routing algorithm might add redundant links if link layer provides an unreliable channel or if QoS constraints from the application layer are particularly tight. Satellite communication systems optimization calls for a vertical design of the air interface protocol stack.

The cross-layer approach requires new interfaces across the layers, which exchange control information beyond the standard ISO/OSI structure to improve the interactions among layers. Cross-layer interfaces can be within, between or beyond adjacent abstraction layers. Although interfaces between adjacent layers are in general preferable, there can be the need for efficient and direct interaction between non-adjacent layers; in general, a layer should be aware of the other layers of the protocol stack. Cross-layer information can be exchanged from higher to lower layers (*top-down approach*) or from lower to higher layers (*bottom-up approach*).

In the classical OSI stack, the exchange of information between adjacent layers is performed through 'send' and 'receive' primitives. In a classical layered approach, non-adjacent layers can communicate only involving intermediate layers. The novelty of the cross-layer approach is to allow the exchange of control information (signaling) among non-adjacent layers [43]. For instance, a 'get function' can be used by higher layer protocols to acquire the internal state of lower layer protocols; moreover, a 'set function' can be adopted by higher layer protocols to change the state of lower layer protocols. Different solutions have been proposed to support the cross-layer exchange of signaling information; an interesting method has emerged from the following papers [44]-[46] where a 'global coordinator' of the different layers is considered allowing to acquire the internal state information from the different protocols to store it in a shared memory and to set the state of the protocols to be adaptable to different events (see Figure 4.2a). The global coordinator may reside in the MAC (i.e., *MAC-centric approach*), in the application layer (i.e., *application-centric approach*) or being an external entity. It should be noted that in a slowly-varying scenario, such as for example the interactive broadband satellite channel with stationary users, the MAC layer could control adaptability (coordinating cross-layer interactions) in an optimal way [47]; this is the case of the MAC-centric approach presented in Figure 4.2b.

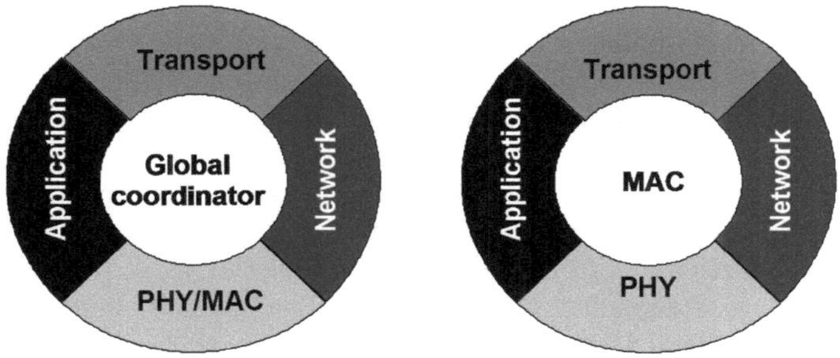

Fig. 4.2: (a) Possible cross-layer air interface based on a global coordinator; (b) Possible MAC-centric cross-layer air interface.

4.4 Cross-layer design: requirements depending on the satellite scenario

4.4.1 Broadband satellite scenario requirements (DVB-S/S2)

Next-generation multimedia broadband satellite networks require the development of key technologies to increase the capacity and efficiency as well as to decrease the total cost for the end-user. Such requirements call for very high throughput, flexibility, multi-beam processing and system adaptivity.

- **Role of Ka band**: Current bent-pipe Ku band satellites create difficulties to develop profitable multimedia satellite models. The current deployment of Ku band spot-beams and frequency re-use will probably be effective for a near-term business model. However, spot-beam coverage, in conjunction with Ka band frequency, can be extremely advantageous. Satellite transponders operating at Ka band frequency permit to achieve a higher G/T and, therefore, higher return channel burst rates. With lower power levels, the price of the terminal significantly decreases. The launch of additional Ka band capacity will greatly affect the multimedia satellite market and will probably lead to more successful models and profitability.

- **Role of DVB-S2**: Typical Ku band broadcasting links are designed with a clear-sky margin of 4 to 6 dB and a service availability target of about 99% of the worst month (or 99.6% of the average year). Since the rain attenuation curves are very steep in the region 99% to 99.9% of the time, many dBs of the transmitted satellite power are useful, in a given receiving location, only for some ten minutes per year. Unfortunately, this waste of satellite power/capacity cannot be easily avoided for broadcasting services, where millions of users, spread over very large geographical areas, receive the same contents at the same time. However, this design methodology devised for broadcasting systems is not optimal for unicast networks. In fact, the point-to-point nature of link connections allows exploiting space and time variability of end-user channel conditions for increasing average system throughput. This is achieved by *Adaptive Coding and Modulation* (ACM) format to best match the user SNIR, thus making the received data rate location- and time-dependent. The inclusion of advanced coding and modulation schemes has been the first objective of the DVB-S2 working group. In particular, ACM has been considered as a powerful tool to increase system capacity, allowing for better utilization of transponder resources and hence providing additional gain with respect to current DVB-S systems. Therefore, ACM is included as normative in DVB-S2 for the interactive application area and optional for *Digital Satellite News Gathering* (DSNG) and professional services. The standardization of the use of ACM by the DVB-S2 standard, introduces therefore an adaptive physical layer, which calls for the development of optimum adaptive resource management strategies to exploit fully ACM potentialities.
- **Applications requirements**: The requirement of increasing bi-directional data rates so that multimedia broadband satellite solutions can be closer to the specifications of terrestrial networks is undoubtedly a core need for any DVB-based or DOCSIS-based network due to the rise in video and large file transfers in enterprises. Future broadband satellite networks should aim to create more symmetry between forward and return links due to a perceived future demand for symmetric applications such as videoconferencing or interactive e-learning. Moreover, satellite solutions must include features and functionalities similar to a terrestrial solution in order to integrate into and coexist with current enterprise infrastructures.

In order to meet application requirements especially of future satellites that implement adaptive physical layer (DVB-S2), a logic reasoning is that cross-layer design is essential to exploit fully new technologies potentialities instead of loosing them by constraining the design to the conventional protocol stack with independent layers.

In what follows, per-layer-based requirements for cross-layer design of broadband satellite systems are presented from the layer 2 perspective.

- **Physical layer requirements**: The DVB-S2 ACM modulator operates at constant symbol rate, since the downlink carrier bandwidth is assumed constant. A sequence of physical layer frames TDM multiplexed is transmitted. Each frame transports a coded block and adopts a uniform modulation format. However, when ACM is implemented, coding scheme and modulation format may change frame-by-frame. Via a return channel, individual *Satellite Terminals* (STs) provide to the *Gateway* (GW) information on the channel status, by signaling the SNIR and the most efficient modulation and coding scheme the ST can support. The ST indications are taken into account by the GW in coding and modulating the data packets addressed to each ST. It is then apparent that the resource management functionalities shall be aware of the physical layer adaptation in order to follow the time variability of capacity.
- **Network layer requirements**: IP-layer QoS provision should be adequately mapped to layer 2 radio resource management protocols. Adequate attention should be also paid to both IntServ and DiffServ approaches. Different multimedia traffic should be provided either with reserved capacity or capacity on demand and QoS guarantees. AF, EF and BE traffic flows of the DiffServ scheme should have an adequate mapping at layer 2. Suitable layer 2 intelligence should be able to perform this important task. In case the broadband satellite sub-network is used as a stand-alone end-to-end network, where the end-to-end QoS can be controlled, a practical solution may be to apply guaranteed QoS to the access network. The implementation of this hybrid solution still needs to be investigated since it requires end-to-end network coordination.
- **Transport layer requirements**: resource management schemes may account for the specific transport layer traffic characteristics, such as TCP, UDP and multicast/broadcast. Note that in this scenario (i.e., broadband satellite communications for fixed users) a memoryless channel has to be considered that causes random packet losses, impacting the performance of the transport layer. Few examples are provided below.
  - The ECN (*Explicit Congestion Notification*) signaling for TCP traffic could be exploited at layer 2 to modify some traffic shaping functions or policing schemes.
  - The TCP congestion window (estimating the network congestion level) could be used at layer 2 to adaptively reserve capacity for TCP-based traffic; such approach could improve the QoS experienced for TCP-based applications and could also improve the multiplexing efficiency of such traffic flows (throughput). Note that the congestion window behavior plays a fundamental role in TCP-based satellite communications due to the very high round-trip propagation delays.
- **Application layer requirements**: different traffic types (e.g., real-time traffic and non-real-time traffic) should have specific SLAs and a monitoring action should be jointly performed with layer 2 in order to modify adaptively the service priority.

4.4.2 Mobile satellite scenario requirements (S-UMTS)

The mobile user scenario adds specific criticalities in the management of resources due to the dynamically changing propagation conditions. Such circumstances made even more crucial the need of cross-layer protocol design. The management of air interface resources (layer 2) must be improved to exploit dynamically updated information exchanged with all the other layers and, in particular, OSI layers 1, 3, 4 and 7. In fact, the congestion of the scarcely available satellite air interface resources as well as the congestion of the related fixed network are too critical aspects that must be taken into due account when designing the air interface protocol stack and, in particular, layer 2 resource management protocols.

Focusing on cross-layer information available at layer 2, we can consider the following contributions coming from other (even non-adjacent) layers:

- **Physical layer requirements**: radio channel conditions should be continuously estimated. In particular, signal strength, BER or PER estimations should be made available to implement multi-mode (i.e., modulation and coding) adaptivity and the selection of appropriate formats and priority levels at layer 2. These capabilities are supported by a possible satellite extension of the *High Speed Downlink Packet Access* (HSDPA) standard, as discussed in Chapter 5.
- **Network layer requirements**: in the IP traffic management, user mobility should be adequately taken into account. Hence, layer 2 protocol should provide a prioritized management for traffic coming from uses that incur in handover phases (this may be very important and time-critical in the presence of non-GEO satellites). In addition to this, mechanisms for IP-layer QoS provision should be adequately mapped to layer 2 radio resource management protocols, as already described in the previous sub-Section (see requirements for network layer in sub-Section 4.4.1).
- **Transport layer requirements**: resource management schemes should be improved to account for the suitable rules for specific transport layer traffic, such as TCP, UDP and multicast/broadcast. Note that in this scenario correlated packet losses are experienced that may affect the transport layer behavior; typically, a multi-state channel model (e.g., good/bad model) should be considered. For details on requirements, please refer also to the related part in sub-Section 4.4.1.
- **Application layer requirements**: different traffic types (e.g., real-time traffic and non-real-time traffic) should have specific SLAs and a monitoring action should be jointly performed with layer 2 in order to modify adaptively the service priority.

4.4.3 LEO satellite scenario requirements

LEO satellite networks are deployed as an enhancement to terrestrial wireless networks in order to provide broadband services to users regardless of their location. They provide significant benefits including wide area coverage, unique broadcast capability, ability to meet different QoS requirements, the possibility to communicate with hand-held devices and low access cost. At the same time, these networks present protocol designers with an array of important challenges, including handover procedures, mobility and location management.

Two broadband transport technologies, ATM (*Asynchronous Transfer Mode*) and IP, are proposed for future broadband LEO satellite networks. In the recent literature most publications are oriented towards the ATM-based LEO satellite scenario. For these reasons, such scenario is described in details later on.

In case of IP-based LEO satellite networks, with IP-routing implemented on board, the satellite network can seamlessly integrate with the terrestrial Internet. Another advantage is the IP QoS support without any required interworking with terrestrial IP QoS mechanisms. Multicast application provision is also well supported by using on-board router. However, routing in mobile satellite IP networks is considered a complex issue, because, one cannot simply use terrestrial Internet routing for on-board routing. The mobile IPv6 protocol, enhanced to support paging and handover, has to be implemented on-board.

ATM is a basic transport mechanism for *Broadband Integrated Services Digital Network* (B-ISDN), broadband Internet access and other technologies. ATM provides high transmission rates, bandwidth-on-demand, compatibility with previous existing protocols and guaranteed QoS. ATM-based LEO satellite networks are expected to support a wide range of multimedia services and applications and to provide their users with appropriate QoS based on the strong end-to-end QoS mechanisms offered by the ATM technology. However, the limited bandwidth of the satellite channel, satellite rotation around the Earth and the mobility of end-users make QoS provisioning and mobility management a challenging task. The following list provides a description of the requirements to support QoS in ATM-based LEO satellite systems.

- **Common LEO system requirements**: The main resources in LEO networks are the satellite radio bandwidth and the buffer capacity of the on-board ATM switches. Because the total link capacity has to be divided among several carriers, and given the limited buffer capacity of the ATM switch, advanced resource reservation cross-layer mechanisms have to be developed. They have to ensure fair bandwidth sharing and provide users with the negotiated QoS guarantees as end-users roam in the system. At the same time, the network and the end-systems have to be protected from congestion. One of the most important QoS parameters for LEO satellite networks is the *Call Dropping Probability* (CDP), quantifying the likelihood that an on-going connection will be forcedly terminated due to an unsuccessful handover attempt. Moreover, *Call Blocking Probability* (CBP) quantifies the chance that a new call request is denied entry into the system for lack of available resources [48]. Cross-layering is aimed to optimize bandwidth allocation, and to provide for low CDP for reliable handovers and acceptable CBP for new calls, while maintaining high resource utilization.

- **ATM layer requirements**: ATM-based LEO satellite networks should be able to meet different QoS requirements at the ATM layer. These requirements are stated in terms of the objective values of the network performance parameters, as specified in ITU-R Recommendation S.1420 [49]. Some of the QoS parameters may be offered on a per-connection basis and are negotiated between the end-system and the network. Other QoS parameters cannot be negotiated.
- **MAC layer requirements**: The most important resource management function is bandwidth allocation. The main constraint is the bandwidth available to all users on the satellite uplink. Unlike a fixed ATM network, the satellite can only control the bandwidth in the downlink towards the end-system. Thus, dynamic bandwidth allocation should be developed in order to meet QoS guarantees for the various *Virtual Channels* (VCs), as defined in the traffic contracts. Moreover, it is necessary to ensure the utilization of the unused bandwidth by connections with no explicit guarantees, as a BE service. Additionally, the MAC protocol should provide support for the ATM service categories. Only a QoS-aware MAC protocol is able to comply with the QoS requirements of different ATM service categories and the ATM signaling. MAC for ATM via satellite is also faced with the fact that an ATM cell does not have a dedicated field for the service parameters. In ATM, the service parameters of a connection are announced to the ATM switches along with the VPI/VCI value during the connection setup. Thus, the service parameters of the ATM cells belonging to a certain connection can be identified only through its VPI/VCI value. Consequently, the MAC layer needs some kind of lookup table with the service parameters of the ATM connections and the corresponding VPI/VCI values, if QoS of different ATM service categories has to be supported. This determines a special design of the protocol stack [50].
- **Network layer requirements**: The most important resource management function is CAC. The CAC algorithm operates at the call level in the network. It defines the procedure performed by the network during the call set-up phase to determine if the connection request can be accepted without infringing on existing commitments. If the request exceeds the available bandwidth, the role of the CAC is to deny the connection. In this case, we say that the connection is blocked. CAC schemes should be improved and mapped to layer 2 radio resource management protocols. A detailed analysis of CAC schemes is provided in Chapter 6.

4.5 Conclusions

In this Chapter we have provided a comprehensive literature review of existing cross-layer design approaches. From the literature review and taking into consideration the particular characteristics of the satellite scenario, a set of requirements has been identified for resource management with cross-layer design. These requirements have been shown to be different for the different scenarios from broadband to mobile and from GEO-based to LEO-based systems. The need of a cross-layer air interface design has been discussed and a couple of possible cross-layer architectures presented.

References

[1] L. Castanet, A. Bolea-Alamanac, M. Bousquet, "Interference and Fade Mitigation Techniques for Ka and Q/V Band Satellite Communication Systems", in *Proc. of Internat. Workshop of COST Actions 272 and 280 on Satellite Communications - From Fade Mitigation to Service Provision*, ESTEC, Noordwijk, The Netherlands, May 2003 [available online: http://www.cost280.rl.ac.uk/documents/WS2%20Proceedings/documents/pm-5-002.pdf].

[2] V. Kawadia, P. R. Kumar, "A Cautionary Perspective on Cross-Layer Design", *IEEE Wireless Communications*, Vol. 12, No. 1, pp. 3-11, February 2005.

[3] A. Maharshi, L. Tong, A. Swami, "Cross-Layer Designs of Multichannel Reservation MAC under Rayleigh Fading", *IEEE Transactions on Signal Processing*, Vol. 51, No. 8, pp. 2054-2067, August 2003.

[4] J.-P. Ebert, A. Wolisz, "Combined Tuning of RF Power and Medium Access Control for WLANs", *Mobile Networks and Applications, Special Issue on Mobile Multimedia Communications*, Vol. 6, No. 5, pp. 417-426, September 2001.

[5] Q. Liu, S. Zhou, G. B. Giannakis, "Queuing with Adaptive Modulation and Coding over Wireless Links: Cross-Layer Analysis and Design", *IEEE Transactions on Wireless Communications*, Vol. 4, No. 3, pp. 1142-1153, May 2005.

[6] L. Alonso, R. Agusti, "Automatic Rate Adaptation and Energy-Saving Mechanisms based on Cross-Layer Information for Packet-Switched Data Networks", *IEEE Communications Magazine*, Vol. 42, No. 3, pp. S15-S20, March 2004.

[7] P. Lettieri, M. B. Srivastava, "Adaptive Frame Length Control for Improving Wireless Link Throughput, Range and Energy Efficiency", in *Proc. of IEEE INFOCOM '98*, San Francisco, CA, Vol. 2, pp. 564-571, March/April 1998.

[8] R. Ludwig, A. Konrad, A. D. Joseph, R. H. Katz, "Optimizing the End-to-End Performance of Reliable Flows over Wireless Links", *Wireless Networks*, Vol. 8, No. 2/3, pp. 289-299, March 2002.

[9] H. Boche, M. Wiczanowski, "Optimal Scheduling for High Speed Uplink Packet Access - a Cross-Layer Approach", in *Proc. of the* 59th *IEEE Vehicular Technology Conference* (VTC 2004 Spring), Milan, Italy, Vol. 5, pp. 2575-2579, May 2004.

[10] A. Fu, E. Modiano, J. Tsitsiklis, "Optimal Energy Allocation and Admission Control for Communications Satellites", *IEEE/ACM Transactions on Networking*, Vol. 11, No. 3, pp. 488-500, June 2003.
[11] D. Noble, M. Satyanarayanan, D. Narayanan, J. E. Tilton, J. Flinn, K. R. Walker, "Agile Application-Aware Adaptation for Mobility", in *Proc. of the* 16th *ACM Symposium on Operating System Principles*, ACM, 1997.
[12] A. Alwan, R. Bagrodia, N. Bambos, M. Gerla, L. Kleinrock, J. Short, J. Villasenor, "Adaptive Mobile Multimedia Networks", *IEEE Personal Communications*, Vol. 3, No. 2, pp. 34-51, April 1996.
[13] H. Liu, M. El Zarki, "Adaptive Source Rate Control for Real-Time Wireless Video Transmission", *Mobile Networks and Applications*, Vol. 3, No. 1, pp. 49-60, June 1998.
[14] W. Yuan, K. Nahrstedt, S. V. Adve, D. L. Jones, R. H. Kravets, "Design and Evaluation of a Cross-Layer Adaptation Framework for Mobile Multimedia Systems", in *Proc. of the SPIE/ACM Multimedia Computing and Networking Conference* (MMCN), 2003.
[15] A. Sali, A. Widiawan, S. Thilakawardana, R. Tafazolli, B. G. Evans, "Cross-Layer Design Approach for Multicast Scheduling over Satellite Networks", in *Proc. of the* 2nd *International Symposium on Wireless Communication Systems* (ISWCS2005), Siena, Italy, pp. 701-705, September 05-09, 2005.
[16] M. van der Schaar, S. Shankar, "Cross-Layer Wireless Multimedia Transmission: Challenges, Principles, and New Paradigms", *IEEE Wireless Communications Magazine*, Vol. 12, No. 4, pp. 50-58, August 2005.
[17] W. Kumwilaisak, Y. T. Hou, Q. Zhang, W. Zhu, C.-C. Jay Kuo, Y.-Q. Zhang, "A Cross-Layer Quality-of-Service Mapping Architecture for Video Delivery in Wireless Networks", *IEEE Journal on Selected Areas in Communications*, Vol. 21, No. 10, pp. 1685-1698, December 2003.
[18] G. Pau, D. Maniezzo, S. Das, Y. Lim, J. Pyon, H. Yu, M. Gerla, "Cross-Layer Framework for Wireless LAN QoS Support", in *Proc. of the IEEE International Conference on Information Technology Research and Education* (ITRE), 2003.
[19] J. Chen, T. Lv, H. Zheng, "Joint Cross-Layer Design for Wireless QoS Content Delivery", *IEEE International Conference on Communication*, 2004.
[20] R. Kravets, P. Krishnan, "Application-driven Power Management for Mobile Communication", *Wireless Networks*, Vol. 6, No. 4, pp. 263-277, July 2000.
[21] P. Agrawal, S. Chen, P. Ramanathan, K. Sivalingam, "Battery Power Sensitive Video Processing in Wireless Networks", in *Proc. of the IEEE PIMRC*, Boston, 1998.
[22] M. Marchese, M. Mongelli, "Rate Control Optimization for Bandwidth Provision over Satellite Independent Service Access Points", in *Proc. of IEEE Globecom 2005*, St. Louis, MO, USA, pp. 3237-3241, November 28 - December 2, 2005.
[23] M. Marchese, M. Mongelli, "On-Line Bandwidth Control for Quality of Service Mapping over Satellite Independent Service Access Points", *Computer Networks*, Vol. 50, No. 12, pp. 1885-2126, August 2006.
[24] ETSI, "Satellite Earth Stations and Systems (SES). Broadband Satellite Multimedia, Services and Architectures", *ETSI Technical Report*, TR 101 984 V1.1.1, November 2002.
[25] ETSI, "Satellite Earth Stations and Systems (SES). Broadband Satellite Multimedia, IP over Satellite", *ETSI Technical Report*, TR 101 985 V1.1.2, November 2002.

[26] C. G. Cassandras, G. Sun, C. G. Panayiotou, Y. Wardi, "Perturbation Analysis and Control of Two-Class Stochastic Fluid Models for Communication Networks", *IEEE Transactions on Automatic Control*, Vol. 48, No. 5, pp. 770-782, May 2003.

[27] M. Baglietto, F. Davoli, M. Marchese, M. Mongelli, "Neural Approximation of Open-Loop Feedback Rate Control in Satellite Networks", *IEEE Transactions on Neural Networks*, Vol. 16, No. 5, pp. 1195-1211, September 2005.

[28] F. Davoli, M. Marchese, M. Mongelli, "Resource Allocation in Satellite Networks: Certainty Equivalent Approach versus Sensitivity Estimation Algorithms", *International Journal of Communication Systems*, Vol. 18, No. 1, pp. 3-36, February 2005.

[29] N. Iuoras, T. Le-Ngoc, "Dynamic Capacity Allocation for Quality-of-Service Support in IP-Based Satellite Networks", *IEEE Wireless Communications Magazine*, Vol. 12, No. 5, pp. 14-20, October 2005.

[30] N. Celandroni, F. Davoli, E. Ferro, A. Gotta, "Long-Lived TCP Connections via Satellite: Cross-Layer Bandwidth Allocation, Pricing and Adaptive Control", *IEEE/ACM Transactions on Networking*, Vol. 14, No. 5, pp. 1019-1030, October 2006.

[31] N. Celandroni, F. Potortì, "Maximising Single Connection TCP Goodput by Trading Bandwidth for BER", *International Journal of Communication Systems*, Vol. 16, pp. 63-79, January 2003.

[32] P. Chini, G. Giambene, D. Bartolini, M. Luglio, C. Roseti, "Dynamic Resource Allocation based on a TCP-MAC Cross-Layer Approach for Interactive Satellite Networks", in *Proc. of IEEE International Symposium on Wireless Communication Systems 2005* (ISWCS 2005), ISBN 0-7803-9206-X, Siena, Italy, September 5-9, 2005.

[33] M. Sooriyabandara, G. Fairhurst, "Dynamics of TCP over BoD Satellite Networks", *International Journal of Satellite Communications and Networking*, Vol. 21, No. 4-5, pp. 427-449, July-October 2003.

[34] J. Border, M. Kojo, J. Griner, G. Montenegro, Z. Shelby, "Performance Enhancing Proxies Intended to Mitigate Link-Related Degradations", IETF RFC 3135, June 2001.

[35] S. Floyd, V. Jacobson, "Random Early Detection Gateways for Congestion Avoidance", *IEEE/ACM Transactions on Networking*, Vol. 1, No. 4, pp. 397-41, August 1993.

[36] W. Stanislaus, G. Fairhurst, J. Radzik, "Cross Layer Techniques for Flexible Transport Protocol Using UDP-Lite over a Satellite Network", in *Proc. of IEEE International Symposium on Wireless Communication Systems 2005* (ISWCS 2005), ISBN 0-7803-9206-X, Siena, Italy, pp. 706-710, September 5-9, 2005.

[37] M. Ibnkahla, Q. M. Rahman, A. I. Sulyman, H. A. Al-Asady, J. Yuan, A. Safwat, "High-Speed Satellite Mobile Communications: Technologies and Challenges", in *Proc. of the IEEE*, Vol. 92, No. 2, pp. 312-339, February 2004.

[38] F. Alagöz, B. R. Vojcic, D. Walters, A. AlRustamani, R. L. Pickholtz, "Fixed versus Adaptive Admission Control in Direct Broadcast Satellite Networks with Return Channel Systems", *IEEE Journal on Selected Areas in Communications*, Vol. 22, No. 2, pp. 238-249, February 2004.

[39] N. Celandroni, F. Davoli, E. Ferro, A. Gotta, "Adaptive Cross-Layer Bandwidth Allocation in a Rain-Faded Satellite Environment", *International Journal of Communication Systems*, Vol. 19, No. 5, pp. 509-530, June 2006.

[40] N. Celandroni, F. Davoli, E. Ferro, "Static and Dynamic Resource Allocation in a Multiservice Satellite Network with Fading", *International Journal of Satellite Communications and Networking*, Vol. 21, No. 4-5, pp. 469-487, July-October 2003.
[41] N. Iuoras, T. Le-Ngoc, M. Ashour, T. Elshabrawy, "An IP-Based Satellite Communication System Architecture for Interactive Multimedia Services", *International Journal of Satellite Communications and Networking*, Vol. 21, No. 4-5, pp. 401-426, July-October 2003.
[42] T. Le-Ngoc, V. Leung, P. Takats, P. Garland, "Interactive Multimedia Satellite Access Communications", *IEEE Communications Magazine*, Vol. 41, No. 7, pp. 78-85, July 2003.
[43] Q. Wang, M.-A. Abu-Rgheff, "Cross-Layer Signalling for Next-Generation Wireless Systems", in *Proc. of the IEEE Wireless Communications and Networking Conference* (WCNC), New Orleans, USA, March 16-20, 2003.
[44] M. Conti, J. Crowcroft, G. Maselli, G. Turi, "A Modular Cross-Layer Architecture for Ad Hoc Networks", *Chapter 1 in Handbook on Theoretical and Algorithmic Aspects of Sensor, Ad Hoc Wireless, and Peer-to-Peer Networks*, Jie Wu (editor), CRC Press, New York, 2005.
[45] G. Carneiro, J. Ruela, M. Ricardo, "Cross-Layer Design in 4G Wireless Terminals", *IEEE Wireless Communications Magazine*, Vol. 11, No. 2, pp. 7-13, April 2004.
[46] V. Vardhan, D. G. Sachs, W. Yuan, A. F. Harris, S. V. Adve, D. L. Jones, R. H. Kravets, K. Nahrstedt, "GRACE: A Hierarchical Adaptation Framework for Saving Energy", Computer Science, University of Illinois *Technical Report* UIUCDCS-R-2004-2409, February 2004.
[47] M. A. Vázquez Castro, G. Seco Granados, "Cross-Layer Packet Scheduler Design of a Multibeam Broadband Satellite System with Adaptive Coding and Modulation", to appear on *Transactions on Wireless Communications.*
[48] P. Todorova, S. Olariu, H. N. Nguyen, "A Two-Cell Lookahead Call Admission and Handoff Management Scheme for Multimedia Satellite Networks", in *Proc. of the Thirty-Sixth Annual Hawaii International Conference on System Sciences* (HICSS - 36), Big Island of Hawaii, USA, January 6-9, 2003.
[49] ITU-R Recommendation S.1420: "Performance for Broadband Integrated Service Digital Network Asynchronous Transfer Mode via Satellite", 1999.
[50] H. Bischl, M. Werner, A. Dreher, L. Richard, E. Lutz, J. Bostic, H. Brandt, P. Todorova, F. Krepel, M. Emmelmann, "ATM-Based Multimedia Communication via NGSO-Satellites", *International Journal of Satellite Communications and Networking*, Vol. 23, No. 1, pp. 1-32, January/February 2005.
[51] M. Methfessel, K. F. Dombrowski, P. Langendörfer, H. Frankenfeldt, I. Babanskaja, I. Matthaei, R. Kraemer, "Vertical Optimization of Data Transmission for Mobile Wireless Terminals", *IEEE Wireless Communications*, Vol. 9, No. 6, pp. 36-43, December 2002.

Part II

Cross-Layer Techniques for Satellite-Dependent Layers

5

ACCESS SCHEMES AND PACKET SCHEDULING TECHNIQUES

Editors: Giovanni Giambene[1], Cristina Párraga Niebla[2], Victor Y. H. Kueh[3]

Contributors: Kostantinos Avgeropoulos[4], Wei Koong Chai[3], Giovanni Giambene[1], Samuele Giannetti[1], Du Hongfei[3], Victor Y. H. Kueh[3], Cristina Párraga Niebla[2], Veronica Pasqualetti[1], Aduwati Sali[3], Orestis Tsigkas[4]

[1]CNIT - University of Siena, Italy

[2]DLR - German Aerospace Center, Institute of Communications and Navigation, Wessling, Germany

[3]UniS - Centre for Communication Systems Research, University of Surrey, UK

[4]AUTh - Aristotle University of Thessaloniki, Greece

5.1 Introduction

The dual objectives of achieving efficient satellite resource utilization and acceptable user QoS levels require a consistent, controllable and flexible *Radio Resource Management* (RRM) scheme. The interest is here in managing packet data traffic of multimedia nature in mobile satellite systems. Complexity is added by the presence of multimedia traffic classes with differentiated QoS requirements and for the dynamically-varying channel conditions with (possible) consequent adaptations at the physical layer.

The MAC layer is the 'place' in the protocol stack where RRM techniques operate. In fact, the achievable resource utilization efficiency and the resulting

QoS are governed by MAC protocols that are used in the uplink case to manage the transmissions of dispersed terminals to an Earth station through the satellite and that are also employed in downlink to schedule the different transmissions from the Earth station to the terminals. Hence, the two essential components of the MAC layer are: access protocols and scheduling techniques. These are also the main targets of this Chapter.

The studies carried out in this Chapter are related to Scenario 1 for what concerns S-UMTS (see Chapter 1, Section 1.4); however, the last part of this Chapter refers to a TDMA-like air interface.

5.2 Uplink: access schemes

Since early 1960's, satellite access protocols have attracted the attention of various researchers. These protocols control the access of a station to the transmission medium. For terrestrial networks, where the transmission medium could be a coaxial cable or a twisted pair, several MAC protocols such as Ethernet, Token Rings and Token Buses have matured. However, these protocols are not suitable for satellite networks. Although the functionalities required and the users' QoS requirements are similar, the design of a satellite access protocol is more complicated and restrictive due to its operating environment. In brief, there are five reasons why many access protocols designed for terrestrial networks are not suitable for satellite ones [1]:

- The long propagation delay constrains the performance of access protocols.
- Satellite and terrestrial links have very different characteristics.
- Hardware modifications to controllers used in space are almost impossible and hence, satellite access protocols need a simple control mechanism.
- In contrast with terrestrial networks where topological changes are slow, satellite networks are characterized by topological changes and network reconfigurability in case of failures is mandatory.
- Power limitation in satellite networks is much stringent and therefore, the use of buffer memory, transponder capacity and processing power are more restrictive.

In the access protocol design phase, there are several factors to be taken into account. One of them is the type of applications that would traverse the satellite network. The traffic pattern the satellite network is expected to support is also a main input to the design process. As new network technologies and applications emerge, access protocols also evolve accordingly. Generically, there are five main access protocol categories:

- *Fixed Assignment* (FA),
- *Random Access* (RA),
- Fixed rate demand-assignment,
- Variable rate demand-assignment and

- Free assignment.

Fixed assignment protocols were the initial access protocols being used in commercial systems. However, because they were inefficient, newer proposals were demand-assignment protocols. The main application at that period was telephony and thus, fixed demand-assignment was proposed. Later, the need to support packet-switched data network has led to the introduction of random access protocols to satellite networks in early 1970's. Although improvements for the protocols in this class have been proposed for satellite, their low upper bound utilization has encouraged researchers to seek for alternatives. The result is the use of variable demand-assignment protocols. Based on the buffer state, users compute and send resource requests. The requested resource will be allocated for a finite period, usually in terms of a number of frames. With the increasing need to support multimedia traffic, the access protocol has to be able to manage traffic flows (i.e., traffic classes) with distinct QoS requirements. As a response, hybrid protocols have been proposed, combining diverse resource allocation mechanisms for different traffic types. For instance, to support real-time inelastic traffic, fixed demand assignment coupled with additional admission control could be used while for elastic data, a combination of variable demand-assignment and free assignment (e.g., a sort of round-robin allocation) could be the right choice.

In the following sub-Sections we examine random access protocols for S-UMTS. We begin by describing the current proposals for random access in S-UMTS and continue with an overview of PRMA-like schemes. Finally, we examine how PRMA can be adopted by S-UMTS and which cross-layer approach can be adopted to optimize the access protocol performance.

5.2.1 Random access in UMTS and application to S-UMTS

The S-UMTS air interface is currently defined by ETSI in technical specifications 101.851-1 to 101.851-4 [2]-[5]. These specifications do not define the type of satellite system (GEO or non-GEO) to be used, although the focus is towards GEO systems. Attention is given however to the consistency between the terrestrial and the satellite part of the system in terms of air interface design. For this reason, the general design and channel structure of the satellite air interface follows that of T-UMTS, modified appropriately in order to accommodate the special characteristics of satellite communications (long delay, Doppler effect, propagation loss, etc.). Table 5.1 below presents the physical channels used in S-UMTS and describes how these are mapped to transport channels, which in turn provide services to the higher layers.

The only common uplink physical channel available in S-UMTS is the *Physical Random Access Channel* (PRACH), which is mapped one-to-one to the *Random Access Channel* (RACH) at the transport level. In one cell, several RACHs/PRACHs can be configured. The *Physical Common Packet Channel* (PCPCH), the other common uplink channel available in T-UMTS, is not

Physical Channel	**Direction**	**Description**
DPDCH	Both	Carries the DCH transport channel
DPCCH	Both	Layer 1 control information for DPDCH
PRACH	Uplink	Carries the RACH transport channel
P-CPICH	Downlink	Phase reference for downlink channels
S-CPICH	Downlink	Phase reference for dedicated downlink channels
P-CCPCH	Downlink	Carries the BCH transport channel
S-CCPCH	Downlink	Carries the FACH and PCH transport channels
SCH	Downlink	Synchronization (spot search)
AICH	Downlink	Acquisition indicators (random access results)
PICH	Downlink	Paging indicators
MICH	Downlink	MBMS indicators

Physical Channels	
DPDCH	Dedicated Physical Data Channel
DPCCH	Dedicated Physical Control Channel
PRACH	Physical Random Access Channel
P-CPICH	Primary Common Pilot Channel
S-CPICH	Secondary Common Pilot Channel
P-CCPCH	Primary Common Control Physical Channel
S-CCPCH	Secondary Common Control Physical Channel
SCH	Synchronization Channel
AICH	Acquisition Indicator Channel
PICH	Paging Indicator Channel
MICH	MBMS Indicator Channel

Transport Channels	
DCH	Dedicated Channel
RACH	Random Access Channel
BCH	Broadcast Channel
FACH	Forward Access Channel
PCH	Paging Channel

Table 5.1: Transport and physical channels.

supported in the satellite air interface. RACH is characterized by open-loop power control and a collision risk in every transmission. RACH is crucial for the operation of the UMTS air interface, since it is used not only for initial channel access to the network (e.g., call origination, paging response and registration messages), but also for sending short data bursts (e.g., *Short Message Service*, SMS), as investigated in the following simulative study.

In 3GPP specifications [6], the PRACH transmission is based on a *Slotted-ALOHA* (S-ALOHA) approach with fast acquisition indication. The *User Equipment* (UE) can start the random access procedure at the beginning of a number of a well-defined time intervals, called access slots, by sending a preamble burst, as detailed below. There are 15 access slots on a 2-frame structure (totally, 20 ms duration) and they are interspaced by 5120 chips.

The PRACH transmission consists of two parts: *preamble* and *data message*. Since, the preamble can be transmitted one or several times (due to possible collisions), we can affirm that the structure of the random access burst is composed by one or several preambles and one message (see Figure 5.1). The random access procedure to transmit the preamble is defined in [2]-[5],[7]. The preamble part has a length of 4096 chips and the message part has a length of 10 or 20 ms. Only when the preamble is successfully detected the UE can transmit the message part with a power related to the detected preamble and with a channelization code corresponding to the signature selected to transmit the preamble.

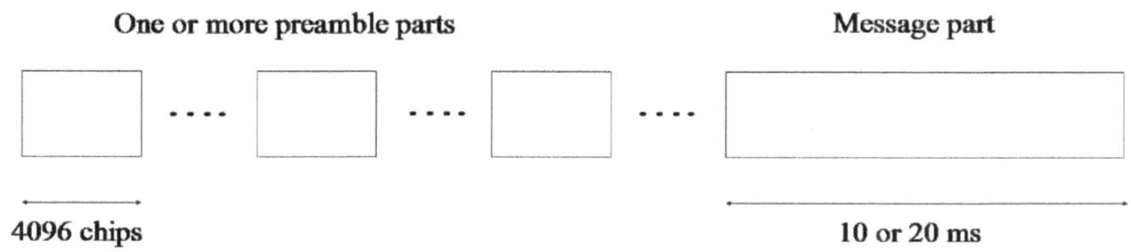

Fig. 5.1: Structure of the message transmission on RACH.

To construct the preamble, the UE uses two components: the preamble *scrambling code* (there are 8192 such codes available) and the preamble *signature code* (16 signatures to choose from, obtained as a repetition of a Hadamard codeword). These codes, sequences of chips with values +1 or −1, are combined to determine the complex preamble transmission code. More details can be found in [4].

The 10 ms message is split into 15 slots, each of 2560 chips (each slot of these has half duration with respect to access slots). The message consists of two parts: the data part and the control part, which are transmitted simultaneously (see Figure 5.2) using different channelization (spreading) codes that both depend on the signature used to construct the preamble part. The control part has a *Spreading Factor* (SF) of 256 and the data part can have different spreading factors in the set {32, 64, 128, 256}. The content of the data bits depends on the higher layers. The 8 pilot bits of the control part are used to support channel estimation for coherent detection and the *Transport Format Combination Indication* (TFCI) bits are used to indicate the spreading factor and the format of the data part.

Access Service Class (ASC) represents a certain PRACH partition (i.e., sub-channels and signature codes, as explained below) and an associated access persistency value (i.e., a probabilistic check to determine whether a preamble transmission can be attempted in the current access frame). There

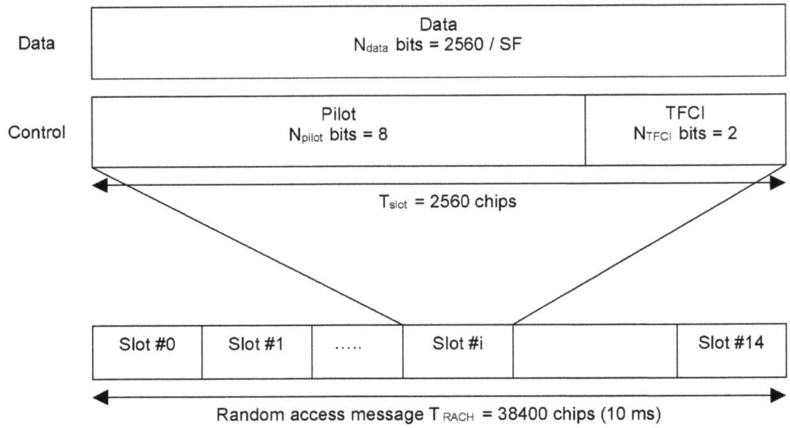

Fig. 5.2: Structure of the RACH message part (slots are here shorter than the access slots; in this case, a slot contains 2560 chips so that 15 slots correspond to 10 ms).

are 8 ASCs, numbered from 0 (highest access priority) to 7 (the lowest access priority) [8]. ASC 0 shall be used for emergency calls. A PRACH sub-channel defines a sub-set of the access slots. There are a total of 12 sub-channels. Typically, every 8 frames the allocation pattern of the different access slots to the different sub-channels repeats. The higher layers communicate to the physical layer the available signatures and sub-channel groups for each ASC.

There are at most 16 PRACH channels per cell; each of them corresponds to a different preamble scrambling code. On a given access slot of a PRACH, up to 16 simultaneous transmissions are possible by using distinct (orthogonal) signatures codes. A PRACH channel is defined by the following parameters: preamble scrambling code, spreading factor for data part, available signatures for each ASC, available sub-channels (i.e., slots) for each ASC and power control information. Available sub-channels and signature codes are broadcast through the BCH channel. When there is data to be transmitted, the UE performs PRACH selection randomly. Then, MAC selects the appropriate ASC for the traffic type to be managed. Consequently, an access slot and a signature are randomly selected among those available for the selected ASC. In the PRACH access mechanism, the main difference with respect to the classical S-ALOHA system is that, besides the time of the transmission, the UE also randomly chooses the signature and the scrambling code that will be used to transmit the preamble.

Once the preamble is sent, the UE waits for an acquisition indication (a sort of acknowledgment message) sent by the Node-B on the *Acquisition Indicator Channel* (AICH), a downlink physical channel that is received in the entire cell or part of the cell in case of sectorization. This transmission may fail for various reasons (interference from other terminals, fading, etc.). If an acquisition indication is not received by the time the UE response timer expires (τ_{pa}), the UE schedules a new transmission attempt on the ASC resources. Note that this timer must be set to a value greater than the estimated round trip delay. In the GEO satellite scenario, this timer can be set to either 280 or 560 ms (the actual selection is made by upper layer procedures) depending on the fact that the satellite is regenerating or not [2].

The system can provide dynamic persistency by publishing a dynamic persistency value through the *Broadcast Channel* (BCH). This value should be determined on the basis of an estimate of the number of contending UEs.

The flow chart in Figure 5.3 describes the random access protocol on the RACH channel. For further details the interested reader should refer to 3GPP specifications [5].

The message transmission is performed with a scrambling code that is one-to-one mapped to the scrambling code used for the preamble.

The remainder of this sub-Section is devoted to the performance evaluation of RACH in a GEO bent-pipe scenario. A C++ simulator has been implemented with a slightly simplified access procedure with respect to that in Figure 5.3 (i.e., no power ramping has been considered; only one PRACH has been simulated). We refer to a GEO bent-pipe satellite scenario, where the Node-B that manages the RACH protocol is on the Earth: the UE exchanges messages with the Node-B via the GEO satellite. In this study the Earth station provides a feedback to the UE about its transmission attempts. Hence, there is a round-trip propagation delay of about 560 ms to know the outcome of this transmission (τ_{pa} timer has been set accounting for such propagation delay).

In order to evaluate whether the access attempt has been successful or not, we have to consider collision events and the uplink interference conditions typical of CDMA transmissions. An access (i.e., preamble transmission) is considered successful if the following conditions are fulfilled [9]:

1. No other UE selects the same access slot and the same signature code on the same PRACH (otherwise there is a collision event; the capture effect is not considered in this case).
2. The received *Signal-to-Interference Ratio* (SIR) at the satellite exceeds a given threshold, SIR_t.

The above SIR issues (point ♯2) can be taken into account in the access phase by assuming a maximum number of transmissions (MaxUE) that can be tolerated in the same access slot for interference reasons. Hence, when there are n concurrent access attempts with $n >$ MaxUE, there is a too high interference level (i.e., SIR $< SIR_t$) so that all n transmission attempts (using different signature codes) are unsuccessful. We can consider that MaxUE is proportional to $\frac{1}{SIR_t}$.

The simulator numerical settings are detailed below:

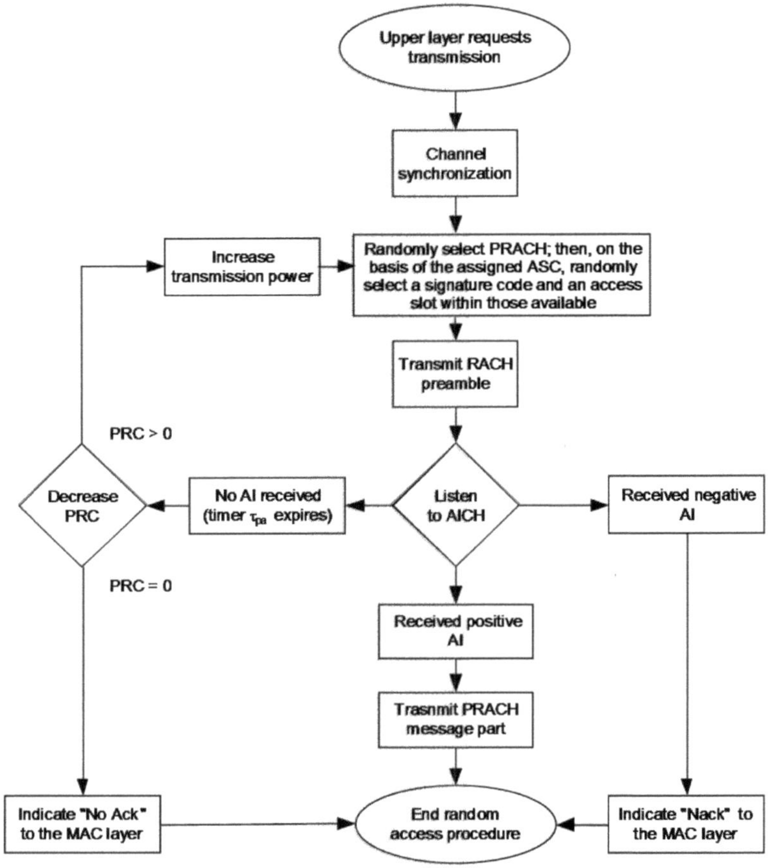

Fig. 5.3: Random access process on the PRACH channel (PRC, *Power Ramping Control*, denotes a mechanism to increase the transmission power of the access burst in subsequent attempts).

- We have considered a GEO bent-pipe satellite scenario with round-trip propagation delay of 560 ms.
- Only one PRACH has been simulated (i.e., one scrambling code is used).
- We consider two different cases for the interference conditions concerning the preamble transmission: MaxUE = 6 (mild interference conditions) and MaxUE = 3 (severe interference conditions). More appropriate MaxUE values could be determined with a complex analysis of the interference conditions deriving from the simultaneous transmissions of different preambles on the same access slot with different signature codes and the same scrambling code. Such a study is beyond the scope of this work.

- All the signature codes can be used by an ASC. While, different ASCs are distinguished by a different set of used sub-channels. In particular, the numbers of sub-channels distributed among the ASCs are as follows: ASC0 = 8, ASC1 = 3, and ASC2 = 1. Hence, the highest-priority ASC0 has a greater number of resources (i.e., sub-channels), thus guaranteeing lower collision and interference probabilities. Note that in this study, all the 12 sub-channels are used.
- We refer here to a case with persistency probability equal to one: the transmission of the preamble is soon attempted or reattempted by randomly select the resources.
- A source (i.e., UE) generating a message does not generate another message until the previous one has been transmitted. Hence, a source can be in the OFF state (waiting for the generation of a new message) or in the ON state (waiting for message transmission).
- There are 10 sources per ASC. The OFF state sojourn time is exponentially distributed with *mean message arrival rate* denoted with λ. As soon as the source leaves the OFF state, a procedure is started to transmit a 10 ms message.
- After the successful transmission of the preamble, message transmission requests are served according to the priority order of the related ASC. A 'virtual' message transmission queue corresponds to a PRACH (messages from ASC0 are prioritized with respect to ASC1, etc.). These message transmissions use a suitably shifted scrambling code with respect to the scrambling code of the preamble transmission that also combines this code with a signature code. We neglect interference between simultaneous message and preamble transmissions related to the same PRACH. Hence, preamble transmissions and message transmissions use separated resource spaces. Of course the message part can be received at the Node-B with errors according to a certain *Frame Error Rate* (FER) value.
- Simulation runs have a duration of 500 s.

We evaluate through simulations both the *mean preamble delay* (from the arrival of the message for the S-RACH transmission to the instant when the terminal receives the acknowledgment -AICH message- that the random access is successful) and the *mean message delay* (from the instant when the AICH message is received to the instant when the message transmission completes). The *total mean message delay* (from message arrival to message transmission) is the sum of the two above mean delay components. Results are shown in Figure 5.4 considering both the cases MaxUE = 6 and MaxUE = 3. The ideal preamble delay (lower bound) only contains a frame duration and a round trip delay. As expected, the mean preamble delay increases with the mean arrival rate λ and reduces with the MaxUE value. Moreover, the mean preamble delay increases from ASC0 to ASC1 and to ASC2 (i.e., the higher priority ASC0 permits to achieve lower mean preamble delay values). As expected, the mean message delay increases with the mean arrival rate λ and is practically insensitive to the MaxUE value variation (the message transmission on PRACH can be described as a simple queuing system -M/D/1-like queue with state-dependent arrival rate- with no interference with preamble transmissions, as previously assumed). Moreover, the mean message delay for ASC0 is lower than that for ASC1 that, in turn, is lower than that for ASC2.

Note that for all the ASCs, the mean preamble values are not so different, thus proving the robustness of the preamble access protocol: the time-code space is a sufficiently wide resource space also for the ASCs with lower number of assigned sub-channels. The random access scheme for preamble, based on different sub-channels and signature codes, has an intrinsic stability since it uses a form of special capture effect due to the codes. In addition to this, the mechanism that a source in the ON state cannot generate a new message, allows reducing the load of random preamble attempts and the load of messages to be transmitted on the PRACH 'virtual' queue. This mechanism further provides stability to both the random access phase and the subsequent message transmission queue.

As a final consideration, we may note that these results prove that the total message delay is high in a GEO bent-pipe scenario. A possible improvement has been proposed for the GEO satellite case in [9] where the message transmission immediately follows the preamble transmission.

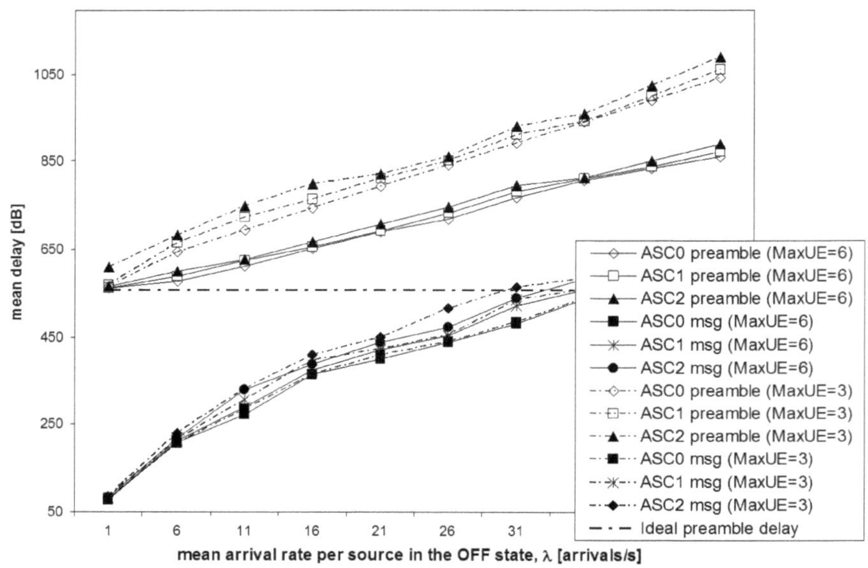

Fig. 5.4: PRACH performance in the presence of traffic on three ASCs with differently allocated resources and two cases for MaxUE values.

5.2.2 The Packet Reservation Multiple Access (PRMA) protocol

PRMA is a random access mechanism based on TDMA and S-ALOHA. Since its initial proposal in [10] it has attracted the attention of both research community and industry, especially because of its efficiency when handling real-time traffic. PRMA can be viewed as a *Dynamic TDMA* (D-TDMA) protocol where time slots are allocated to the users on demand. It is targeted mostly for voice and data traffic [11]-[14].

PRMA voice *User Terminals* (UTs) use speech activity detectors so that the channel is accessed only when there is voice traffic to be sent. This is important, since a voice channel is active for less than 50% of the time in a telephony dialog, which means that allocating slots statically for the entire call would waste resources.

As in all time division mechanisms, a PRMA carrier is divided into time slots of duration T_s that are grouped into frames of duration $T_f = T_s N$. Each slot has two states: *available* and *reserved*.

The figure below shows the state diagram for a UT in the simplest case where a UT is allowed to reserve only one slot at a time [more complex state diagrams result when more than one reserved slot per UT is allowed in a frame: $N-1$ or $2(N-1)$ states are added, depending on the mechanism used to reserve additional slots]. The UT starts in the *silent* state. When a talkspurt begins, the UT moves to the *contending* state where it attempts to reserve one slot in order to transmit the voice data. Random access transmissions are only allowed in available slots and occur according to a permission probability scheme. We assume that a UT monitors the state of the slots (using a downlink control channel) and therefore knows which of them are available. If a transmission is correctly received by the base station (no collisions), the transmitting UT is notified via a downlink control channel (this channel is often broadcast and can be used by UTs for slot state monitoring). In this case, the UT moves to the *active* state and the slot becomes reserved. This means that only the reserving UT is allowed to transmit in that slot in subsequent frames. When the talkspurt ends, the UT releases the slot by sending a special signal and moves again to the silent state while the slot becomes available. If the random access burst is not correctly received, usually due to a collision with other UTs that transmit their random access burst in the same slot, the base station informs the UTs that a collision has occurred and the UT remains in the contending state and schedules a retransmission attempt. The UT behavior in the access phase is depicted in the diagram in Figure 5.5.

During the access phase, a packet can be dropped (front-end clipping phenomenon): if the voice packet transmission delay D (i.e., the time between the packet generation and the packet successful transmission) exceeds a certain limit (30-40 ms), D_{max}, the packet is dropped and the UT will attempt to transmit the next one following the same procedure. Of course, the probability that a packet gets dropped is an important performance parameter and must be kept very low (lower than 1%) for guaranteeing a good voice

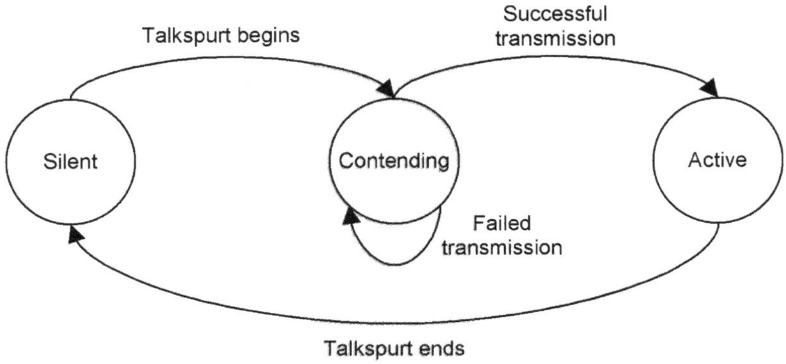

Fig. 5.5: State diagram of the PRMA protocol.

quality.

As shown in [10], PRMA outperforms the classical S-ALOHA protocol in terms of packet dropping probability and is therefore more preferable. It is also flexible enough to accommodate data and voice traffic. Moreover, there have been proposals where a UT can reserve more than one slot per frame to accommodate more demanding real-time traffic. There are certain issues however that are critical for PRMA performance, some of them are even more important in the case where it is used for satellite systems. These issues are:

- *Frame and slot duration, channel bandwidth and voice codecs.* In our discussion above, we mentioned that in order to transmit a talkspurt the UT reserves one slot per frame. This assumes that the channel bandwidth, the slot and frame durations and the codec used must be coordinated in order to receive the required voice quality at the receiver. This means that if the channel bit-rate is R_c and the codec voice bit-rate is R_s, then the maximum number of slots per frame is

$$N_{\max} = \frac{T_f R_c}{R_s T_f + L_h} \tag{5.1}$$

  where L_h is the length of each packet header.
- *Scheduling retransmissions and resolving collisions.* We can assume that as soon as a talkspurt begins, the UT selects the next available slot to transmit the random access burst in order to make a reservation as soon as possible. If there is a collision and all UTs that participated in the collision select another available slot in deterministic manner (e.g., they all select the next available slot), then they will enter a collision deadlock since all of them will select exactly the same slot to transmit. To avoid such deadlocks, a probabilistic collision resolution mechanism must be employed. In the simplest case, each UT may decide to transmit with a probability p, known

as *permission probability* or *persistence*. Selecting an appropriate value for this probability is vital in order to achieve a fast collision resolution mechanism and to guarantee a stable protocol behavior.

- *Contending-to-active state transition time*. Obviously, there is a strict requirement for the time it takes for a UT to move from the contending to the active state. If this time exceeds the packet deadline, then the first packet of the talkspurt must be dropped reducing the voice quality at the receiver (front-end clipping phenomenon). The exact limit depends on the codec used, but it usually amounts to a few tens of milliseconds.
- *Round Trip propagation Delay* (*RTD*). As already discussed, the base station is responsible for transmitting the results of a random access attempt to UTs. This means that after a UT transmits its random access burst, it must wait and listen for the result of its transmission (success or failure) on the downlink channel for a duration at least equal to RTD. In other implementations, the base station does not reply after a failed transmission and the UTs assume that they failed after not receiving a response within a given timeout. This means that in every transmission (or retransmission) the round trip delay is directly added to the contention phase. While this is not an issue in terrestrial systems with very small RTD values, it is quite critical for satellite systems. To cope with this problem, a modified PRMA protocol, called *PRMA with Hindering States* (PRMA-HS) has been proposed in [11]. In this PRMA version, the UT employs a more aggressive behavior in the contending state by continuously reattempting random access transmissions during RTD, without stopping for waiting the base station reply. It has been proved that while this approach increases the contention load with possibly useless re-transmissions, it still outperforms the classical PRMA scheme in mobile satellite systems.
- *Available slots versus collision probability*. In the classical PRMA protocol, we described above, the number of available slots (i.e., the number of unreserved slots that are available for contention) is variable. This means that as more slots become reserved the probability that two or more UTs transmit their random access bursts in the same available slot (collision probability) increases. There are cases where this phenomenon is not desirable. Therefore, there have been proposals in which a separate channel is used for contention (for example, this channel may simply consist of a certain amount of minislots in a reserved portion of the frame, thus significantly reducing the variations on the collision probability. There is obviously a trade-off here, as these contention-dedicated resources may cause a waste of bandwidth.

5.2.3 Adopting PRMA-like schemes in S-UMTS

GEO systems cannot adopt PRMA since their long RTD (max 280 ms in the case of a regenerating satellite; max 560 ms for a bent-pipe satellite) makes the state transition time from the contending to the active state to exceed the limit posed by the codec (i.e., the voice packet deadline, typically of few tens of ms). For these systems a simple reservation scheme may be used where a reservation per call is made. In LEO systems, however, RTD is much smaller (between 5 and 30 ms) and PRMA techniques are applicable. A feasibility study for the adoption of PRMA in the LEO case is made in [11]-[14], including the selection of the permission probability p and the frame duration T_f.

It should be noted that there is a substantial difference between the S-UMTS air interface and the air interface assumed by classical PRMA. PRMA relies solely on time division, whereas S-UMTS can be characterized as a hybrid CDMA/TDMA system. Therefore, CDMA/TDMA variations of PRMA must be considered such as the one proposed in [15], where UTs select a code in addition to a time slot in order to transmit their access bursts in a very similar fashion as we previously described.

The flow chart shown below in Figure 5.6 is an example of how S-UMTS channels can be used in order to adopt a PRMA-based scheme. We assume that a UT will require to use a *Dedicated Channel* (DCH) consisting of one *Dedicated Control Channel* (DCCH) and one or more *Dedicated Traffic Channels* (DTCHs) (DCCH and DTCH are logical channels) depending on the upper layer requirements. These requirements can be stated in the message part of the RACH burst. If the burst is not received properly, the UT schedules a retransmission using the permission probability p, which is announced by the system using the BCH channel, as specified in [5]. Note also that by using different channels for contention (RACH) and data transmission (DCH) we keep the collision probability constant and independent of the already assigned DCH channels. This separation of contention and data channels constitutes a substantial difference from classical PRMA schemes.

Note that in the presence of different traffic classes sharing the same RACH access channel, different permission probability values should be used to take into account the traffic urgency and other priority requirements.

As a conclusion, we may observe that S-UMTS as well as T-UMTS can adopt PRMA-like schemes without dramatic alterations to the air interface, since the already-available transport channels can be utilized by higher layers to implement PRMA. Due to this, variations of PRMA, such as the PRMA-HS mentioned earlier, can be also adopted in S-UMTS in order to improve the overall system performance. Anyway, it should be reminded that PRMA may only be used in LEO satellite systems.

5.2.4 Stability analysis of access protocols

The behavior of S-ALOHA-like protocols, such as the protocol used for PRACH or the PRMA-like variants used for S-UMTS, calls for a suitable design of the access protocol parameters.

As for the PRACH access protocol, 3GPP MAC specifications do not

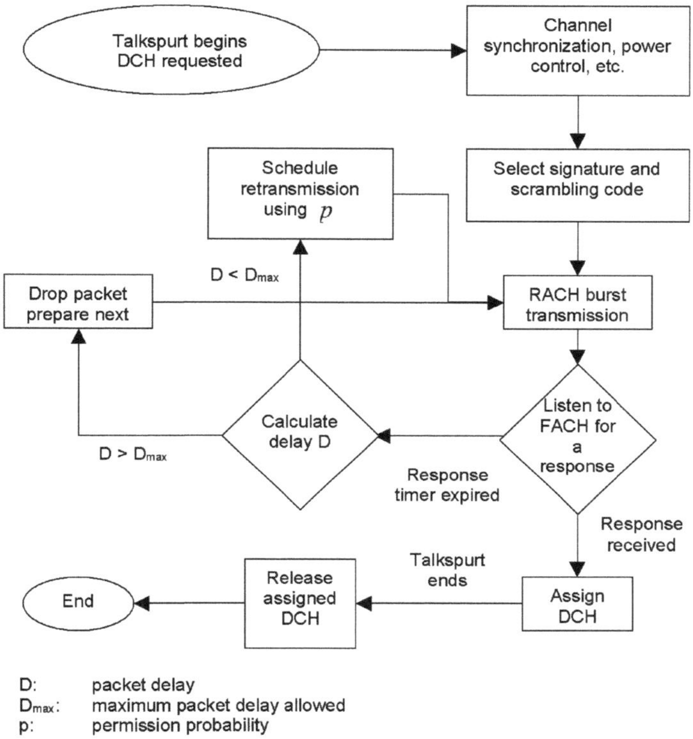

Fig. 5.6: PRMA-like access protocol (voice traffic).

provide a specific scheme to determine explicitly the ASC configuration for different traffic classes. However, access control could be coordinated by the satellite Earth station in order to define dynamically the access characteristics. This should be implemented by means of the feedback BCH signal. The studies made in [13],[16],[17] could be exploited to optimize both the access delay (for the different traffic classes) and the energy consumption during the access phase.

Also PRMA-like protocols need suitable settings for the control parameters. In particular, the permission probabilities can be used to modify the backlog period (after a contention) or to refrain a terminal from attempting transmissions. The problem is that a too aggressive protocol may lead to protocol bi-stability (i.e., too many collisions occur so that the throughput of correctly carried out requests goes to zero). This is a critical problem especially when many UTs contend for the same (access) resources [13],[17]. It is therefore important to adopt an explicit cross-layer scheme that dynamically adjusts transmission probability values on the basis of different aspects, such as the characteristics of each traffic class, the radio channel behavior and traffic load conditions. Analytical studies as those carried out in [13] can provide the appropriate framework for the cross-layer (adaptive) design of the access protocol parameters.

5.3 Downlink: scheduling techniques

5.3.1 Survey of scheduling techniques

The nature of the scheduling mechanisms employed on network links greatly impacts the QoS levels that can be provided by a network. The basic function of the scheduler is to arbitrate between packets that are ready for transmission on the link. Based on the scheduling algorithm, as well as the traffic characteristics of the flows multiplexed on the link, certain performance measures can be obtained. These can then be used by the network to provide end-to-end QoS guarantees.

First In First Out (FIFO) is not only the simplest scheduling policy, but also the most widely deployed one in the Internet today. As its name suggests, FIFO (or else *First Come First Served*, FCFS) serves packets according to their arrival order. This scheduling policy does not provide any guarantees to end-users.

Fixed priority mechanisms between two or more classes aim to provide the lowest possible delay for the highest priority class. The link multiplexer maintains a separate queue for each priority. The scheduler sends the data from the highest priority class before sending data for the next class. A packet in a lower priority queue is served only if all the higher priority queues are empty. As each queue is served in an FCFS manner, fixed priority schedulers are almost as simple as the FCFS scheduler with the added complexity of having to maintain queues. While this scheduling policy offers service differentiation, care should be taken in order not to starve lower priority classes. Moreover, it should be noted that fixed priority mechanisms do not readily allow end-to-end performance guarantees to be provided on a per-class basis.

Weighted Round Robin (WRR) [18] aims to give a weighted access to the available bandwidth to each class, ensuring a minimum allocation and distribution. The scheduler services each class in a round-robin manner according to the weights. If one or more classes are not using their full allocation, the unused capacity is distributed to the other classes according to their weights. A class can achieve a lower effective delay by giving it a higher weighting than the traffic level it is carrying.

Class-Based Queuing (CBQ) or *Hierarchical Link Sharing* (HLS) [19] is a more general term for any mechanism that is based on the class. Each class is associated with a portion of the link bandwidth and one of the goals of CBQ is to guarantee roughly this bandwidth to the traffic belonging to the class. Excess bandwidth is shared in a fair way among the other classes. There is no requirement to use the same scheduling policy at all levels of a link sharing hierarchy.

Generalized Processor Sharing (GPS) [20] is an idealized fluid discipline with a number of very desirable properties, such as the provision of minimum service guarantees to each class and fair resource sharing among the classes. End-to-end guarantees on a per-class basis can be provided if the traffic characteristics of the classes are known. Due to its powerful properties, GPS has become the reference for an entire class of GPS-related packet-scheduling disciplines, and relatively low cost implementations have started reaching the market. *Weighted Fair Queuing* (WFQ) [21] and its variants similarly aim to distribute available bandwidth over a number of weighted classes by using a combination of weighting and timing information to select which queue has to be served. The weighting effectively controls the ratio of bandwidth distribution between classes under congestion. However, it has been shown [22] that the tight coupling between rate and delay under GPS in the deterministic setting leads to sub-optimal performance and reduced network utilization.

The *Earliest Deadline First* (EDF) is a dynamic priority scheduler, with an infinite number of priorities. The priority of each packet is given by its deadline. EDF has been proven to be optimal [23] in the sense that, if a set of tasks is schedulable under any scheduling discipline, then the set is schedulable under EDF as well. EDF scheduling in conjunction with per-class traffic shaping permits the provision of end-to-end delay guarantees.

The *Service Curve-based Earliest Deadline first* policy (SCED) [24] is based on service curves, which serve as a general measure for characterizing the service provided to a user. Rather than characterizing service by a single number, such as minimum bandwidth or maximum delay, service curves provide a wide spectrum of service characterization, specifying the service by means of a function. It is shown that the SCED policy has greater capability to support end-to-end delay-bound requirements than other known scheduling policies.

Scheduling techniques for wireless systems

The approaches presented above are designed according to specific goals in terms of fairness and service requirements, without taking into account the transmission media. At present, the success of wireless networks pushes towards the design of scheduling techniques that, not only are aware of the characteristics of the transmission channel, but might also take some profit of this knowledge to achieve better performance.

Wireless systems are characterized by time-varying and location-dependent link states conditioned by interference, fading and shadowing. As a result, wireless channels are error-prone. This aspect has been considered in the literature from different perspectives in order to design scheduling techniques suited for wireless environments.

A first approach consists in the emulation of an error-free channel by deferring transmissions of user terminals experiencing bad channel conditions, and compensating them when their channels are again in a good state. Among the users in a good channel state, a scheduler suited for wired systems is typically considered. Examples of this approach are the *Idealized Wireless Fair Queuing*, the *Channel Condition-Independent Fair Queuing*, the *Server-Based Fairness Approach* and the *Wireless Fair Service scheduler*. These techniques are described below referring to a channel with BAD and GOOD states, interpreting them as error channel and error-free channel, respectively.

The *Idealized Wireless Fair Queuing* (IWFQ) simulates an error-free channel by applying a compensation model on top of the WFQ scheduler [25]. A start tag and a finish tag are associated to each packet, as for WFQ. The flows are serviced according to increasing service tags of the flows perceiving error-free channels. The compensation model operates as follows: if a flow receives service in one round, its service tag is increased by a factor l (lead bound); furthermore, for each round that a flow experiences a BAD channel, its service tag is decreased by a factor b_i (lag bound). This way, flows, which are in error channel during some time, are able to capture the resources as soon as they experience error-free channel, since their service tag is very low. The drawback is that leading flows, i.e., those with higher service tags, might be starved for long periods and therefore QoS bounds cannot be guaranteed and the service degradation is abrupt.

Similar to IWFQ, the *Channel Condition-Independent Fair Queuing* (CIF-Q) simulates an error-free channel by applying a compensation model on top of the *Stochastic Fairness Queuing* (STFQ, proposed in [26] as an enhancement of WRR). The compensation model applied here avoids abrupt service degradation. A *lag* parameter l is assigned to each flow, which is positive if the flow is lagging and negative if the flow is leading. In principle, flows are scheduled according to STFQ; however, if a flow i in error channel has allocated resources, the scheduler looks for other backlogged flows that perceive an error-free channel. If a flow j is found that fulfills this requisite, the flow i gives way to flow j, and their lag parameters are updated: l_i is incremented and l_j is decreased [25]. Hence, flow i still receives a fraction of its service, yielding to a graceful service degradation.

In *Server-Based Fairness Approach* (SBFA), a specific amount of transmission bandwidth is reserved for compensation purposes only. This is achieved by creating a virtual flow called *Long-Term Fairness Server* (LTFS) that will be used to manage the compensation. If a flow cannot be served because it experiences BAD channel conditions, the corresponding packet is queued in the LTFS. The scheduler treats the LTFS flow the same way as any other flow for the channel allocation. The share of bandwidth corresponding to LTFS is determined by a weight relative to the total bandwidth (as in a WRR approach). Since the lag of a flow is not bounded and the packets in the LTFS flow are served according to a FIFO policy, no packet delay bounds can be guaranteed.

Applying the *Wireless Fair Service* scheduler, each flow i has a lead bound of $l_{i,max}$ and a lag bound of $b_{i,max}$. Each leading flow relinquishes a portion of its lead $l_i/l_{i,max}$ for lagging flows. On the other hand, each lagging flow gets a fraction of the aggregated relinquished resources that is proportional to its lag: $b_i/\sum_{i \in S} b_i$, where S is the set of backlogged flows [25]. In practice, the leading flows free their resources in proportion to their lead and those resources are fairly distributed among the lagging flows. This approach achieves fairness, as well as delay and bandwidth guarantees.

The above scheduling techniques assume a simplified two-state channel model representing an error state and an error-free state. A more realistic model is to consider that each channel state is associated with a certain error probability, which allows for more flexibility in scheduling decisions. Based on this assumption, several scheduling techniques have been proposed in the literature, driven by the comparison of the channel quality level experienced by the user terminals having backlogged packets. Detailed examples of these techniques are reported below referring to the UMTS scenario.

Packet scheduling in UMTS

In case of CDMA cellular systems, the resources are the bandwidth, the codes, the RLC buffers at the RNC node and the UE, and the transmit power. In UMTS, the packet scheduler works in close-cooperation with the other resource management functions, in particular the admission control and the load (congestion) control entities [27]. Scheduling is part of the congestion control function, namely it is a form of *reactive* resource management, as opposed to the *proactive* characteristic of admission control. The packet scheduler can decide the allocated bit-rates and the length of the allocation among users. In W-CDMA, this can be done in several ways, in a code or time division manner or power scheduling-based.

In the code division approach, a large number of users can have a low bit-rate channel available simultaneously. When the number of users wanting capacity increases, the bit-rate, which can be allocated to a single user, decreases. In time division scheduling, the capacity is given to one user or only to a few users at each time instant. A user can have a very high bit-rate, but can use it only very briefly. When the number of users increases in the time division approach, each user has to wait longer for transmission. Power-based scheduling may be employed in response to the condition of the radio link between sender and receiver. If the power devoted to a code is kept fixed, the possible supported rate for a given transmission quality (interpreted in this context in terms of E_b/I_0) increases for GOOD and decreases for BAD channel conditions. Likewise, if the information rate is kept constant, maintaining the same transmission quality is obtained by employing two different levels of transmit power (i.e., the concept of power control).

The most common packet schedulers for UMTS are described below. The

Maximum C/I Scheduler serves in each resource allocation interval the flow with the best *Carrier-to-Interference ratio* (*C/I*) [28]. This approach is unfair, provided that flows corresponding to users located in the coverage edge have in general poor *C/I* performance and are starved in general, experiencing uncontrolled long delays.

The *C/I Proportional Scheduler* (*C/I* PS) also serves the flow with the best channel quality among backlogged flows. The main difference to the Maximum *C/I* Scheduler is that once a flow gets the resource, it is served until its queue is empty. This method does not guarantee fairness and QoS; it just maximizes channel efficiency and in turn network throughput. Furthermore, users with poor channel conditions might remain in a waiting status for a long time, experiencing very high delays.

Finally, more advanced scheduling techniques have been proposed, that operate on the basis of trade-off criteria (throughput vs. fairness) and even profit from developments in the fields of digital modulation and forward error correction. Examples of these approaches are the *Proportional Fair scheduler* and an enhanced version of it, named *Exponential Rule scheduler*.

The *Proportional Fair* (PF) scheduling algorithm has been originally developed to offer an appealing trade-off between user fairness and cell capacity in terrestrial *High Speed Downlink Packet Access* (HSDPA) as well as in CDMA/HDR [29]-[31] (see also the following sub-Section). With this approach, the server retrieves information about the instant quality of the downlink channel (*Channel Quality Indicator*, CQI). According to this CQI measure, the server calculates for each flow in every scheduling round the *Relative Channel Quality Index* (RCQI) that is a trade-off measure between the maximum throughput that the flow can achieve (according to the modulation and coding rate it can afford) and the service it got in the past. The scheduler serves in each resource allocation interval the flow with the highest RCQI value. The maximum achievable throughput by a flow is determined by the highest modulation and highest coding rate that can be applied according to the experienced channel conditions, reported in the CQI. The RCQI parameter provides a trade-off between channel efficiency and fairness, avoiding that users with good enough channel are starved due to the presence of users with better channel conditions. However, delay bounds cannot be guaranteed with this approach.

The *Exponential Rule* scheduler introduces enhancements to the PF scheme that aim at balancing the weighted delay of all backlogged flows when the differences of weighted queue delay among users become significant [31]. This is achieved by adding a multiplicative exponential parameter to the RCQI metric. The exponential function is dependent on the weighted instantaneous delay compared to the cumulative delay. If a significant increase on the delay is detected, the function gets a high value (due to its exponential profile) that increases the final value of the RCQI metric, thus giving high priority to that user in front of the others. In addition to the trade-off between fairness and transmission efficiency, this scheduling technique provides also guarantees in terms of delay bounds.

Although satellite systems can be considered as a specific case of wireless systems, additional effects might have impact on the scheduling performance, such as the propagation delay and channel state dynamics different from the terrestrial case. These issues are considered in the following sub-Sections.

5.3.2 Scheduling techniques for HSDPA via satellite

Overview on terrestrial HSDPA

HSDPA is a step beyond the W-CDMA air interface, in order to improve the performance of downlink multimedia data traffic according to the increasing demand for high bit-rate data services. For that purpose, the main targets of HSDPA are to increase user peak data rates, to guarantee QoS and to improve spectral efficiency for downlink asymmetrical and bursty packet data services, supporting a mixture of applications with different QoS requirements [28].

The HSDPA concept is based on an evolution of the *Downlink Shared Channel* (DSCH), denoted as *High Speed-DSCH* (HS-DSCH). DSCH time-multiplexes the different users and is characterized by a fast channel reconfiguration time and a packet scheduling procedure, which is very efficient for bursty and high data rate traffic in comparison with DCH. HS-DSCH introduces several adaptations and control mechanisms that enhance peak data rates, and spectral efficiency for bursty downlink traffic.

The HS-DSCH structure is based on a *Transmission Time Interval* (TTI) whose duration is selected on the basis of the type of traffic and the amount of users supported (in the order of 2 ms). In comparison with the typically longer TTIs of W-CDMA (10, 20 or 40 ms), the shorter TTI in HSDPA allows for lower delays between packets, multiple retransmissions, faster channel adaptation and minimal wasted bandwidth.

Two fundamental CDMA features are disabled in HS-DSCH, i.e., fast power control and *Variable Spreading Factor* (VSF), being replaced by other features such as *Adaptive Coding and Modulation* (ACM), multi-code operation, *Fast L1 hybrid ARQ* (FL1-HARQ) and fixed spreading factor equal to 16 [28]. The fixed spreading factor allows the allocation of 15 codes in each TTI (the 16th code is used for signaling purposes) that can be assigned to either the same UE to enhance its peak data rate or several UEs code-multiplexed in the same TTI.

Furthermore, in order to achieve low delays in the link control, the MAC layer functionality corresponding to HS-DSCH (namely MAC-hs) is placed in the Node-B (instead of the RNC, where the MAC layer functionality corresponding to DSCH is typically located). This solution allows the scheduler to work with the most recent channel information, so that it is able to adapt the modulation scheme and coding rate to better match the current channel conditions experienced by the UE. However, this solution introduces some changes in the interface protocol architecture, as depicted in Figure 5.7 [32].

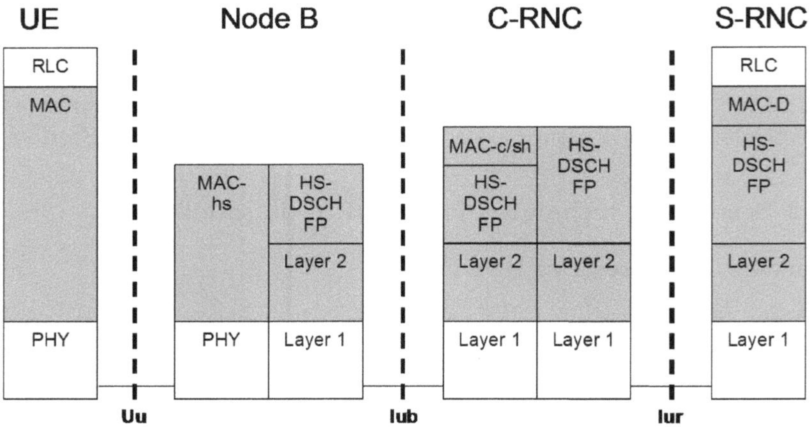

Fig. 5.7: Interface protocol architecture of HSDPA.

The adaptability of HS-DSCH to physical channel conditions is based on the selection of a coding rate, a modulation scheme, and the number of allocated codes to the scheduled UE in each TTI. In particular, the HS-DSCH encoding scheme is based on the Release'99 rate-1/3 turbo encoding, but adds rate matching with puncturing and repetition to obtain a high resolution on the effective code rate, ranging approximately from 1/6 to 1. To facilitate very high peak data rates, the HSDPA concept has added 16QAM on top of the existing QPSK modulation available in Release'99. A modulation scheme and code rate combination is denoted as *Transport Format and Resource Combination* (TFRC). Under very good channel conditions, the selection of highly efficient TFRCs combined with the allocation of several orthogonal codes to the scheduled UE (multi-codes operation), allow the UE to receive theoretically up to 10 Mbit/s [28]. However, this might be constrained by the UE capabilities, due to the limitation of receiving several parallel codes [33].

The packet scheduler can be considered as the central entity of the HSDPA design. In the HSDPA protocol stack architecture, the packet scheduler is located in the MAC-hs at the Node-B. The tasks corresponding to the MAC-hs layer in the UE and in the Node-B are summarized in Table 5.2.

According to a certain packet scheduling algorithm, the HS-DSCH transport channel is mapped onto a pool of physical channels, *High Speed Physical Downlink Shared Channels* (HS-PDSCHs), to be shared among all the HSDPA users in a time-multiplexed way.

The scheduler governs the distribution of the available radio resources in the cell among the UEs, i.e., it selects which UE is scheduled in the next TTI and which settings should be used (TFRC and number of parallel codes), supported by the link adaptation functionality. The scheduler relies on channel state information sent from each UE in order to perform its functions. The UE

MAC-hs in UE	MAC-hs in Node-B
Generation of ACK and NACK responses to received packets	MAC PDU flow control
Routing of packets to the correct reordering queue based on queue identifier	Scheduling and priority handling
Reordering of PDUs	Request of retransmission if NACK received
Removing of MAC-hs header and padding bits	Selection of appropriate transport format and resource combination

Table 5.2: MAC-hs functions in UE and Node-B.

is requested by the RNC to send periodically a specific CQI on the uplink *High Speed Dedicated Physical Control Channel* (HS-DPCCH). The periodicity is selected from the set {2, 4, 8, 10, 20, 40, 80, 160} ms. The CQI provides the following information related to the currently experienced channel conditions by the UE [7]:

- TFRC mode (most efficient modulation scheme and coding rate that can be used);
- Maximum number of parallel codes that can be used by the UE;
- Specification of a transport block size (i.e., the transport layer PDU) for which the UE would be able to receive data with a guaranteed FER lower than or equal to 10%, after first transmission.

There are different CQI tables for several UE categories. Table 5.3 shows an example [34]. If CQI indicates that the quality is degrading, the scheduler can choose a less ambitious TFRC that will cope better with the poor channel conditions.

Implications of the satellite component in HSDPA

The HSDPA concept and architecture have been designed for terrestrial environments. In a satellite scenario, the allowed complexity on board of the satellite, the selected constellation (LEO, MEO, GEO) and a different propagation environment condition the applicability of the HSDPA concept as it is defined and the feasibility of the promised peak data rates.

One of the major advantages of HSDPA with respect to the W-CDMA interface is the location of the scheduling function at the Node-B, allowing for shorter delays and better adaptability to time-varying channel conditions. However, the location of the different network entities, such as Node-B or RNC, is not uniquely determined in a satellite-based UMTS system. Depending on the available complexity on the satellite, part of the functionalities typically located at the Node-B or at the RNC in a UMTS network can be

CQI value	Modulation and coding	Number of codes used per TTI	Bits per TTI (transport block size)
1	QPSK 1/3 (on each code 960 bits are sent in a TTI)	1	137
2		1	173
3		1	233
4		1	317
5		1	377
6		1	461
7		2	650
8		2	792
9		2	931
10		3	1262
11		3	1483
12		3	1742
13		4	2279
14		4	2583
15		**5**	**3319**
16	16QAM 1/3 (on each code 1920 bits are sent in a TTI)	5	3565
17		5	4189
18		5	4664
19		5	5287
20		5	5887
21		5	6554
22		5	7168
23		7	9719
24		8	11418
25		**10**	**14411**
26		12	17237
27		15	21754
28		15	23370
29		15	24222
30		15	25558

Table 5.3: Example of CQI mapping in transport block size for TTI = 2 ms (terrestrial standard); the highlighted CQIs are those considered for simulations referring to a GOOD/BAD channel model.

executed on board or not. In the case of a bent-pipe satellite, all medium access control mechanisms must be located at the Gateway station or the network control center. In any case, the large distances involved in a satellite system disable the HSDPA capabilities of fast retransmissions and quick adaptation to physical channel variations, thus scaling the performance that link adaptation mechanisms can achieve.

In GEO satellite systems, retransmissions take too long time. Therefore, FER upper bounds should be adequately much lower in order to reduce statistically the number of required packet retransmissions.

Furthermore, the behavior of the channel is not comparable to the terrestrial mobile channel: deeper and longer fades are expected in the satellite case, in contrast to the fast fades of the terrestrial mobile channel.

All the issues discussed above condition the performance of the packet scheduler, which is the main entity of the HSDPA concept. With the purpose of testing the performance of different scheduling techniques in a simplified satellite-HSDPA scenario, the following assumptions have been made:

- A multi-beam GEO bent-pipe satellite has been considered.
- All *Radio Access Network* (RAN) functionalities corresponding to the network part are located at a Gateway station, as can be observed in Figure 5.8.
- The propagation delay between Gateway station and UE is approximately 280 ms, i.e., round-trip propagation delay is about 560 ms.
- Each UE performs channel estimation. The result is sent back (in the form of CQI) to the Gateway station.
- CQI information transmission interval is extended to 40 ms in order to save power at the UEs (this is not that critical, considering that the impact of round-trip delay in the acquisition of channel state information should be more dominant than this larger periodicity).
- During the time interval between two CQI updates, channel conditions are considered constant by the scheduler for a given UE.
- The TTI duration of the terrestrial HSDPA is kept also in the satellite case in order to reduce packet delays and to have fine scheduling time granularity.
- A GOOD/BAD Markovian channel model is considered at the physical layer for the sake of simplicity. Accordingly, one CQI value from Table 5.3 is selected for each channel state: CQI = 15 for the BAD state and CQI = 25 for the GOOD state. Note that the channel variation dynamics in a satellite environment are slow; in particular, a mean GOOD (BAD) sojourn time of 6 s (2 s) has been considered.
- Code-multiplexing of different users in the same TTI is not applied in this simplified study, i.e., only one UE is served in each TTI. According to this assumption, the task of the scheduler is to select the UE to be served in each TTI. The service got by the scheduled UE depends on the transport block size determined by the CQI currently supported by the UE (see Table 5.3).

On the basis of the assumptions above, the channel state information that the UE transmits to the Gateway station is outdated when received at the Gateway due to the high propagation delay. To cope with this, either higher margins in the selection of the CQI value to be sent shall be considered or delay compensation strategies shall be applied that permit to predict what will be the channel evolution by the time that the CQI information reaches the Gateway. For the interested reader, some delay compensation techniques

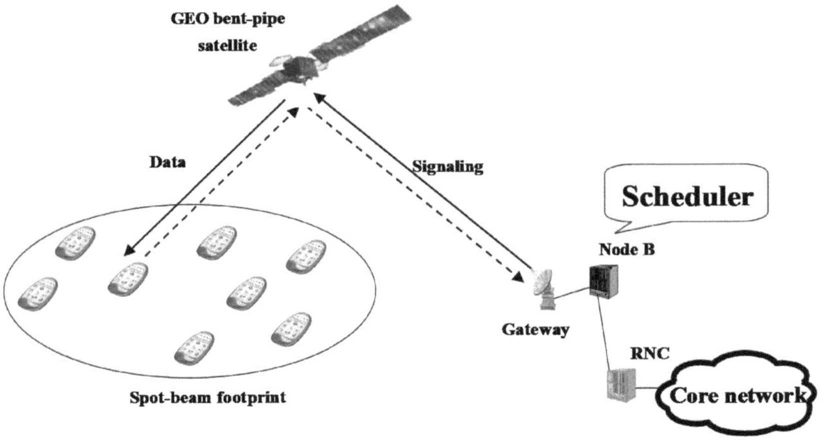

Fig. 5.8: S-HSDPA network architecture.

are proposed in [35] for Ku and Ka band satellite links.

If no countermeasures are adopted, channel state transitions cause temporal misalignments between the current channel state and the considered CQI by the Gateway station. In particular, in the presence of a transition from BAD to GOOD, the system uses a more conservative mode than necessary for 560 ms plus maximum 40 ms ([1]). This does not affect the service quality (i.e., no packet losses are caused), but the resource utilization is not optimal. On the other hand, in the presence of a transition from GOOD to BAD, the system does not adequately protect transmissions during 560 ms plus maximum 40 ms, so that the transmitted data during this period is lost due to channel impairments with high probability. For the sake of simplicity, it is assumed that FER = 1 during misalignment periods from GOOD to BAD channel states.

The traffic scenarios may also affect the resource utilization performance achieved by any scheduling technique. If the traffic generated by the scheduled UE in the next TTI is not sufficient to fill the transport block assigned for the transmission, part of the capacity remains unused and certain inefficiency is experimented.

PHY-aware scheduling approaches for HSDPA over satellite

The design of suitable scheduling techniques for HSDPA-like transmissions in a satellite environment must consider the several degrees of freedom imposed

[1] Depending on the CQI transmission timing with respect to the current state transition (GOOD to BAD or vice versa), the delay to receive a packet with an updated TFRC ranges from 560 to 600 ms.

by the HSDPA interface in addition to the specific characteristics of satellite links. HSDPA has a sort of hybrid TDM/CDM air interface, where packet scheduling can be done in two dimensions: *time* and *code*. The code dimension allows for two flavors of resource management strategies: code-multiplexing and multi-code operation. By means of code-multiplexing, several UEs can be scheduled in the same TTI, thus enhancing the resource utilization. Using the multi-code operation, the throughput of one UE can be improved on a TTI basis by allocating several codes to it.

Based on the CQI information periodically sent by each UE, the scheduler can find out the achievable throughput by each UE in the next TTI by checking a look-up table like Table 5.3; this scheme is considered here like an explicit cross-layer technique that envisages the dynamic interaction of physical and MAC layer. The throughput achievable by each UE is determined by the most efficient applicable modulation and coding rate, the maximum number of codes that can be allocated to the UE and the transport block size that can be used. It should be noted that the effective code rate must be calculated taking into account the *Cyclic Redundancy Check* (CRC) code of 24 bits that is added to each transport block before encoding and additional puncturing and repetition, which yield a number of physical layer bits equal to:

- 960 bits $\times$ number of assigned codes, if QPSK modulation is used;
- 1920 bits $\times$ number of assigned codes, if 16QAM modulation is used.

The effective code rate is given by:

$$\text{Code rate}\,(CQI) = \frac{24 + (\text{transport block size})}{b_{code} \times (\text{number of codes})} \tag{5.2}$$

where $b_{code} = 960$ bits for QPSK modes and $b_{code} = 1920$ bits for 16QAM modes.

For the method adopted in HSDPA to pass from transport to physical layer, the interested reader should refer to [36].

The availability of channel state information and the relation between channel state and achievable throughput by a UE adds new dimensions for optimization to the scheduling problem. Typically, a scheduler manages the share of resources among flows accessing the media according to some fairness or QoS criterion. However, in a system that supports ACM and code-multiplexing on top of time-multiplexing (and multi-code operation), the scheduler operation becomes even more complex. Several approaches can be adopted in the design of scheduling techniques, depending on the optimization goals. We consider here some of those schemes already introduced in sub-Section 5.3.1.

A first approach is to ignore the additional degrees of freedom of HSDPA and to schedule the backlogged traffic according to either a fairness criterion or driven by QoS constraints. In this case, algorithms such as EDF can be applied. However, this approach does not exploit the flexibility of HSDPA to use efficiently the available resources, since the channel state corresponding to each flow is transparent to the scheduler.

A second approach is to maximize the transmission efficiency by scheduling the flows that can achieve the highest throughput in the current TTI, i.e., those flows that are associated to better channel conditions, which is the strategy applied by the *opportunistic scheduler* [37]. However, this approach does not guarantee QoS, since those UEs with worse channel conditions shall be blocked for long periods, even if their channel state is good enough for transmission.

A third approach is to schedule the flows according to a hybrid criterion that combines fairness and transmission efficiency maximization in a trade-off manner. This concept has been proposed for scheduling in HSDPA in terrestrial environments under the name of PF scheme [30],[31] (see sub-Section 5.3.1).

A scheduler has been considered to manage downlink transmissions (HS-DSCH) that is in the Node-B (Earth Station) according to the architecture in Figure 5.8. In particular, the scheduler is at MAC-hs level and it is assumed to have different queues for the different UEs. Each queue (at IP level) contains the multimedia traffic corresponding to one UE. Suitable priority indexes are considered to serve these queues; these indexes are related to either the EDF scheduler or the PF scheme. In what follows, the performance achieved by these schedulers are compared in the presence of video streaming and Web traffic [38],[39]. The assumptions previously made (see the previous part on "Implications of the satellite component in HSDPA") are considered for this simulation study.

EDF scheduler

This scheduling technique, described in sub-Section 5.3.1, serves packets according to their urgency. The EDF scheme is quite appropriate for the management of real-time traffic flows that are characterized by deadlines. Such scheme requires the dynamic management of the buffer for each traffic class when packets with different deadline values have to be served.

To implement the EDF criterion it has been considered that the priority index for the generic k-th UE in the current n-th TTI interval, $P_k[n]$, is given by the ratio between the transmission delay of its oldest IP packet, $d_k[n]$, and the packet deadline, $T_{deadline}$:

$$P_k[n] = \frac{d_k[n]}{T_{deadline}} \qquad k = 1, 2, \ldots, N \tag{5.3}$$

where N denotes the number of UEs per spot-beam.

The above priority index does not permit to prioritize the real-time video traffic with respect to the interactive Web traffic. This approach could degrade the video performance in the presence of significant Web traffic load. To cope with this, a differentiation in the priority index in equation (5.3) is needed. In particular, equation (5.3) is used for video traffic so that a video IP-packet has an increasing priority up to (almost) 1 when the packet is close to its deadline and risks to be dropped. Moreover, a modified priority index is used for Web IP-packets that saturates to 0.9 when these packets are close to (or exceed) their *virtual* deadline:

$$P_k[n] = \min\left\{0.9, \frac{D_k[n]}{\mathrm{T_{deadline}}}\right\} \qquad k = 1, 2, \ldots, N. \tag{5.4}$$

Hence, very urgent video packets will be served with highest priority than any Web packet. In what follows, the scheme where the priority index (5.3) is used for both video and Web traffic flows will be denoted as EDF; whereas, the name *Prioritized-EDF* (P-EDF) is applied to the scheme where the priority index (5.3) is used for video flows and the priority index (5.4) is adopted for Web traffic flows.

PF scheduler

This strategy serves the UE with largest RCQI, which represents the ratio between the maximum data rate currently supported by each UE (according to its CQI and by using a look-up table like Table 5.3) and the 'average' service that the UE got in the past, according to a suitable sliding window. On the basis of [30], the RCQI value corresponding to the k-th UE can be computed as follows.

$$RCQI_k[n] = \frac{R_k[n]}{T_k[n]} \qquad k = 1, 2, \ldots, N \tag{5.5}$$

where n is related to the time measured in TTI units, $R_k[n]$ is the bit-rate supported by the k-th UE in the n-th TTI interval (depending on its current CQI) and $T_k[n]$ represents the average throughput achieved by the k-th UE up to the present TTI (according to a defined memory length).

$R_k[n]$ and $T_k[n]$ can be computed as follows [30]:

$$R_k[n] = \min\left\{CQI_k[n], \frac{B_k[n]}{\mathrm{TTI}}\right\} \tag{5.6}$$

$$T_k[n] = \left(1 - \{B_k[n] > 0\} \cdot \frac{1}{N_k}\right) \cdot T_k[n-1] + \frac{1}{N_k} \cdot R_k'[n-1] \tag{5.7}$$

where $CQI_k[n]$ denotes the maximum bit-rate supported by the k-th UE at the current time, calculated as the throughput that is allowed by the CQI in the next TTI interval (according to a look-up table like Table 5.3). $B_k[n]$ represents the amount of data waiting for transmission in the Node-B buffer of the k-th UE at current time; $\{B_k[n] > 0\}$ is either 1 or 0 depending on whether the Boolean expression is right or not. N_k represents the memory of the averaging filter (which has been set to 1000 TTI units), and $R_k^{'}[n-1]$ denotes the bit-rate used for the transmission to the UE during the $(n-1)$-th scheduling interval. It is assumed that $T_k[1] = CQI_k[1]$.

According to the assumptions made on the GOOD/BAD channel (i.e., CQI = 15 for the BAD state and CQI = 25 for the GOOD state) and on the basis of Table 5.3, we have the corresponding bit-rate capacities:

- $CQI_k[n] = R_{bad} = 3319$ bits/TTI $\approx$ 1.6 Mbit/s in the BAD state;
- $CQI_k[n] = R_{good} = 14411$ bits/TTI $\approx$ 7.2 Mbit/s in the GOOD state.

Note that in the PF case an explicit cross-layer scheme has to be adopted since scheduling takes into account the dynamic variation of the radio channel conditions for the UEs.

The software simulator presented in [38],[39] has been used to evaluate the performance of S-HSDPA transmissions, using both EDF and PF techniques in order to manage video streaming and Web traffic downlink flows.

S-HSDPA performance: simulation results

In order to evaluate the performance of S-HSDPA transmissions when using EDF, P-EDF and PF schedulers, the following metrics have been considered:

- Efficiency in the utilization of radio resources, η;
- Percentage of IP-video packets lost due to deadline expiration, P_{drop};
- Percentage of IP packets lost due to GOOD-to-BAD channel misalignment, $P_{loss\_channel}$ (without considering packet retransmissions);
- Mean delay for the transmission of an IP-Web packet, $Delay_{Web}$.

Let C denote the mean capacity considering the GOOD/BAD channel previously described [39] and the related CQI values associations in Table 5.3. Hence, the resource utilization efficiency η (<1) can be measured as follows:

$$\eta = \frac{\text{Mean aggregated transmitted bit-rate}}{C} \quad . \tag{5.8}$$

P_{drop} is obtained as the ratio between the number of IP-video packets that are lost due to deadline expiration (the deadline has been set to 150 ms) and the number of generated IP-video packets. $P_{loss\_channel}$ is computed as the ratio of the number of IP packets that are lost at the receiver due to the GOOD-to-BAD channel misalignment (considering both video and Web traffic together) and the number of transmitted IP packets.

In the following graphs, the above different performance metrics are plotted as a function of the *system load*, ζ, that is defined as:

$$\zeta = \frac{\text{Mean aggregated generated bit-rate}}{C} \quad [\text{Erl}]\,. \tag{5.9}$$

Of course, $\eta \leq \zeta$ due to the fact that not all the generated bits are transmitted (some of them may be dropped due to deadline expiration in the case of video traffic).

The following simulation results have been obtained considering an equal number of video and Web traffic sources; both video and Web sources produce the same mean bit-rate (variable parameter) according to the formulas detailed in [39]. Each simulation run corresponds to 4×10^4 s. Moreover, we have used $T_{deadline}$ = 150 ms for video packets and $T_{deadline}$ = 2 s for Web packets ([2]).

Figure 5.9 shows the P_{drop} behavior as a function of ζ for PF, EDF and P-EDF schedulers, when the total number of traffic sources is equal to 30. All these scheduling schemes employ the physical layer adaptability, but PF and EDF achieve extremely poor P_{drop} performance since they do not include a strategy to provide a strong prioritization of video traffic with respect to the Web one. Whereas, P-EDF attains a low P_{drop} value that permits to fulfill the P_{drop} requirement ($\leq$ 1%) up to ζ about equal to 1 Erlang.

Figure 5.10 shows the $Delay_{Web}$ behavior as a function of ζ for PF, EDF and P-EDF schedulers in a scenario where the total number of traffic sources is equal to 30. As expected, EDF and PF schemes allow the lowest $Delay_{Web}$ values; the high $Delay_{Web}$ values with the P-EDF scheme are due to the strong prioritization of video traffic that entails higher transmission delays for Web traffic.

In the following graphs, the performance comparison is focused on P-EDF and PF techniques. Figure 5.11 presents the comparison of η as a function of ζ for PF and P-EDF in a scenario where the total number of traffic sources is equal to 30. It can be observed that P-EDF allows a better efficiency than PF since it permits to achieve a lower P_{drop} value. Finally, Figure 5.12 provides $P_{loss\_channel}$ results as a function of ζ for PF and P-EDF for cases with total number of traffic sources equal to 30. The obtained results show that all the $P_{loss\_channel}$ values are quite close and around 7%, a limit loss value that could be still tolerated by some recent video codecs, such as H.263 used in UMTS.

It should be noted that the PF scheduler in some way may provide a more frequent service to UEs in the GOOD state than P-EDF. Hence, with PF, it could be more probable to schedule a UE that is changing its state from GOOD to BAD (thus incurring in packet losses due to channel state

[2] Video packets exceeding the deadline are dropped; while, Web packets exceeding their deadline are sent anyway since they are related to interactive traffic.

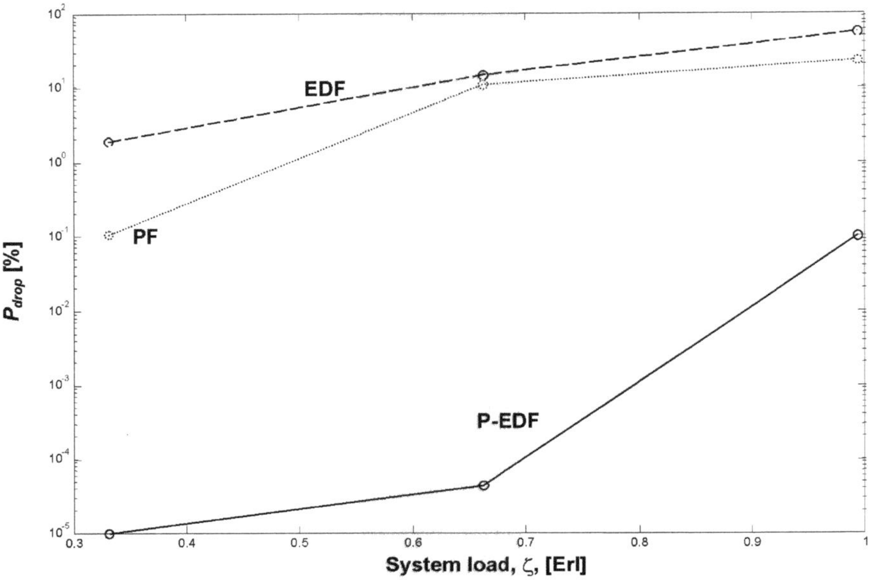

Fig. 5.9: S-HSDPA results in terms of P_{drop} for video packets.

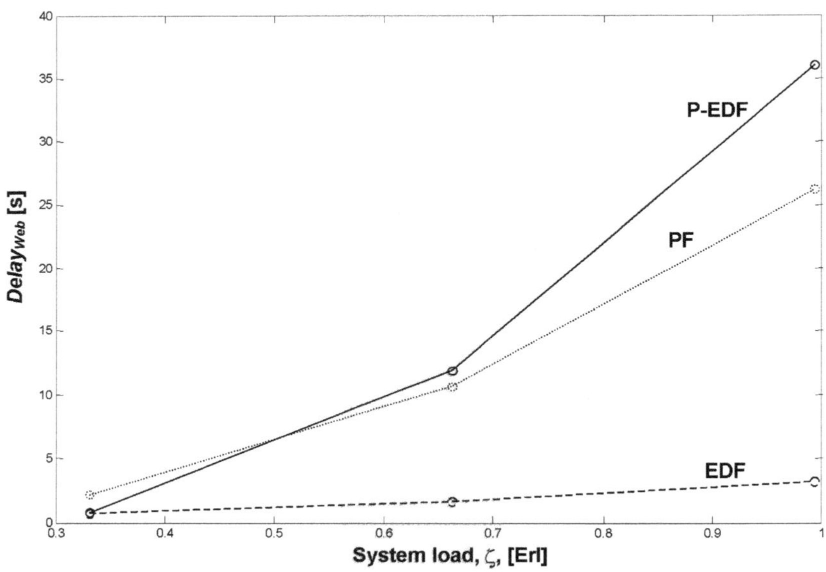

Fig. 5.10: S-HSDPA results in terms of $Delay_{Web}$ for Web packets.

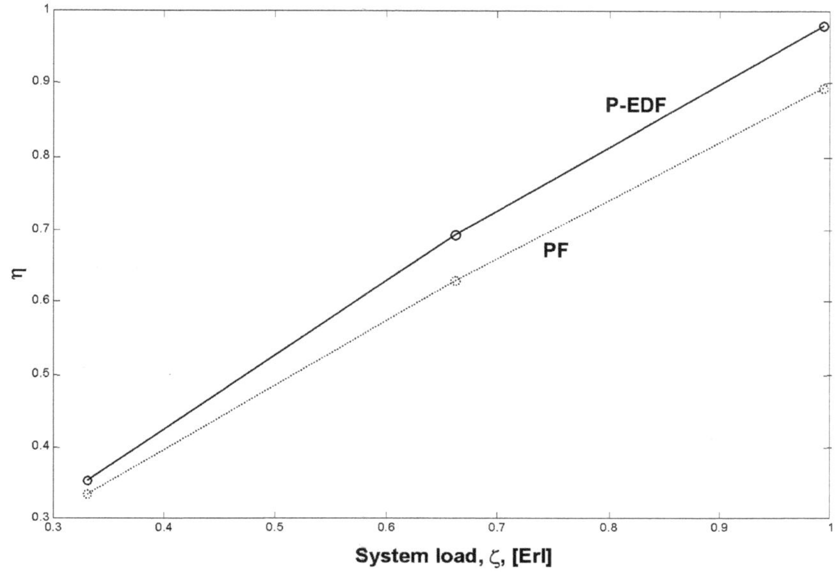

Fig. 5.11: Resource utilization comparison as a function of system load for P-EDF and PF schemes.

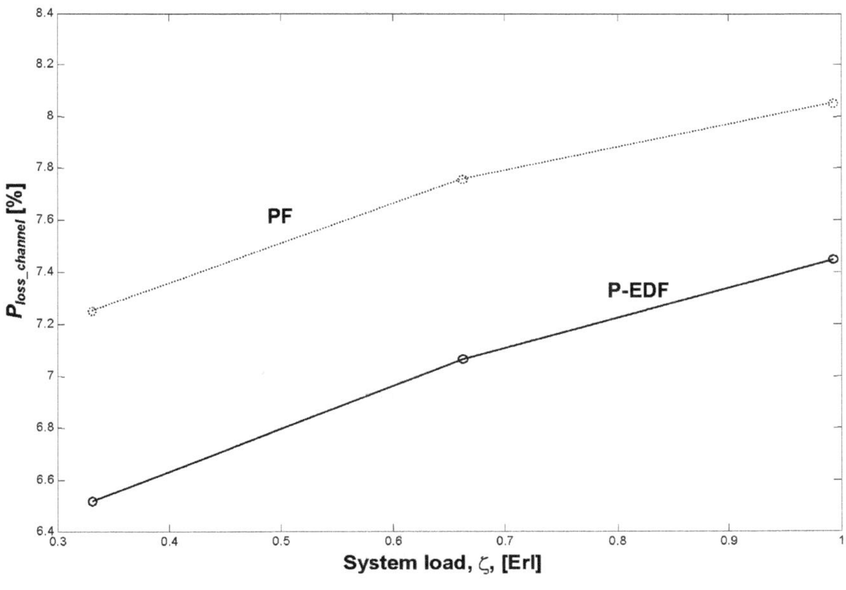

Fig. 5.12: $P_{loss\_channel}$ as a function of system load for P-EDF and PF schemes.

misalignment). These are the reasons why PF is characterized by a higher $P_{loss\_channel}$ value than P-EDF. Moreover, the higher the number of UEs, the higher the probability to schedule a UE when a transition occurs from GOOD to BAD.

In conclusion, the simulation results reported here prove that S-HSDPA is feasible, provided that suitable scheduling functions and traffic flow prioritization are employed.

5.3.3 Scheduling techniques for broadcast and multicast services in S-UMTS

With the increasing use of high-bandwidth applications in 3G mobile systems, especially with a large number of users receiving the same high data rate services, efficient information distribution is essential. Thus, broadcast and multicast are techniques to decrease the amount of transmitted data within the network and to use resources more efficiently. In particular, broadcast/multicast is a method for transmitting datagrams from a single source to several destinations. Due to the broadcast nature of the radio channel, this method is efficient for sessions sharing the same (or even common) contents. If the nature of the offered service lends itself to spatial and temporal bundling of the demands into one transmission, the benefit of multicast and broadcast is that data are sent just once by the network and transmitted to users, located in the same cell, over a single common channel without clogging up the air interface with multiple transmissions of the same data, as caused by multiple usage of unicast sessions.

Due to the broadcast nature and ubiquitous coverage, satellite systems may become a very efficient complement to terrestrial mobile networks, removing their asymmetric load and providing them with far more point-to-point equivalent capacity for far less investment cost.

Design requirements

Requirements of broadcast and multicast services delivery and impact on packet scheduler design

Even though the broadcast and multicast delivery mode is able to give many benefits for certain application areas such as inherently 'non-interactive' applications, e.g., video/audio streaming and file downloading applications in the presence of a high user density (stadiums, trade shows, etc.), there are still many challenging issues to be solved such as the resource management for providing the QoS constraints with the same conditions for all members in the same group.

UMTS allows a user or an application to negotiate the characteristics of the service at connection set-up. The network may check whether sufficient resources are available, and returns the results to the application, which can accept or deny the connection request according to a CAC scheme. After admission of the connection request, the network should keep the performance of the connection as contracted. This is also the case of broadcast and multicast users. By admitting the connection request, the access network has to make a choice for the type of the radio access bearer taking into account several conditions, like attributes of the requested service, number of group members in the cell, current load conditions etc. In contrast to unicast, i.e., point-to-point service delivery, the network has to select the type of the transport channel (i.e., a common channel or a dedicated one). For instance, if there is only one multicast member in the cell, it is not worth to use a common channel since a common channel needs additionally a return link dedicated channel for maintaining the quality of the connection, i.e., measurement control/report, power control and the error correction due to its unidirectional nature. In other words, the usage of a common channel is not always more effective than that of dedicated channels. Therefore, well-defined criteria for selecting the transport channel type among others (e.g., the minimum number of members in the multicast group, momentary load condition, current/predictable channel condition, QoS constraints of the session and so on) are necessary in order to utilize optimally system capacity. Since the number of members in a multicast session can be dynamically changing, there should be another criterion for the appropriate timing when a *Radio Access Bearer* (RAB) re-assignment will be necessary. Such criterion will certainly affect the scheduling assignment. Another issue to consider is on the method the transmission power should be (re)assigned to reflect the group dynamics of a multicast session, since users can join or leave a multicast group at any time. Controlling the transmission power in a UMTS network is crucial in maximizing the capacity that the network supports. This is due to the fact that UMTS uses the CDMA technology, which is interference-limited. In order to get a feedback channel for the power control, several methods can be considered, such as: use of an additional bi-directional DCH between each multicast member and the base station (i.e., Node-B) or usage of the RACH, as specified in UMTS.

After the assignment of a certain RAB to the multicast session, the network should maintain the contracted performance throughout the session. In practice, it is considerable that the network has to maintain not only this multicast session, but also other multicast sessions as well as other unicast sessions, which have their requirements in terms of delay, throughput, jitter, priority and so on. Moreover, especially for the satellite network, it is also considerable that the group members are distributed with great distances from each other. Hence, the selection of an appropriate *Transport Format* (TF) has a strong impact on the performance of connections, not only the multicast session itself, but also on the other active sessions due the generated interference level. According to the W-CDMA channel sharing technique, for each TTI, we have to decide how to accommodate datagrams over channels by choosing an optimal, or sub-optimal, TF combination, for the currently active sessions. This TF selection has to be done dynamically according to the changing load conditions, the number of multicast members and the radio propagation condition. Of course, the performance experienced by the most of group members cannot be worsened by a minor number of them.

Reference scenario and impact on packet scheduler design

The provision of multimedia services in broadcast and multicast mode has been regarded as a key for the efficient use of the precious wireless resources, and is currently under standardization within the *Multimedia Broadcast Multicast Services* (MBMS) framework [2] in 3GPP. However, serious concerns are expressed as to whether T-UMTS can cope with the additional requirements of MBMS delivery on top of the other point-to-point T-UMTS services due to the spectrum limitations and very limited means to improve the spectrum efficiency. On the other hand, satellites are a promising platform for MBMS delivery due to their unique wide area coverage capabilities.

Considering that broadcast and multicast traffic flows are asymmetric in nature, the baseline satellite system architecture under consideration is effectively unidirectional [40], as shown in Figure 5.13. It relies on the existing 3G mobile network *point-to-point* (p-t-p) service capability for the return link to manage and to control the delivered services, for example for access to content decoding keys and retrieval of multimedia content blocks corrupted on the satellite forward link. The space segment consists of a GEO satellite that features a transparent payload with multiple beams. This choice provides the desired flexibility in updating/enhancing the system throughout its life and is accompanied by reduced technology and investment risks. In build-up areas such as in urban and indoor environments, terrestrial repeaters/gap-fillers can be introduced to enhance the signal availability. They are designed to be smoothly co-sited with 3G base stations (i.e., Node-Bs) to prevent additional installation costs [41].

The UE+ considered here is a multi-mode terminal (i.e., satellite and terrestrial 2G/3G radios), with frequency band extension. It is able to perform parallel idle mode, i.e., maintaining either GSM activity or UMTS activity during S-MBMS reception. The basic type does not have a dedicated receiver for S-MBMS and is then required to switch from UMTS terrestrial to satellite reception. The hub includes 3G RAN equipment (i.e., RNC) and 3G core network functions. It collects incoming media services from the *Broadcast Multicast-Service Center* (BM-SC) and generates the W-CDMA waveform and redirects the signal to the satellite feeder link. The BM-SC provides functions for S-MBMS user service provisioning and delivery; for example, it controls user access to services, authorizes and initiates bearer services within the network, schedules and transmits MBMS data across the network.

Given that there is no real-time interaction between the user and the satellite RAN in the considered baseline architecture, the operation of the packet scheduler is therefore different than in the previous S-HSDPA case.

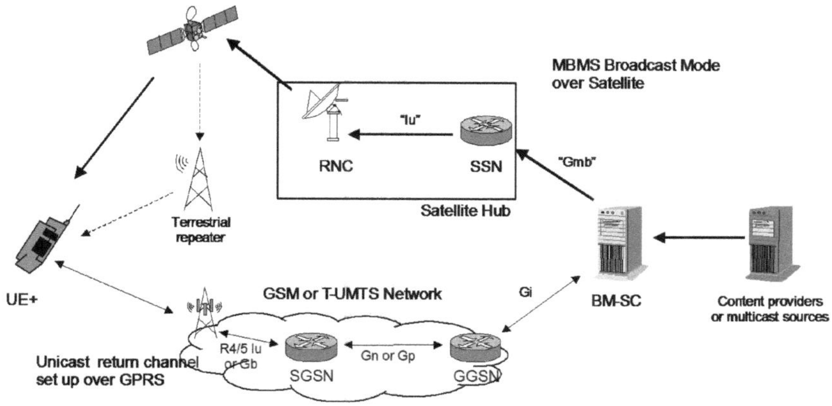

Fig. 5.13: S-MBMS architecture and its interworking with a terrestrial network.

The packet scheduler in the unidirectional satellite system has to decide on allocations without knowledge of the state of individual channels, i.e., channel state-dependent scheduling is not possible. In any cases, even if such information were available, it would have to be exploited in a complex way due to the point-to-multipoint nature of the services, i.e., the decisions regarding the scheduling of a single service data flow need to consider the state of multiple links corresponding to all the users that have activated the service in each (multicast) group.

The role of the packet scheduler in S-MBMS is not that dominant in determining the system throughput as in the T-UMTS case. Nevertheless, the scheduler is still responsible for two important tasks that are executed with a period equal to the TTI of the radio bearers [42]:

- Time multiplexing of flows with different QoS requirements into fixed physical channels, in a way that can satisfy these requirements.
- Adjusting the transmit power of the physical channel carrying the data flows on the basis of the required reception quality of the service (in terms of the target FER) under the constraint that the total available power for all the physical channels within a beam is fixed.

The packet scheduling strategy can be generally conceptualized into two steps, as described in Figure 5.14.

These two steps effectively constitute the discipline of the packet scheduler.

Functional design of packet scheduler for multicast traffic

Service prioritization

In MBMS, each service is one-to-one mapped onto an *MBMS point-to-multipoint Traffic Channel* (MTCH), a logical channel, which is then mapped

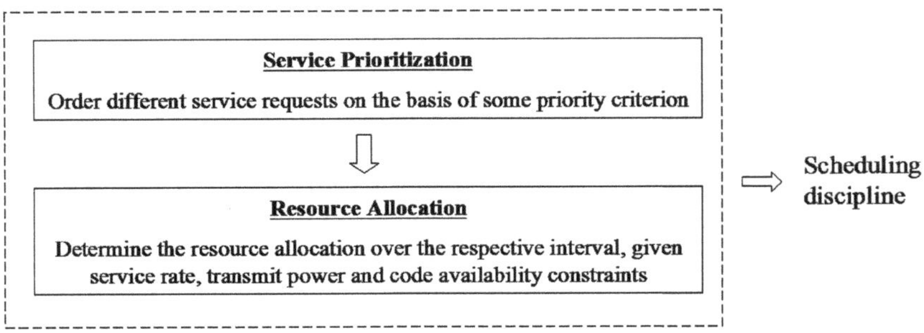

Fig. 5.14: Packet scheduling procedure.

onto the FACH transport channel. At the physical level, the *Secondary Common Control Physical Channel* (S-CCPCH) can carry one or more FACH(s). The incoming service requests are ordered according to some priority criterion. In selecting the respective criteria, the service attributes are considered, which are normally mapped onto the traffic handling priorities, as defined by the UMTS QoS classes. Note that the prioritization can be more or less dynamic; in a more dynamic prioritization, the relative priority of the different channels may change in each resource allocation interval (this is normally the TTI), depending for example on the maximum delay tolerated by a service or the number of packets buffered.

We firstly describe a semi-dynamic prioritization performed at two levels. The first prioritization is static: the scheduler orders the services according to their QoS classes (streaming, background) and the type of service delivery (streaming, hot download, cold download), i.e., streaming service MTCHs have higher priority than hot download service MTCHs, while hot download MTCHs have higher priority than cold download service MTCHs, with both download type of services belonging to the background class. Essentially, this means that an explicit cross-layer design approach has been adopted herein, whereby the upper layer information regarding the service attributes are signaled down to the packet scheduler. In fact, QoS attributes are regarded as the parameters from the application layer, which are used in the scheduling entity, so that QoS-based scheduling can be considered as a cross-layer approach. The second level of prioritization is related to the treatment of MTCHs featuring the same level of priority, i.e., when there are two or more MTCHs services having the same priority level. This prioritization is more dynamic and two alternatives can be envisaged:

- The first one is based on the rotation of the serving order of the MTCHs at each one of the three 'groups' (streaming, hot download, cold download) determined from the first prioritization level. Separate lists are maintained for each of these 'groups', whereby MTCHs are served according to their current order in the list: the MTCH at the top of the list is served first, then the second one, etc. When an MTCH is served, it is removed from the head of the list and is placed at the end of it, i.e., in a round-robin manner.

- The second scheme is based on the *Service Credit* (*SCr*) concept, which extends the idea of tokens from the leaky bucket algorithm to CDMA packet-switched mobile communication systems. The *SCr* of a service accounts for the difference between the actual offered bit-rate (by the scheduler) and the requested bit-rate, i.e., the guaranteed bit-rate for this service. Hence, a service obtaining a higher bit-rate than requested has $SCr < 0$, while a service obtaining a lower bit-rate than requested has $SCr > 0$. In each TTI, the *SCr* for a service is updated as follows:

$$\begin{aligned} SCr\,[n] = SCr\,[n-1] + (Guaranteed\_rate/TB\_size) \\ - Transmitted\_TB\,[n-1] \end{aligned} \tag{5.10}$$

  where $SCr[n]$ is the service credit at the current TTI, n, and is measured in number of transport blocks per TTI; $SCr[n\text{-}1]$ is the service credit in the previous TTI; "*Guaranteed_rate*" is the number of bits per TTI that would be transmitted at the guaranteed bit-rate; "*TB_size*" is the number of bits in the *Transport Block* (TB) considered, and *Transmitted_TB*$[n\text{-}1]$ is the number of successfully transmitted TBs in the previous TTI.
  Obviously, this dynamic prioritization scheme is directly applicable to streaming services, which feature a guaranteed rate attribute; however, it may be expanded to download services even if they are not explicitly characterized by the guaranteed bit-rate attribute (see Figure 5.14).

Rather than performing service prioritization in a semi-dynamic way, a more efficient packet scheduling algorithm performs service prioritization dynamically, depending on the waiting time/queuing delay experienced by packets in each MTCH/FACH at the beginning of each TTI. Resource is then allocated to respective physical channels (i.e., S-CCPCH) according to the priority assigned to each MTCH/FACH flow as long as their power and load condition can be satisfied. This scheduling scheme is named *Delay Differentiation Queuing* (DDQ) [43]. It is worth noticing that the packet scheduling algorithm remains under the assumption of one-to-one mapping from logical channels (MTCHs) to transport channels (FACHs).

DDQ is not a priority queue and is based on the *Hybrid Proportional Delay* (HPD) scheduling scheme [44], which is widely used in the differentiated service networks. It assumes that there are QoS ratios between different QoS priority classes. In each TTI, the serving indexes will be calculated for each queue. These serving indexes are obtained based on the average waiting delay for all the packets currently in the queue, the average queuing delay for all the packets that have left the queue before this TTI, the packet arrival rate and the QoS priority ratio index.

The mathematical formulation of DDQ can be expressed as follows. Let α_i be QoS class factor, which is essentially a time-independent parameter designated for each queue i. Let $\delta_i(n)$ be the average queuing/waiting delay at current n-th allocation instant (i.e., n-th TTI) for each queue i. This measure describes the delay states of all packets passing through the respective queue, including both the packets which are currently in the queue and those packets which have already left the queue. The delay index is calculated for each queue i in each TTI as in equation (5.11):

$$\delta_i[n] = \frac{\sum_{j=0}^{N_q} W_{i,j}^q[n] + \sum_{j=0}^{N_d} W_{i,j}^d[n]}{N_q + N_d} \tag{5.11}$$

where $W_{i,j}^q[n]$ is the waiting delay for the j-th packet currently in queue i; N_q is the number of packets in the queue; $W_{i,j}^d[n]$ is the queuing delay for the j-th packet, which has left queue i before this TTI (i.e., current time slot n); N_d is the number of packets that have been served and left the queue before this TTI.

For the service flow of the FACH queue i at the current time slot (i.e., TTI for UMTS) n, the priority is defined as:

$$P_i[n] = \alpha_i \, \delta_i[n] \quad . \tag{5.12}$$

Consequently, the serving orders are calculated and assigned to each FACH according to (5.12) at the beginning of each TTI.

With the above approaches of semi-dynamic and dynamic service prioritization in mind, the dynamically changing priorities of MTCHs indicate the serving order of FACHs and S-CCPCHs for each TTI. It must also be noted that it is generally assumed that only services with similar characteristics and QoS requirements are multiplexed together to the same transport channel.

Resource allocation

Once all the services to be transmitted are prioritized, the next step is the allocation of resources to them. This phase consists of bit-rate and transmit power assignments within the specific resource allocation interval (i.e., TTI). The data rate assignment consists in the selection of the *Transport Format Combinations* (TFCs), which directly determine the per FACH transport block size, namely how much data from each transport channel mapped to the physical channel will be forwarded to the physical layer in TTI. For each active physical channel (S-CCPCH), the exact TFC is selected from the *Transport Format Combination Set* (TFCS), which is passed during the admission of a new service as well as its mapping on a specific bearer. This TFC selection step is of paramount importance since the capacity allocated to each service is strongly related to the QoS perceived by the end-users, and, therefore, the selection of the TFC has to take into consideration constraints in terms of service requirements (e.g., minimum guaranteed rate, maximum tolerated delay) as well as system-level constraints (system load, transmit power per beam).

As for the power allocation, the transmit power setting for the S-CCPCH is based on the required reception quality of the active service flows mapped to S-CCPCH, which in our case is defined in terms of the most demanding target FER among these service flows. The calculated power is only allocated as long as it is within the constraint of the total available power for all the physical channels, which is fixed within a beam. In the resource allocation phase, the per S-CCPCH TFC selection and power allocation are made in parallel.

As illustrated in Figure 5.15, the description of the DDQ packet scheduling scheme can be summarized as follows:

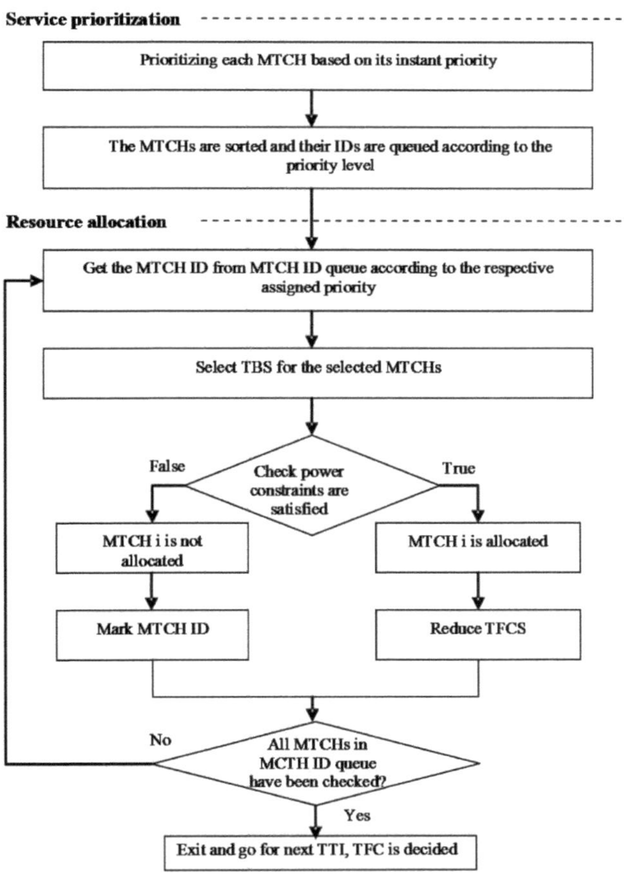

Fig. 5.15: Flowchart of DDQ scheme.

- For all S-CCPCHs, the packet scheduler tries to serve the MTCHs according to the priorities dynamically allocated to them in the particular TTI. The higher priority MTCH queues will be served ahead of the lower priority MTCH queues. For those MTCH queues having the same priority class, the queue with the longest packet queue will be served first.
- For each MTCH l, mapped on FACH j and on S-CCPCH i, the packet scheduler scans the TFCS of the physical channel to find all the different TBS sizes that could be used. A sorted list of all candidate TBS sizes, in decreasing order, is created.
  - The scheduler first seeks to allocate the maximum TBS size to the first FACH. This is the case when the sum of data at the MTCHs queues is greater than the maximum supported TBS size for this FACH in the TFCS; the allocation of data (transport block) that each MTCH can transmit is based on the priority of each MTCH mapped to this FACH, with the highest priority channel assumed to be given the maximum share.
  - Otherwise, if the sum of data from all the MTCHs queues is less than the maximum supported TBS size for this FACH, the selected TBS size is the minimum available in the TFCS that can serve this sum of queued data.
- For each S-CCPCH, the packet scheduler checks the power required on the basis of the BLER requirement of the active service flow. These power allocation decisions involve the search in lookup tables (BLER versus E_b/N_t) to determine the transmitted power for each S-CCPCH, satisfying both power and load constraints.

The packet scheduler will then derive a reduced TFCS out of the initial one for the S-CCPCH i, including only those TFCs that feature the selected TBS size for FACH j. Further allocations in the same TTI for another MTCH/FACH mapped on the same S-CCPCH will have to consider this reduced TFCS. As for the power allocation, the power required to satisfy the active service flow with the most demanding target BLER is selected, as long as the total transmit power per beam is not exceeded; otherwise, this service is not scheduled.

These procedures are repeated recursively until all the FACHs mapped to each S-CCPCH are assigned.

Performance evaluation

In order to demonstrate the performance of the packet scheduling schemes proposed for broadcast and multicast services over S-UMTS, simulations have been carried out for a wide range of scenarios by using a simulator developed under the ns-2 environment. Specifically, the DDQ packet scheduling algorithm has been evaluated via simulations in a typical S-MBMS scenario, and compared with the *Multi-Level Priority Queuing* (MLPQ) scheduling scheme described in [42]. The main characteristics of MLPQ are that it always processes packets starting from those non-empty queues having the highest priority first, with queues having the same priority served in a round-robin fashion. As a result, packets in the lower-priority queues may suffer from a considerably longer queuing delay. Moreover, according to this scheduling policy, there is no differentiation made between queues with the same QoS ranking. Therefore, this is not an efficient mechanism in differentiated QoS multimedia services provisioning with respect to both efficiency and fairness. Rather than prioritizing queues in a strict method, other essential QoS metrics should also be considered in the scheduling discipline design.

The following typical scenario with 3 S-CCPCHs each of 384 kbit/s has been simulated:

- S-CCPCH 1: 64 kbit/s download (FACH 1), 256 kbit/s streaming (FACH 2), 64 kbit/s streaming (FACH 3);
- S-CCPCH 2: 256 kbit/s streaming (FACH 4), 128 kbit/s streaming (FACH 5);
- S-CCPCH 3: 384 kbit/s download (FACH 6).

The above scenario is summarized in Table 5.4.

S-CCPCH	**1**	**2**	**3**
Bit-rate [kbit/s]	384	384	384
Streaming [kbit/s]	256×1; 64×1	256×1; 128×1	-
Download [kbit/s]	64×1	-	384×1

Table 5.4: Simulation multiplexing scenario (FACHs to S-CCPCHs).

Here we assume one-to-one mapping between MTCHs to FACHs, while multiplexing only occurs from transport channel to physical channel. Therefore, FACHs transport channel to physical channel multiplexing scenario is specified in the simulation as in Table 5.4.

DDQ and MLPQ performance results are compared via simulation metrics, such as mean delay, mean jitter and channel utilization.

Analysis of delay and delay variation

As illustrated in Figure 5.16, by using the DDQ packet scheduling algorithm, the download multimedia services (i.e., FACH 1 and FACH 6) experience much less mean delay compared with MLPQ. It is noted that the significant reduction in delay of lower-class services does not result in a dramatic performance degradation for the higher-class counterparts (i.e., FACH 2 to FACH 5). These results demonstrate that DDQ provides the download service the highest possible degree of utilizing those spare resources remaining after

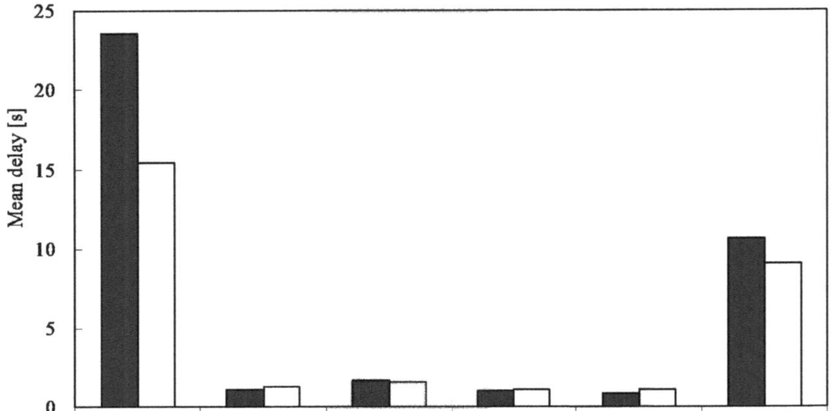

	FACH 1	FACH 2	FACH 3	FACH 4	FACH 5	FACH 6
■ MLPQ	23.570605	1.068729	1.666252	0.996016	0.865440	10.616220
□ DDQ	15.395824	1.265035	1.593606	1.110081	1.062310	9.055439

Fig. 5.16: Mean packet delay at RLC buffers for different packet scheduling algorithms.

streaming applications have been serviced, given that the detrimental affect is not posing significant degradation on the QoS target of streaming users.

Figure 5.17 shows the mean jitter experienced by each individual service when employing MLPQ and DDQ packet scheduling. Obviously, DDQ features much lower jitter for both streaming service and download service than MLPQ, especially for lower-class and lower date rate users. Since the unidirectional streaming service in S-MBMS is quite sensitive to delay-variation (jitter), this result proves that DDQ packet scheduling provides a way to balance all FACH queues in order to get the minimum delay variation for streaming services.

Analysis of channel utilization ratio

Figure 5.18 shows the average S-CCPCH physical channel utilization for both MLPQ and DDQ. Both schedulers managed to achieve throughput values close to the optimum. For instance, the S-CCPCH channel utilization ratios are 97.8%, 96.2%, 85.4% respectively under MPLQ scheduling; whilst they achieve 98.4%, 96.2%, 86.4% respectively under DDQ scheduling. Therefore, DDQ manages to obtain a slight channel utilization improvement on those S-CCPCHs carrying background traffic.

To summarize, the DDQ algorithm achieves the following advantages over the MPLQ scheduling scheme:

- Dynamic proportional delay-driven prioritization;

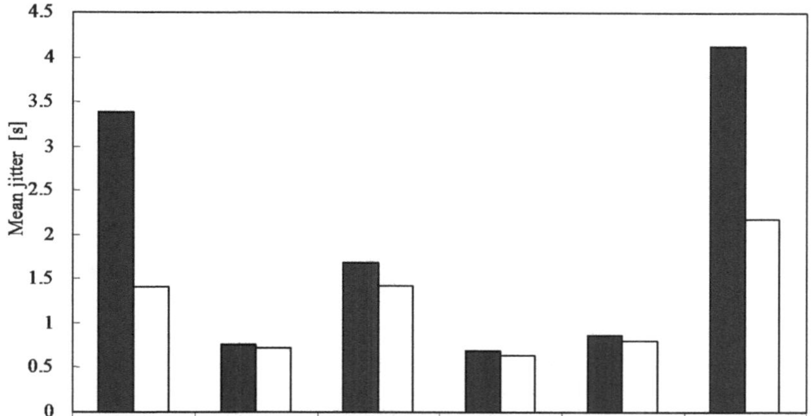

	FACH 1	FACH 2	FACH 3	FACH 4	FACH 5	FACH 6
■ MLPQ	3.386130	0.751228	1.682824	0.691275	0.873156	4.119604
□ DDQ	1.407796	0.713677	1.418577	0.632931	0.795434	2.170275

Fig. 5.17: Mean packet jitter at RLC buffers for different packet scheduling algorithms.

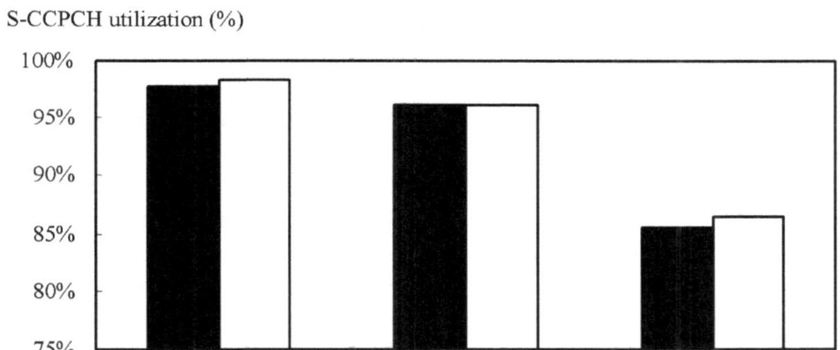

	S-CCPCH 1	S-CCPCH 2	S-CCPCH 3
■ MLPQ	0.978351	0.961814	0.854248
□ DDQ	0.983770	0.961734	0.864324

Fig. 5.18: S-CCPCH utilization for MLPQ and DDQ.

- Highest utilization for the background class without posing significant degradation on streaming class;
- Significant improvement on mean delay and mean jitter performance;
- Better overall system utilization.

5.3.4 Packet scheduling with cross-layer approach

Due to the nature of wireless transmissions, satellite communications suffer from strong variations of the received signal power due to shadowing and multipath fading. Shadowing of the satellite signal is due to obstacles in the propagation path (buildings, trees, bridges, etc). Whereas multipath fading occurs because the satellite signal is received not only via the direct path, but also being reflected from objects in the surrounding area. Due to different propagation distances, the multipath signals may add destructively and leads to deep fades.

Due to these variations, the most critical part in satellite communications is the communication link between the satellite and the user terminal (i.e., downlink). The downlink availability could be the limiting factor for the performance of the overall system. Thus, a scheduler employing an explicit cross-layer technique is proposed, where signaling interactions from the physical layer are employed so that the scheduler is aware of the channel state of the users. In this cross-layer design, a multicast packet scheduler is developed that relies on the prediction of the wireless channel conditions to improve the performance of downlink transmissions via satellite for a TDMA-based air interface.

The domain architecture for the multicast service under consideration is illustrated in Figure 5.19. The entities in the service provider will provide the interface between RAN and external packet data networks. The scheduler is at the Earth station and a GEO satellite relays the multicast information to all users through *Multicast Terminals* (MTs) and *Terminal Equipment* (TE). A *reliable* multicast transport protocol is assumed to guarantee delivery and congestion control mechanism. A unicast return link will be required for acknowledgments.

Based on a TDMA framework, the system under consideration supports scheduled access on both downlink and uplink. Downlink capacity is organized into fixed 80 ms MF-TDMA frames that are composed of a sequence of fixed 20 ms time slots. *Channel State Information* (CSI) of each user, which is the information from physical (PHY) layer, is considered in the decision mechanism whether to transmit or not the next multicast packet. The CSI parameter is averaged over a total of N frames to become a conditional parameter for the next transmission. CSI parameter is updated periodically. The update through uplink bearer is contained within a 200 kHz sub-band, which is further divided in frequency, and time slots and each slot may contain a burst from a user.

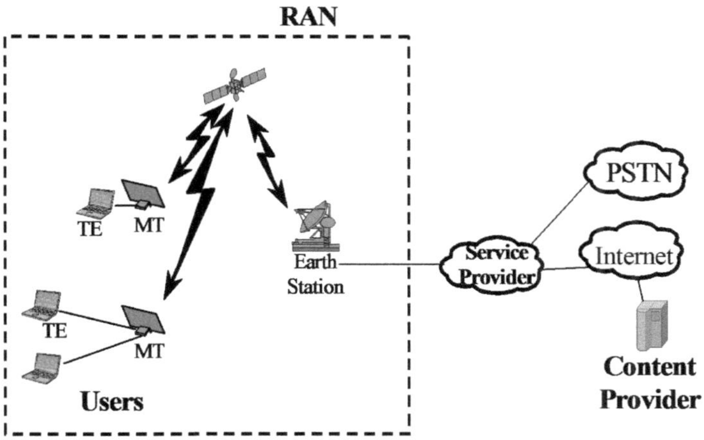

Fig. 5.19: Domain architecture for a TDMA-based satellite system.

It is assumed that the CSI update provides a reasonably accurate prediction of the slowly varying elements (i.e., valid at least within one round-trip-delay) of the channel conditions. The satellite link is modeled using the Lutz's two-state model with variation of fade duration for open and shadowed environments [45]. The following results indicate that a significant performance improvement is possible by adopting a cross-layer design approach in a fading environment.

Description of scheduler's task

Prior to assigning the slot, which is the resource allocation step, the different multicast services need to be prioritized. In our scheme, the prioritization is performed at two levels. The first prioritization is static: the scheduler orders the services according to their QoS classes (streaming and best effort), i.e., streaming service is assigned higher priority than best effort. The second level of prioritization is based upon the cross-layer information provided with CSI for services featuring the same level of priority. This prioritization is more dynamic and confined only for best-effort traffic. The algorithm is described as follows:

- For all incoming multicast packets, the packet scheduler aims to serve the packets according to priorities dynamically allocated to them. Streaming traffic packets have higher priority to access to time slots at all times.
- For the remaining slots, if best effort traffic packets arrive, the scheduler scans the CSI intended for the multicast group. The acquisition of CSI will be performed for each user in the intended multicast group. The update of channel condition is acquired in every 20 ms, according to the slot and

burst definition. The channel state information is contained in the bearer control signaling data unit. In the evaluation of scheduler's performance, we consider TDMA channels in which each frame is divided into fixed control and data sub-frames. The user channel information is updated through bearer control signaling data unit. The data sub-frame length is the block length which is a *data transmission unit* size of 125 bytes and it fits into time slots each holding one packet.
- The packet scheduler will check for estimated E_b/N_0 values of each user i in the multicast group, Γ_i. A reference E_b/N_0 threshold, γ_T, is compared with Γ_i and the number of users satisfying this reference will be the decision making parameter for slot allocation.
- If within one slot, more than one packets of the same priority arrives, then the scheduler will check for the packet with higher percentage of satisfied users. If the number of satisfied users from a particular multicast group is above a certain threshold, the slot is allocated to that packet. If not, then the packet will be delayed and retried for the next slot, provided that the next slot is not intended for higher priority traffic.

For more details, the interested reader is referred to [46].

Performance evaluations

The following study assumes stationary users and slowly-varying channels in satellite links where fade duration holds within one CSI update. The results are here based on perfect channel predictions; we assume no channel estimation loss occurs. This assumption might be impractical in a satellite environment where the propagation delay is high, but the results with this assumption permit to have a good indication of the effectiveness of this scheme to achieve a high reliability multicast transmission.

In this study, two different scenarios have been examined. The first one, the single environment scenario, assumes that all users are subject to identical channel conditions (the single environment model uses an elevation angle α = 80° and values for μ and σ calculated for urban areas with K factor of 7, where K represents the Ricean factor which is defined as the ratio of the dominant component to the scatter contribution [45]). The proposed technique aims at reducing the number of retransmissions that stem from bad channel conditions. Figure 5.20 [46] exhibits some rather interesting results, where parameter *zeta* is defined as follows: a packet is retransmitted only if the percentage of users in the multicast group that experience *Packet Loss Rate* (PLR) higher than a defined PLR threshold is greater than a percentage, denoted by parameter zeta. It should be pointed out that PLR significantly diminishes, by endowing the multicast packet scheduler with CSI (cross-layer approach). Moreover, PLR is hardly affected by an increase in the multicast group size, whereas the greater the parameter *zeta* is, the lower PLR is.

Figure 5.21 [46] illustrates the probability that at least one user will request retransmission versus the multicast group size. Apparently, as the

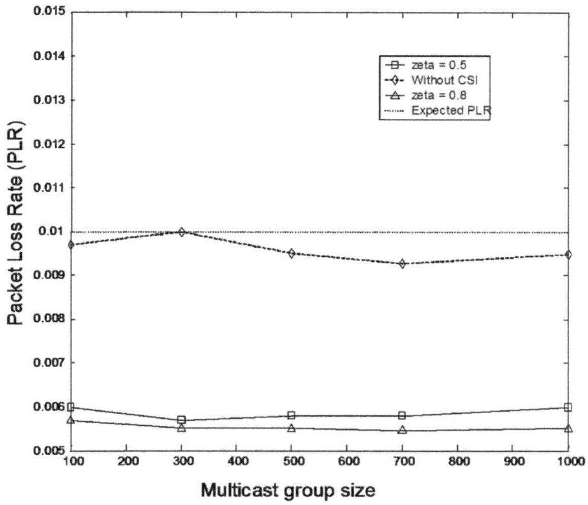

Fig. 5.20: Packet loss rate versus multicast group size. See reference [46]. Copyright ©2005 IEEE.

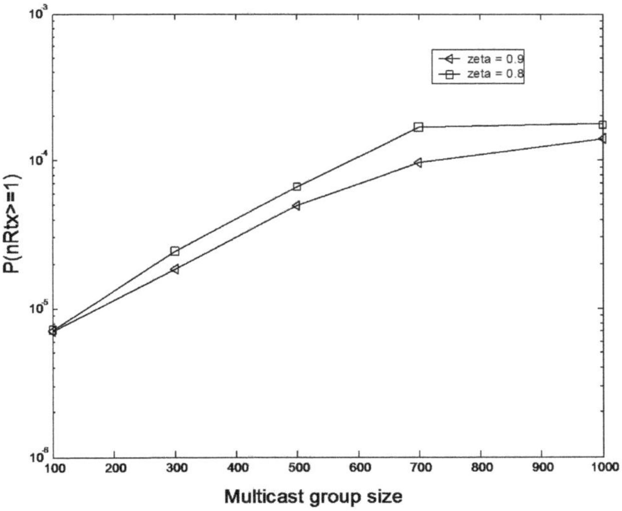

Fig. 5.21: Probability of at least one user in a multicast group requesting retransmission (failure rate) versus multicast group size. See reference [46]. Copyright ©2005 IEEE.

size of the multicast group increases, so does the probability of retransmission. Furthermore, this probability can be reduced by increasing parameter *zeta*. What is more important is that the greater the size of the multicast group, the higher the average packet delay, as illustrated in Figure 5.22. The value of *zeta* used to obtain the results in Figure 5.22 is 0.9.

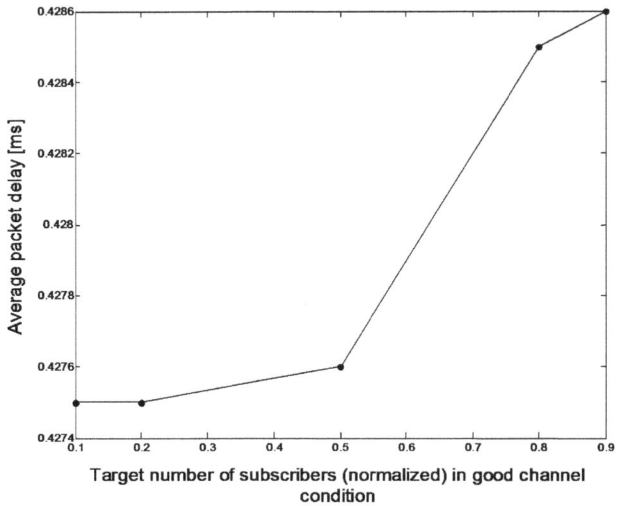

Fig. 5.22: Average packet delay versus target number of users (normalized) in good channel condition.

At this point, multi-channel environments are simulated based on the parameters in Table 5.5.

Area	**Rician K factor**	α (°)	μ	σ	**% Users**
Suburban	0 dB	20	−1.69	2.70	20
Urban	3 dB	20	−13.90	3.06	40
Urban	7 dB	80	1.75	0.80	30
Suburban	10 dB	60	0.14	0.40	10

Table 5.5: Simulation parameters for the multi-environment scenario.

In this case, multicast users are subject to different channel conditions, and the empirical models presented in [45] were deployed. A user requires retransmission only if the difference between γ_{ref}, which is the reference E_b/N_0 from the AWGN channel model to achieve a target PLR of 10^{-2}, and $\Gamma(t)$, which is the E_b/N_0 value of the signal received from this user, is greater than a given E_b/N_0 threshold, γ_T:

$$\gamma_{ref} - \Gamma(t) > \gamma_T \quad \text{[dB]}. \tag{5.13}$$

Figure 5.23 [46] depicts the probability that at least one user will request retransmission versus E_b/N_0 threshold. Evidently, the retransmission probability decreases as E_b/N_0 threshold increases. It should also be noted that the retransmission probability decreases as parameter *zeta* increases. As far as the average packet delay is concerned, it becomes clear from Figure 5.24 that as E_b/N_0 threshold increases, the mean delay increases since the multicast packet scheduler refrains from transmitting packets to multicast groups typically experiencing bad channel conditions [46].

This approach has been shown to reduce unnecessary transmission of best-effort traffic and hence reduces unnecessary bandwidth usage and retransmission requests. However, in achieving relatively good channel utilization for a multicast group, higher average packet delay is expected in the cross-layer scheduler. The average packet delay can be regulated according to the power threshold a user is estimated to receive at the downlink transmission. It is also important to note that this approach consumes an amount of resources in performing the channel prediction algorithm. The accuracy of the channel quality is highly dependent on the channel model used.

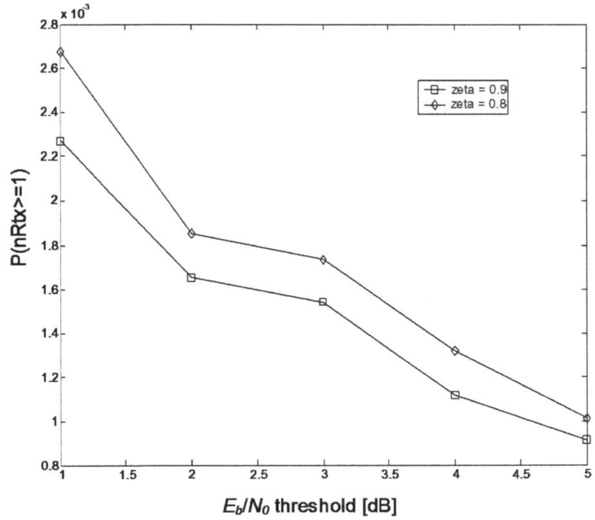

Fig. 5.23: Probability of at least one user in a multicast group requesting retransmission (failure rate) versus E_b/N_0 threshold. See reference [46]. Copyright ©2005 IEEE.

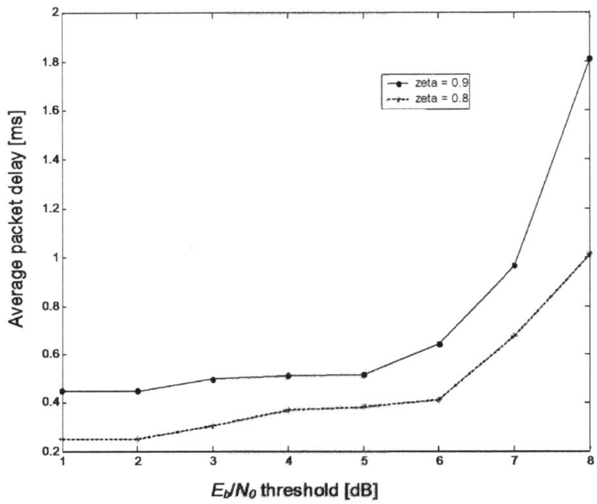

Fig. 5.24: Average packet delay versus E_b/N_0 threshold. See reference [46]. Copyright ©2005 IEEE.

5.4 Conclusions

Satellite communications have a potential market in providing high downlink bit-rate services and in supporting multicast services on broad areas of the Earth. These are the reasons why this Chapter has focused on HSDPA and MBMS provision via a GEO bent-pipe satellite. In both cases suitable network architectures and radio resource management techniques have been investigated to support such services in an appropriate and efficient way.

For HSDPA, the results showed by the *Proportional Fair* scheduler are sub-optimal for mixed traffic classes, since it does not provide any QoS differentiation among diverse applications. The study of proposed enhancements to the PF scheduler to support QoS differentiation, such as the *Exponential Rule*, should be addressed in the future for the satellite case. Furthermore, the impact of the round trip time in the acquisition of channel state information has been shown in the form of packet losses in intervals of misalignments between current channel state and information available at the Gateway. In particular, the simulation results in a simplified scenario (using a GOOD/BAD channel model) show non-negligible losses due to the use of outdated information in the selection of the best suited TFRC for transmission. Hence, if it is desired to reduce the number of retransmissions, delay compensation strategies or larger margins in the selection of TFRCs should be adopted. Furthermore, a more complex channel model should be considered in order to take into account the channel variation dynamics typical of S Band (S-UMTS band).

For the provision of broadcast and multicast services, it has been shown that packet scheduling is an important element within the RRM framework. Aiming at a more efficient provision of heterogeneous QoS-differentiated MBMS services over S-UMTS, novel packet scheduling algorithms have been proposed. These algorithms take into account the impact of important performance factors reflecting service QoS demands in order to provide traffic differentiation and overall system performance optimization. To tackle the deteriorating effect of changing propagation environments in multicast transmissions, channel estimation can fill the void whilst obtaining the current channel state. Statistical channel models can be used to represent channel variations to be exploited by packet scheduler for its decisions. For traffic with strict delay bound, a negotiation between delay and channel states can be facilitated by a cost function where a trade-off between delay and throughput is expected.

References

[1] Y. Cao, V. O. K. Li, "Scheduling Algorithms in Broadband Wireless Networks", in *Proc. of the IEEE*, Vol. 89, No. 1, January 2001.

[2] ETSI, "Satellite Earth Stations and Systems (SES); Satellite Component of UMTS/IMT2000; G-family; Part 1: Physical Channels and Mapping of Transport Channels into Physical Channels (S-UMTS-A 25.211)", TS 101 851-1.

[3] ETSI, "Satellite Earth Stations and Systems (SES); Satellite Component of UMTS/IMT2000; G-family; Part 2: Multiplexing and Channel Coding (S-UMTS-A 25.212)", TS 101 851-2.

[4] ETSI, "Satellite Earth Stations and Systems (SES); Satellite Component of UMTS/IMT2000; G-family; Part 3: Spreading and Modulation (S-UMTS-A 25.213)", TS 101 851-3.

[5] ETSI, "Satellite Earth Stations and Systems (SES); Satellite Component of UMTS/IMT2000; G-family; Part 4: Physical Layer Procedures (S-UMTS-A 25.214)", TS 101 851-4.

[6] 3GPP, "Physical Channels and Mapping of Transport Channels onto Physical Channels (FDD)", TS 25.211, 2005.

[7] 3GPP, "Physical Layer Procedure (FDD)", TS 25.214 V6.3.0 (2004-09).

[8] ETSI, "Universal Mobile Telecommunications System (UMTS); Medium Access Control (MAC) Protocol Specification (3GPP)", TS 125.321 V3.11.0 (March 2002).

[9] V. Y. H. Kueh, A. Capellacci, R. Tafazolli, B. G. Evans, "W-CDMA Random Access Channel Transmission Enhancement for Satellite-UMTS", in *Proc. of the* 13th *IEEE International Symposium on Personal, Indoor and Mobile Radio Communications* (PIMRC 2002), Lisbon, Portugal, September 15-18, 2002.

[10] J. Goodman, R. A. Valenzuela, K. T. Gayliard, B. Ramanurthi, "Packet Reservation Multiple Access for Local Wireless Communications", *IEEE Transactions on Communications*, Vol. 37, No. 8, pp. 885-890, August 1989.

[11] E. Del Re, R. Fantacci, G. Giambene, W. Sergio, "Performance Analysis of an Improved PRMA Protocol for Low Earth Orbit-Mobile Satellite Systems", *IEEE Transactions on Vehicular Technology*, Vol. 48, No. 3, pp. 985-1001, May 1999.

[12] G. Benelli, R. Fantacci, G. Giambene, C. Ortolani, "Performance Analysis of a PRMA Protocol Suitable for Voice and Data Transmissions in Low Earth Orbit

Mobile Satellite Systems", *IEEE Transactions on Wireless Communications*, Vol. 1, No. 1, pp. 156-168, January 2002.
[13] G. Giambene, E. Zoli, "Stability Analysis of an Adaptive Packet Access Scheme for Mobile Communication Systems with High Propagation Delays", *International Journal of Satellite Communications and Networking*, Vol. 21, pp. 199-225, March 2003.
[14] R. Fantacci, T. Pecorella, I. Habib, "Proposal and Performance Evaluation of an Efficient Multiple-Access Protocol for LEO Satellite Packet Networks", *IEEE Journal on Selected Areas in Communications*, Vol. 22, No. 3, pp. 538-545, April 2004.
[15] N. Batsios, I. Tsetsinas, F. N. Pavlidou, "Performance Evaluation of CDMA/PRMA Techniques for LEO Constellations", in *Proc. of the Vehicular Technology Conference, 2001*, Vol. 1, pp. 576-580, May 2001.
[16] A. Andreadis, G. Giambene. *Protocols for High-Efficiency Wireless Networks*. Kluwer Academic Publishers, November 2002.
[17] G. Giambene, F. Miano, E. Zoli, "Energy-Efficient Packet Access Scheme for MF-TDMA in non-GEO Satellite Systems", in *Proc. of VTC 2004-S*, Milan, May 17-19, 2004.
[18] M. Katevenis, S. Sidiropoulos, C. Courcoubetis, "Weighted Round-Robin Cell Multiplexing in a General Purpose ATM Switch Chip", *IEEE J. Select. Areas in Commun.*, Vol. SAC-9, No. 8, pp. 1265-1279, October 1991.
[19] S. Floyd, V. Jacobson, "Link-Sharing and Resource Management Models for Packet Networks", *IEEE/ACM Trans. Networking*, Vol. 3, No. 4, pp. 365-386, August 1995.
[20] K. Parekh, R. G. Gallager, "A Generalized Processor Sharing Approach to Flow Control in Integrated Services Networks: the Single Node Case", *IEEE/ACM Trans. Networking*, Vol. 1, No. 3, pp. 344-356, June 1993.
[21] A. Demers, S. Keshav, S. Shenkar, "Analysis and Simulation of a Fair Queueing Algorithm", *Internet. Res. and Exper.*, Vol. 1, 1990.
[22] L. Georgiadis, R. Guerin, V. Peris, K. N. Sivarajan, "Efficient Network QoS Provisioning Based on per Node Traffic Shaping", *IEEE/ACM Transactions on Networking*, Vol. 4, No. 4, pp. 482-501, August 1996.
[23] L. Georgiadis, R. Guerin, A. Parekh, "Optimal Multiplexing on a Single Link: Delay and Buffer Requirements", *IEEE Transactions on Information Theory*, Vol. 43, No. 5, pp. 1518-1535, September 1997.
[24] H. Sariowan, R. L. Cruz, G. C. Polyzos, "SCED: A Generalized Scheduling Policy for Guaranteeing Quality-of-Service", *IEEE/ACM Trans. Networking*, Vol. 7, No. 5, pp. 669-684, October 1999.
[25] V. Bharghavan, S. Lu, T. Nandagopal, "Fair Queuing in Wireless Networks: Issues and Approaches", *IEEE Personal Communications*, Vol. 6, No. 1, pp. 44-53, February 1999.
[26] P. McKenney, "Stochastic Fairness Queuing", *Journal of Internetworking Research and Experience*, Vol. 2, pp. 113-131, 1991.
[27] H. Holma, A. Toskala (Eds). *WCDMA for UMTS: radio access for third generation mobile communications.* Second edition, John Willey & Sons Ltd, 2002.
[28] T. Kolding, K. Pedersen, J. Wigard, F. Frederiksen, P. Mogensen, "High Speed Downlink Packet Access: WCDMA Evolution", *IEEE Vehicular Technology Society News*, February 2003.

[29] E. Esteves, P. Black, M. Gurelli, "Link Adaptation Techniques for High-Speed Packet Data in Third Generation Cellular Systems", in *Proc. of European Wireless Conference*, Florence, Italy, February 2002.
[30] T. Kolding, "Link and System Performances Aspects of Proportional Fair Scheduling in WCDMA/HSDPA", in *Proc. of the IEEE VTC-Fall 2003*, Orlando, Florida, USA, October 4-9, 2003.
[31] L. Wang, M. Chen, "Comparisons of Link Adaptation Based Scheduling Algorithms for the WCDMA System with High Speed Downlink Packet Access", *Canadian Journal of Electrical and Computer Engineering* (CJECE), Vol. 29, No. 1-2, pp. 109-116, January-April 2004.
[32] 3GPP, "High Speed Downlink Packet Access; Overall UTRAN Description", TR 25.855, Release 5.
[33] H. Ishii, A. Hanaki, Y. Imamura, S. Tanaka, M. Usuda, T. Nakamura, "Effects of UE Capabilities on High Speed Downlink Packet Access in WCDMA Systems", in *Proc. of the* 59th *Vehicular Technology Conference VTC2004-Spring*, Milan, May 17-19, 2004.
[34] 3GPP, "UE Radio Access capabilities", TS 25.306 V6.2.0 (2004-06).
[35] C. Párraga, C. Kissling, "Delay Compensation Strategies for an Efficient Radio Resource Management in DVB-S2 Systems", in *Proc. of IEEE International Symposium on Wireless Communication Systems 2005* (ISWCS 2005), ISBN 0-7803-9206-X, Siena, Italy, September 5-9, 2005.
[36] 3GPP, "User Equipment (UE) Radio Transmission and Reception (FDD)", TS 25.101, Release 6.
[37] X. Liu, E. Chong, N. Shroff, "Opportunistic Transmission Scheduling with Resource-Sharing Constraints in Wireless Networks", *IEEE Journal on Selected Areas in Communications*, Vol. 19, No. 10, pp. 2053-2064, October 2001.
[38] G. Giambene, S. Giannetti, V. Y. H. Kueh, C. Párraga, "Packet Scheduling Techniques for HSDPA and MBMS Transmissions in Satellite UMTS", in *Proc. of ICSSC 2005*, Rome, Italy, September 26-28, 2005.
[39] G. Giambene, S. Giannetti, V. Y. H. Kueh, C. Párraga, "HSDPA and MBMS Transmissions via S-UMTS", *COST 290*, TD(06)013, 5th MCM, Delft, The Netherlands, February 9-10, 2006.
[40] K. Narenthiran *et al.*, "S-UMTS Access Network for MBMS Service Delivery: the SATIN Approach", *International Journal of Satellite Communications and Networking*, Vol. 22, No. 1, pp. 87-111, January/February 2004.
[41] T. Severijns *et al.*, "The Intermediate Module Concept within the SATIN Proposal for the S-UMTS Air Interface", in *Proc. of IST Mobile Summit 2002*, Greece.
[42] M. Karaliopoulos, P. Henrio, E. Angelou, B. G. Evans, "Packet Scheduling for the Delivery of Multicast/Broadcast Services via S-UMTS", in *Proc. of First International Conference on Advanced Satellite Mobile Systems*, Frascati, Italy, July 2003.
[43] L. Fan, H. Du, U. Mudugamuwa, B. G. Evans, "Novel Radio Resource Management Strategy for Multimedia Content Delivery in SDMB system", in *Proc. of the* 24th *AIAA ICSSC*, San Diego, California, USA, Ref. 2006-5476, June 11-15, 2006.
[44] C. Dovrolis *et al.*, "Proportional Differentiated Services: Delay Differentiation and Packet Scheduling", *IEEE Transaction on Networking*, Vol. 10, No. 1, pp. 12-26, February 2002.

[45] E. Lutz, D. Cygan, M. Dippold, F. Dolainsky, W. Papke, "The Land Mobile Satellite Communication Channel-Recording, Statistics, and Channel Model", *IEEE Transactions on Vehicular Technology*, Vol. 40, No. 2, pp. 375-386, May 1991.

[46] A. Sali, A. Widiawan, S. Thilakawardana, R. Tafazolli, B. G. Evans, "Cross-Layer Design Approach for Multicast Scheduling over Satellite Networks", in *Proc. of IEEE International Symposium on Wireless Communication Systems 2005* (ISWCS 2005), ISBN 0-7803-9206-X, Siena, Italy, September 5-9, 2005.

6

CALL ADMISSION CONTROL

Editors: Stylianos Karapantazis[1], Petia Todorova[2]

Contributors: Stylianos Karapantazis[1], Petia Todorova[2], Franco Davoli[3], Erina Ferro[4]

[1]AUTh - Aristotle University of Thessaloniki, Greece

[2]FhI - Fraunhofer Institute - FOKUS, Berlin, Germany

[3]CNIT - University of Genoa, Italy

[4]CNR-ISTI - Research Area of Pisa, Italy

6.1 Introduction to Call Admission Control

RRM in multimedia satellite networks aims to guarantee the fair distribution of available resources, due to the fact that the total link capacity has to be divided among several users, as well as to fulfill certain pre-negotiated QoS requirements for the lifetime of the connection. RRM is one of the functions that are carried out in the *Data Link Layer* (DLL). A general DLL protocol stack that applies to satellite networks is depicted in Figure 6.1, while Figure 6.2 illustrates the most important RRM entities.

One of the most important resource management functions is *Call Admission Control* (CAC), which comprises the set of functions taken by the satellite network during the phase of connection establishment or connection re-negotiation to decide whether to accept or reject a user's request for a connection. A new user's request can be accepted provided that there are adequate network resources available to guarantee the QoS of both all already-existing connections and the new requested one. Generally, the CAC function results in the blocking of new calls or call dropping in the case of ongoing calls when the bandwidth required for the connection exceeds the available bandwidth. CAC, which turns out to be a crucial function to provide high utilization of network resources, is network-specific and is generally managed by the *Network Control Center* (NCC - recall that a description of the NCC functions is given in Chapter 1, sub-Section 1.4.3). However, in non-GEO satellite systems the CAC function has to be implemented on board of the satellite as well. Nevertheless, it should be mentioned that this approach requires satellites with on-board processing capabilities.

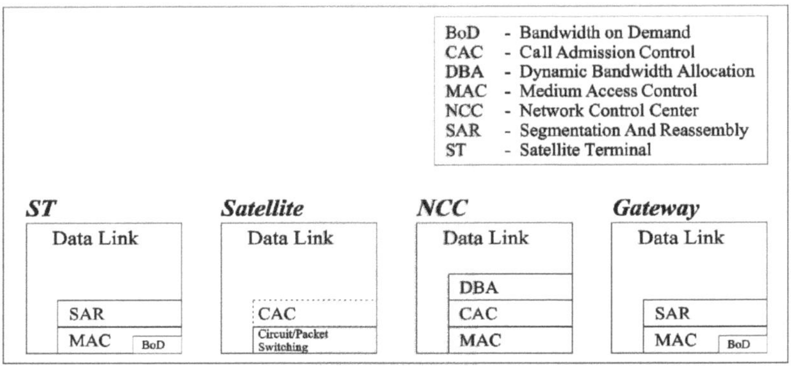

Fig. 6.1: A general protocol stack for the main elements of a satellite network.

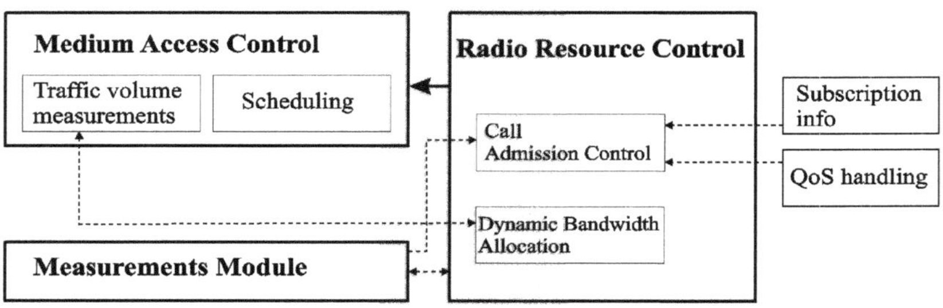

Fig. 6.2: The main RRM entities.

6.2 CAC and QoS management

As noted in [1], the public data network provides a resource that could profoundly impact on high-priority activities of society, like defense and disaster recovery operations. Under stress, however, the public network turns out to be a virtually unusable resource, unless suitable traffic prioritization and CAC are applied to improve its performance. CAC has been extensively studied in the past as a general resource allocation mechanism in various networking contexts. Ross [2] is an excellent reference for CAC mechanisms in general, whereas reference [3] contains a recent survey on this topic in the context of wireless networks.

In the simplest case of resource allocation, a connection is admitted simply if resources are available at the time the connection is requested. This policy is commonly called *Complete Sharing* (CS), where the only constraint on the system is the overall system capacity. In a CS policy, connections that request fewer resource units are more likely to be admitted (e.g., a voice connection will more likely be admitted compared to a video connection). A CS policy does not consider the importance of a connection when resources are allocated. At the other extreme, in a *Complete Partitioning* (CP) policy, every traffic class is allocated a set of resources that can only be used by that specific class. Other solutions are represented by *Trunk Reservation* (TR), where class i may use resources in a network as long as r_i units remain available [4], and *Guaranteed Minimum* (GM) [5],[6], which gives each class its own small portion of resources; once used up, classes can then attempt to use resources from a shared pool. An *Upper Limit* (UL) policy was adopted in [1], and *Virtual Partitioning* (VP) was proposed in [7].

As far as satellite systems are concerned, the architecture of the new satellite systems testifies the interest in ATM, IP and DVB technologies. A general architecture of a satellite system is illustrated in Figure 6.3. An Earth station (Gateway) is in charge of mapping ATM/IP traffic originated from terrestrial terminals over satellite connections, while the NCC performs CAC and DBA functions. The role of the aforementioned functions is to meet the QoS requirements of different service classes, i.e., delay, jitter and packet loss.

A plethora of CAC algorithms were proposed in the literature for terrestrial ATM-based networks. Some of them require an explicit traffic model, while some others require traffic parameters such as peak and average rate. A classification of these schemes is provided in [8] along with the description of their salient features. Nevertheless, it should be noted that while some parameters can be easily specified (for instance, the peak rate), the actual average rate is difficult to estimate, since the source does not know it. Then, the user can declare an upper bound, which, however, results in low bandwidth efficiency. To cope with this issue, measurement-based CAC methods have been proposed. In [9], the authors present a taxonomy as well as a detailed survey of measurement-based CAC techniques. In that study, different measurement-based CAC methods were compared against each other

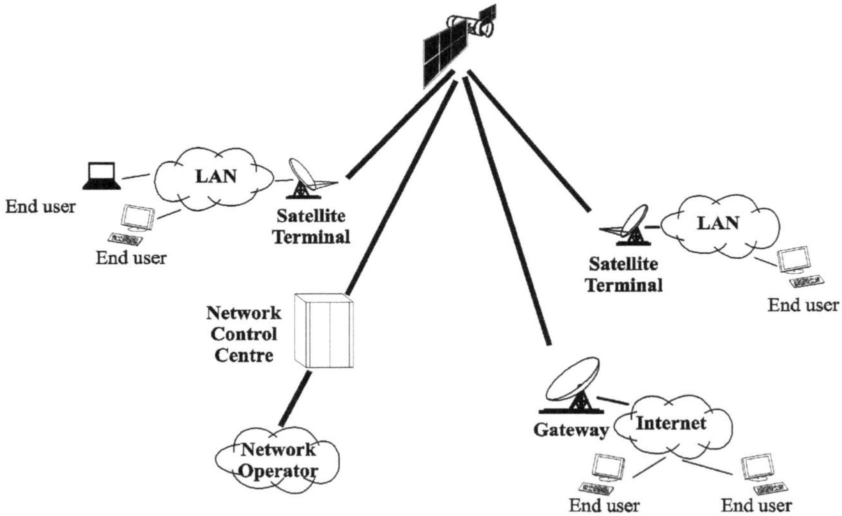

Fig. 6.3: General architecture of a satellite system.

in the light of bandwidth efficiency, *Cell Loss Ratio* (CLR), implementation complexity, scalability and dependency on traffic model. The authors were led to the conclusion that those methods that are based on *effective bandwidth* are the most suitable for high-speed communication systems, since they are simple enough to be implemented in real systems, they attain high bandwidth efficiency and last, but not least, they assume fewer traffic parameters. The rationale behind this category of CAC schemes is rather simple. First, the effective bandwidth for the aggregate connections is measured, namely the equivalent bandwidth needs of ongoing connections. Then, a request for a new connection is accepted provided that the requested bandwidth is smaller than the residual bandwidth, that is, the total link bandwidth minus the effective bandwidth.

Concerning ATM-based satellite networks, they are able to meet different QoS requirements at the ATM layer [10]. These requirements are defined in terms of objective values of the network performance parameters, as specified in ITU-R Recommendation S.1420 [11]. Some of the QoS parameters (*Peak-to-Peak Cell Delay Variation*, *Max Cell Transfer Delay* and *Cell Loss Ratio*) may be offered on a per-call/connection basis and negotiated between the end-system and the network, whereas some other QoS parameters (*Cell Error Ratio*, *Severely Errored Cell Block Ratio* and *Cell Misinsertion Rate*) cannot be negotiated. For each direction of the call/connection, a specific QoS is negotiated, based on a traffic contract between the network and the user. At call set-up time, the user declares the source traffic descriptors and the

QoS class by means of signaling or subscription. The traffic descriptors in the set-up signaling message include a generic list of traffic parameters, specific for each user connection. For each connection request, the CAC function derives the following information:

- The source traffic descriptors, including the traffic characteristics of the ATM source;
- The *Cell Delay Variation Tolerance* (CDVT) value;
- The requested and acceptable values of each QoS parameter, and the QoS class.

In particular, the idea of endowing LEO satellites with on-board ATM switching capabilities (Figure 6.4) combines the advantages of LEO systems, like significantly reduced propagation delay, rendering them suitable for real-time applications, with those offered by ATM, including faster transmission rate, bandwidth on demand, compatibility with existing protocols and guaranteed QoS [12],[13]. By supporting statistical multiplexing, priority queuing and multicasting, ATM technology can accommodate all QoS features requested by the user and therefore, becomes a suitable solution for broadband multimedia communications. However, as LEO satellites' coverage area changes continuously over time, in order to maintain connectivity, end-users must switch from beam to beam and from satellite to satellite, resulting in frequent intra- and inter-satellite handovers.

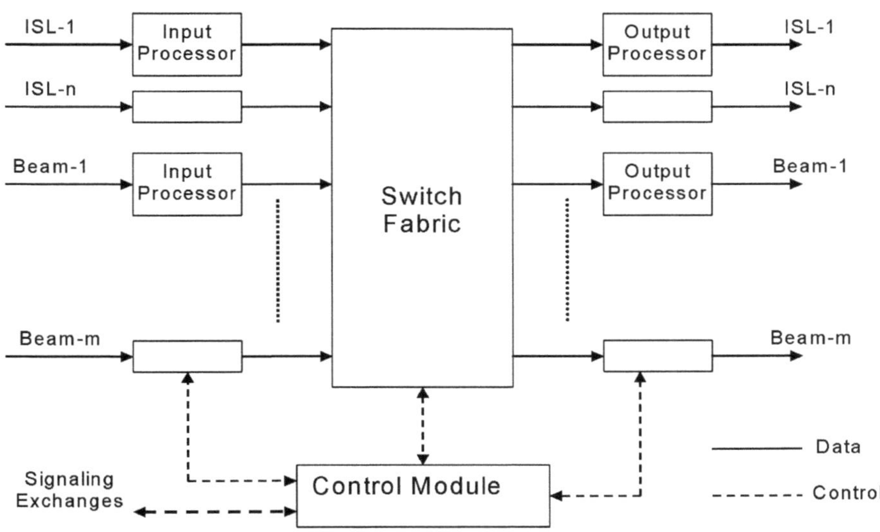

Fig. 6.4: An on-board ATM switching/processing architecture. See reference [12]. Copyright ©2003 IEEE.

The functions of the individual modules in Figure 6.4 are as follows:

- *Switch Fabric*: switching cells from input ports to appropriate output ports.
- *Input Processor*: scheduling, buffer monitoring.
- *Output Processor*: scheduling, buffer monitoring and cell discarding.
- *Control Module*: CAC, handover monitor & control, resource allocation, routing table update, signaling protocol, etc.

The ATM switch uses different input/output ports for the uplink/downlink and for the *Inter-Satellite Links* (ISLs). This is because of the different bandwidth and signaling protocols used.

The functions of the *Control Module* (CM) are shown in Figure 6.5, assuming that signaling and routing table updating are implemented [13]. For intra-satellite handover, the *Handover Monitor & Control* module has to monitor and measure the handover status of all beams belonging to the satellite.

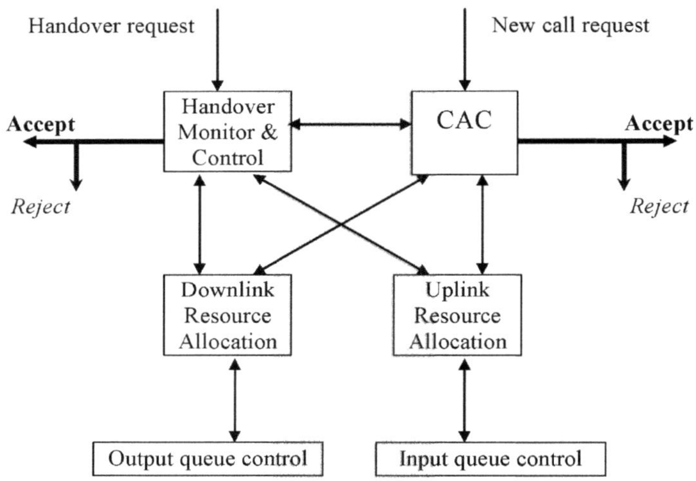

Fig. 6.5: The anatomy of an ATM switch with CAC/handover control module. See reference [13]. Copyright ©2004 IEEE.

It is assumed that the mobile user initiates the intra-satellite handover process based on physical link quality measurements. Then, the mobile user will send a handover request message to the LEO satellite, indicating the new beam identification and the QoS requirements. The satellite CM has to implement the handover/CAC process in order to decide whether or not the new beam could provide the QoS requirements. If the handover is accepted, a handover reply message is sent to the mobile user. At this point buffering processes are needed to guarantee the minimum loss of cells. After the uplink/downlink accesses are finished, buffered cells will be transmitted through new links.

As the Internet has become a ubiquitous communication infrastructure, IP QoS provisioning is a strategic issue in any kind of network. A view that has been gaining considerable interest in the scientific community considers that the *IP Integrated Services* approach (IntServ - service differentiation is focused on individual packet flows) can be used in the wireless access networks (in our case the satellite links) in order to admit or to reject the requests of flows according to the availability of resources and the guarantees provided to other flows. On the other hand, the *IP Differentiated Services* approach (DiffServ - scalable service differentiation, focused on the aggregate of flows) can be employed to avoid complexity and maintenance of per-flow state information in the core network [14],[15]. Both IntServ [15],[16] and DiffServ [17] have been studied for satellite networks; they are considered later in conjunction with CAC schemes.

In general, CAC schemes can be classified into those that offer *Deterministic QoS Guarantees* and those that provide *Statistical QoS Guarantees* [8].

- *Deterministic QoS guarantees*: a new connection is accepted, provided that the worst-case scenario's requirements are met (for instance, the available capacity is greater than the peak rate of the connection). Although this approach represents the simplest solution for traffic management, it tends to over-commit resources, thus resulting in low link utilization.
- *Statistical QoS guarantees*: in this case, the NCC maintains a statistical allocation instead of guaranteeing a peak rate. Losses may occur, but high channel utilization is accomplished. This approach is based on the assumption that having all the connections transmitting at their peak rates at the same time is beyond the realms of possibility, allowing in this way the statistical multiplexing of flows. However, the difficulty of this approach lies in the traffic characterization problem.

An efficient integration of the aforementioned approaches can make up for the weaknesses of each other [18]. In particular, the technique proposed in that study combines the good characteristics of the EDF (*Earliest Deadline First*) scheduler in terms of QoS provisioning with the advantages that stem from statistical multiplexing (a description of the EDF scheduler is provided in sub-Section 5.3.1). Moreover, the mask of the *Dual Leaky Bucket* used for traffic shaping is such that takes into account the statistical variability of the peak rate and the burst size. Note that a *leaky bucket* is simply a finite queue and it can be viewed as a bucket with a small hole in the bottom: no matter at what rate water enters the bucket, the outflow is at a constant rate, when there is any water in the bucket. In other words, a *leaky bucket* is used to smooth out bursts and greatly reduces the chances of congestion. Simulation results showed that the scheme described in [18] improves channel utilization, while providing QoS guarantees at the same time.

Another important issue closely related to CAC and QoS is represented by fade countermeasures for rain-fading attenuation. *Forward Error Correction* (FEC) techniques aim to mitigate channel impairments and diminish BER. Notwithstanding the advantages that accrue from FEC schemes, the price to pay is a decrease in the information bit-rate. Specifically, under the condition of fixed bandwidth availability, these schemes have an impact on the operations of at least two higher layers. These operations are:

- The CAC and bandwidth allocation for guaranteed-bandwidth traffic (indirectly affecting, in turn, the residual bandwidth left to best-effort traffic);
- The performance of the TCP congestion control mechanism.

The impact of fade countermeasures on TCP will be investigated in Chapter 8. Concerning CAC, adaptive control approaches can be adopted in the context of *Constant Bit Rate* (CBR) and *Variable Bit Rate* (VBR) connections, also taking into account the presence of best-effort traffic [19],[20],[21]. In the presence of guaranteed-bandwidth traffic, control actions may take the form of CAC and bandwidth allocation, whose parameters can be determined and, possibly, adaptively tuned on the basis of the fade countermeasures adopted; the ensuing redundancy is applied to ongoing and incoming connections. In particular, the decision on whether to accept or reject a request for a new call is dependent on the measured level of signal attenuation. Moreover, the traffic source rate and FEC rate can be dynamically adjusted in a co-ordinated fashion to satisfy QoS requirements.

6.3 CAC algorithms for GEO satellite systems

GEO satellite systems have been dominant in the telecommunications arena for years and have been the subject of extensive research by virtue of their large coverage area and their intrinsic broadcast/multicast capabilities. The frequency band allocated for satellite services has changed many times over the years. Several proposals for GEO satellite systems suggest the use of Ka band (20-30 GHz). Systems operating at these high frequencies can provide a wide spectrum of multimedia applications to users. Thus, CAC becomes an issue of paramount importance to provide QoS guarantees to calls of different service classes. The *Multi Frequency-Time Division Multiple Access* (MF-TDMA) air interface solution has been adopted by most of the satellite system designers. In the following, a description of CAC algorithms for MF-TDMA GEO satellite systems is given.

6.3.1 CAC schemes for MF-TDMA networks

The study in [22] focuses on resource allocation and CAC in broadband GEO satellite systems. In particular, a GEO satellite with on-board processing capabilities is considered, thus allowing the CAC decision to be taken on board of the satellite. The proposed algorithm, which is called *Dynamic Movable Boundary Strategy*, is geared towards the specific needs of CBR and bursty data traffic and guarantees a minimum number of resources to each service class. In brief, users access the channel in a *Time Division Multiple Access* (TDMA) manner. The frame is divided into three parts: one part dedicated to CBR calls, another one devoted to bursty data traffic, and a third part used as a common resource pool. The CAC decision for CBR traffic, as well as the resource allocation decision, is taken periodically, at the beginning of a time interval called *control period.* This algorithm adapts itself to the network loading conditions by modifying the resource allocation criterion at the beginning of each frame.

A CAC scheme for integrated ATM-satellite systems is proposed in [10]. The proposed algorithm caters for both real-time and non-real-time variable bit-rate traffic by exploiting the statistical multiplexing of traffic sources. In particular, the supported traffic is categorized into four classes. Resources are allocated on a permanent basis for calls of the first traffic class, namely CBR traffic sources, while a semi-permanent allocation based on the statistical multiplexing of traffic sources is employed for calls of the three other traffic classes, that is, for *real-time-VBR* (rt-VBR), *non-real-time-VBR* (nrt-VBR) and *Unspecified Bit Rate* (UBR) traffic. The idea of the algorithm consists in the introduction of the *excess demand probability*, which is the probability that a given number of calls request in a future time more channels than those actually available. A double check is performed before admitting a new call into the system, ensuring that the *excess demand probability* is below a predefined threshold for each traffic class. Specifically, the first check ensures that the *excess demand probability* of all the multiplexed sources is below ε_1, whereas the second check verifies that the *excess demand probability* of real-time traffic (CBR and rt-VBR) is also lower than ε_2, where $\varepsilon_2 < \varepsilon_1$ since real-time traffic is characterized by stringent QoS constraints.

A similar CAC technique is combined in [23] with an in-band signaling scheme in order to combat the adverse effects of the intrinsic propagation delay, which makes the traffic profile different from the one declared. The in-band technique allows requesting resources for semi-permanent connections on a burst basis. In particular, it is adopted by VBR sources and allows the use of a field in the currently transmitted burst in order to notify a new burst arrival, thus obviating the need for signaling exchange between ground and space segments. It should be noted, that in that study only rt-VBR traffic was considered. Such study was extended in [24], where the in-band-signaling scheme was coupled with a resource engagement prolongation technique. When the former is used in conjunction with the latter, the time needed for resource allocation notification is reduced. In brief, if the traffic resource management scheme finds out, when processing a bandwidth request, that resources are still occupied and used by the relevant terminal for the transmission of previous information bursts, then it just lengthens the time interval that those resources remain engaged, thereby diminishing the number of bursts that are lost while waiting for the acknowledgement of the assignment of new resources.

A CAC scheme for DVB-RCS systems is examined in [25]. The scheme presented in that study is coupled with a capacity request scheduling technique with the aim of meeting the QoS requirements of different service classes. In particular, the CAC algorithm employs a preventive congestion control, based on traffic descriptor parameters (that is, peak bit-rate, burstiness, and service category) and decides whether to accept or to reject a new call connection according to the estimation of the *excess demand probability*. The latter sets an upper bound on the burst loss probability.

The concept of *excess demand probability* is also used by the CAC scheme presented in [26], where an integrated terrestrial-satellite system is considered. The CAC scheme consists of two distinct phases: terrestrial admission control and satellite admission control. The authors of that study also propose the use of the IP IntServ architecture in the satellite network and the adoption of a scalable IP DiffServ-like architecture in the terrestrial network. Concerning the satellite admission control, it accepts a given number of calls if the *excess demand probability* is such that a target service quality can be guaranteed.

A CAC scheme geared towards multimedia GEO satellite networks with on-board cross-connectivity, that is, connectivity between any pair of beams, is presented in [27]. It is considered that there exists one Gateway Earth Station associated with each beam. In addition to this, it is assumed that any connection initiated by a user ends in the terrestrial network. Assuming that the QoS requirements of a connection can be met in the home Gateway, then the CAC criterion consists in opting for the destination Gateway that: (*i*) has enough bandwidth to support a connection request, and (*ii*) results in the shortest distance to the connection's terrestrial destination. The amount of resources statically allocated depends on the connection type (i.e., the ATM-based classification of services), the traffic descriptors, and the requested QoS.

In [28], the employment of the IntServ model in a GEO satellite system is examined. Specifically, the authors of that paper study two main classes of service, namely *Guaranteed Services* and *Controlled Load Services*. The former is suited for real-time applications with stringent QoS requirements, whereas the latter provides for adaptive-tolerant real-time traffic (i.e., traffic with loose delay requirements). The satellite CAC supports the statistical multiplexing of traffic over the air interface. A new call is accepted if the network has sufficient bandwidth to satisfy the QoS constraints of the call without degrading the QoS perceived by ongoing calls. Specifically, the authors of that study adopt a technique similar to the one described in [16]. Each flow is characterized by specific parameters that are called token bucket parameters. These parameters are the *token bucket rate* r, the *token bucket size* b, the *peak data rate* p and the *maximum packet size* M. However, what is meant by "*token bucket*"?

Token bucket is an algorithm for traffic shaping, like the *leaky bucket* algorithm, used to regulate the average rate (and burstiness) of data transmission. It simply counts tokens. However, in contrast to the *leaky bucket* algorithm, which does not allow idle terminals to save up permissions to send large bursts later, the *token bucket* algorithm does allow saving, thus permitting some burstiness in the output stream and giving faster response to sudden input bursts. In brief, a counter is increased by one (or a token is added in the bucket) every $1/r$ seconds and decreased by one whenever a packet is sent. When the counter hits zero, no packets can be sent. The token bucket algorithm allows up to b tokens to be added in the bucket. All the token bucket parameters are used by the CAC algorithm in order to estimate the resources that are required for each new flow. Specifically, the source terminal sends a request for a new connection towards the destination. This request serves the purpose of describing the characteristics of the flow in terms of token bucket parameters. Each router (or, in general, each network element) that receives this request computes how it will handle packets of this flow and updates the request by adding this information to it. When the destination receives the request, it can calculate the bandwidth that is required so that the maximum end-to-end delay be below a given threshold by combining the information that each router has added to the request.

Concerning *Guaranteed Services*, the destination (i.e., an edge device located at the border between terrestrial and satellite segments) computes for each flow the bandwidth R and the buffer space B on board the satellite that are required so that the QoS constraints be met. Then, these quantities are sent to a designated Earth station, which decides on whether to accept or reject this new flow. As regards *Controlled Load Services*, a similar CAC procedure is applied. Nonetheless, in this case, the resources that are requested do not guarantee that specific target values in terms of end-to-end delay and packet loss will be met.

The performance of a CAC algorithm that is combined with a variant of the *Resource Reservation Protocol* (RSVP) is assessed in [29]. In that study, the traffic carried by the satellite network is categorized into threes classes, that is, data traffic, multimedia traffic, and control traffic. A pool of channels is available for all classes. However, if all these channels are reserved, the remaining channels can be used only for the transmission of data and control traffic.

A CAC technique for DVB-S/DVB-RCS satellite systems is examined in [18]. In particular, the CAC algorithm that is presented capitalizes on the positive characteristics of the EDF scheduler in order to provide QoS guarantees and attain high channel utilization. The proposed technique is compared with two CAC schemes that are based on the *Deterministic QoS guarantees* and the *Statistical QoS guarantees* approaches.

The authors of [30] study a CAC algorithm for DVB-RCS satellite networks, which is tailored for *Moving Picture Experts Group* (MPEG) traffic sources. MPEG represents a video compression standard for multimedia applications. In essence, MPEG subdivides the video in *Group of Pictures* (GOPs).

The GOP rate changes over time, therefore the CAC scheme described in that study relies on the statistical multiplexing of this kind of traffic. Specifically, the authors propose a statistical multiplexing scheme that is based on discrete bandwidth levels of the GOP rate and compare it to another scheme that relies on the Normal distribution of the aggregate GOP rate [31]. Concerning the latter scheme, the MPEG traffic generated by each source is modeled as a Normal distribution of GOPs with mean rate μ and standard deviation σ. Thereby, supposing that MPEG flows are independent of each other, according to the central limit theorem the aggregate traffic of a set of N multiplexed connections can also be modeled as a Normal distribution.

Albeit that scheme takes some characteristics of the MPEG traffic into account, it cannot account for traffic variations over time. A solution based on a GEO satellite system equipped with on-board processing and on-board switching is investigated in [32], where an integrated CAC and *Bandwidth on Demand* (BoD) algorithm is proposed for a broadband satellite communication system of this kind, loaded with heterogeneous traffic. This algorithm is able to utilize efficiently available bandwidth in order to attain high throughput and maintain a good grade of service for all the traffic types.

Last but not least, an issue of great importance for the designers of satellite systems is the energy allocation. Power is a resource at a premium in satellite systems, therefore a trade-off between consuming and saving energy is always sought. At this point, it should be noted that higher levels of energy consumption translate into higher throughput. Reference [33] derives an optimal threshold policy for the joint problem of CAC and energy allocation, by means of a dynamic programming approach. In particular, as usual in dynamic programming, a *value function* $J_k(a_k,\ r_k,\ d_k)$ is introduced which aims to show how desirable is a satellite with available energy level a_k at time k, given that the current demand is d_k and the current reward is r_k. The term r_k represents the reward for consumption, namely the satellite receives r_k units of reward per unit of energy consumed. This amount of reward depends on distances, atmospheric conditions and financial considerations. The aim is, then, to maximize the *value function* over a consumed energy c_k.

6.3.2 CAC schemes for CDMA networks

Albeit MF-TDMA has been shown to be particularly effective in satellite networks, *Code Division Multiple Access* (CDMA) has emerged as the mainstream air interface solution for the 3rd *Generation* (3G) networks. One scenario that holds considerable appeal involves the integration of *Satellite Universal Mobile Telecommunications System* (S-UMTS) with *Terrestrial UMTS* networks (T-UMTS) [34], thus resulting in a powerful integrated network infrastructure. However, unlike in TDMA/FDMA networks, in CDMA systems users share the same portion of bandwidth at the same time. This is realized by assigning each user a pseudo-random code. A new user can be admitted to the network as long as the *Signal-to-Interference Ratio* (SIR) is adequate for processing at the receiver and the QoS requirements of ongoing calls are met. Thereby, CDMA systems are interference-limited rather than capacity-limited. Despite the vast literature on CAC algorithms for terrestrial CDMA networks, only a handful of studies exists on CAC schemes for satellite CDMA systems.

An interactive SIR-based algorithm for S-UMTS networks is delineated in [35]. The described algorithm aims at finding out if a power equilibrium point can be calculated so that the target SIR of all the ongoing calls and the target SIR of the new call are met. This CAC scheme is applied to the admission of bi-directional, high-demanding services.

The authors of [36] propose a CAC scheme that provides QoS guarantees to integrated voice, videoconference, and data services. The essence of their goals is to maximize the utilization of system resources. The air interface adopted in that work is a combined CDMA/TDMA scheme. The highest priority is given to videoconference calls.

6.4 Handover and CAC algorithms for non-GEO satellite systems

The mind-set of satellite systems' designers over the past decades has been to keep most of the complexity on the ground segment. Notwithstanding, the advantages that stem from this design approach, the growing exigencies for both mobility and ubiquitous access, coupled with advances in technology, led the designers to move satellites closer to the Earth surface in order to enable the provision of delay-sensitive and high bit-rate services. Non-GEO satellite systems attracted considerable attention by virtue of some of the compelling features that are endowed with, such as the low propagation delay and the ability to communicate with handheld terminals. The 1990s were perhaps the public heyday of this type of satellite systems. In that decade, several commercial satellite constellation networks were come to light, while the end of the decade saw the launch and the start of operations of two LEO satellite constellations, namely Iridium and Globalstar, which provide voice service and paging. Nevertheless, the widespread usage of terrestrial cellular systems for the provision of mobile telephony worldwide had usurped many of the "target markets", thus these non-GEO satellite networks have never come to fruition on account of their competitive rather than complementary role with respect to terrestrial cellular systems.

The coverage area of non-GEO satellites, referred to as *footprint*, is divided into slightly overlapping cells, called *spot-beams*. Due to the movement of satellites with respect to the Earth's surface, end-users must switch from spot-beam to spot-beam and from satellite to satellite in order to maintain connectivity. Thus, as in the case of terrestrial cellular systems, the issue of call handover arises in non-GEO satellite constellations as well. Two types of call handover can be distinguished:

- *Intra-satellite handover* (also referred to as *cell handover* or *spot-beam handover*), which refers to the handover of a call between neighboring cells (beams) of the same satellite (Figure 6.6).
- *Inter-satellite handover* (also referred to as *satellite handover*), which relates to the handover of a call between two contiguous satellites (Figure 6.7).

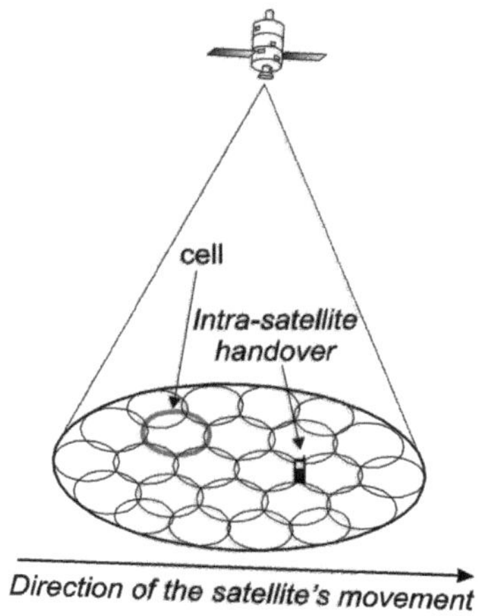

Fig. 6.6: Intra-satellite handover.

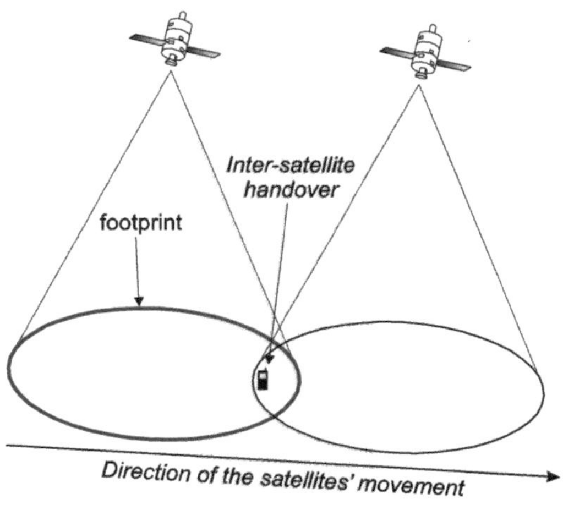

Fig. 6.7: Inter-satellite handover.

It should be pointed out that, in contrast to terrestrial cellular networks, where the handover rate is determined by the motion of the users, in non-GEO satellite systems, the handover rate is determined by the motion of the satellites. In the case of LEO satellites, the ground track speed of satellites is over 5700 m/s (note that users in fast vehicles move with a velocity of 80 m/s at most), hence a satellite is in view for up to 10 min, while the user's sojourn time in a cell can be as short as 1 min.

The handover of a call constitutes a daunting challenge in this kind of satellite systems, since it may result in the forced termination of ongoing calls. To overcome this problem, advanced CAC and handover control techniques are required for improving the QoS performance. While the role of CAC algorithms is to decide whether to accept or reject a new call, handover techniques aim to ensure that the service of a call will not be interrupted when the user moves from one cell (or satellite) to another. In other words, while the aim of CAC techniques is to minimize *Call Blocking Probability* (CBP), handover techniques aim to diminish *Call Dropping Probability* (CDP), also referred to as *Forced Termination Probability.* Unfortunately, any efforts to reduce one of these two probabilities result in an increase in the other one.

6.4.1 Intra-satellite handover and CAC schemes

Several approaches for handover prioritization proposed for terrestrial cellular systems have been studied for the case of intra-satellite handover in non-GEO satellite systems. The techniques that can be found in the literature are based on either *Dynamic Channel Allocation* schemes (that is, any channel can be temporarily assigned to any cell) [37],[38] or *Fixed Channel Allocation* schemes (that is, a set of channels is permanently assigned to each cell) [39]-[49]. CAC and intra-satellite handover schemes are summarized below. The works are referenced on a time-line basis, identifying the seminal works in this field on the one hand, and works that primarily extended previous studies on the other hand.

In [39], a CAC strategy is proposed, along with an *intra-satellite handover* scheme. A metric, called *mobility reservation status*, is introduced that aims to provide the information about the bandwidth that is required by all the active calls in each cell as well as to predict the potential bandwidth requirements by calls in adjacent cells. Supposing that a new call k has been accepted in a given cell (i.e., spot-beam) m, the mobility reservation status of this cell, as well as the mobility reservation status of the next S - 1 cells, is increased by the following quantity:

$$C_{m+i}(k) = \begin{cases} B_k\left(\frac{T_0}{T_{\max}}\right), & i = 0 \\ B_k\left(\frac{T_{\max}}{T_0+i\cdot T_{\max}}\right), & i = 1, \; \dots \;, S-1 \end{cases} \tag{6.1}$$

where B_k is the number of the traffic channels required by this cell, T_0 is the user's dwell time in the source cell m of the call, whereas T_{max} is the maximum time interval that a user can dwell in a generic transit cell. Note that index i from 0 to $S-1$ is used to denote the source cell (index $m+i$, $i=0$) and the next possible transit cells (index $m+i$, $i=1, \dots, S-1$). A new call is admitted into the network only if there are at least B_k available channels in the cell m where the user is located, and at the same time the values of the mobility reservation status of this cell, the previous cell and the next one are below a predefined threshold, called T_{new}. As far as handover requests are concerned, a call is successfully handed over to a new cell provided that the number of available channels in that cell is greater than B_k and its mobility reservation status is below a predetermined threshold, called T_{HO}. Apparently, T_{HO} is greater than T_{new} in order to prioritize handover requests over new call requests.

In [40], an adaptive dynamic channel allocation scheme is examined, which relies on the well-known concept of guard channels, which are channels exclusively used in each cell only to serve handover requests. In particular, the number of guard channels is dynamically adapted based on the estimation of future handover events. In more detail, upon the arrival of a new call request in a cell, the algorithm, by capitalizing upon the deterministic network topology of LEO satellite systems, computes the user's dwell time in that cell. Then it estimates the number of the potential handover requests within this time interval as well as the expected number of channels γ that will be needed to serve these requests. The request will be accepted only if the number of available channels is greater than γ. As regards handover requests, a call is successfully handed over to a new cell as long as there is at least one available channel in that cell.

The study in [41] extends the aforementioned scheme and proposes a *geographical connection admission control* algorithm that aims to guarantee that the forced termination probability will always be below a predefined threshold. This CAC algorithm is based on the estimation of the future CDP of both the new calls and the ongoing ones. Upon the arrival of a new call, these two probabilities are estimated, and the call is admitted into the network provided that these probabilities are below some predefined thresholds.

The techniques presented in [37],[38],[42]-[46] rely on the queuing of handover requests. According to this kind of handover schemes, a handover request is queued for a specific time interval when no channel is available in the next cell. In [37],[38],[42], the queuing time interval is dependent on the overlapping area between contiguous cells.

In [43], a *guaranteed handover service* scheme was proposed. According to that technique, a handover request can be queued up to a time interval equal to the user's sojourn time in the cell, that is, as soon as a handover occurs, a handover request is sent to the next transit cell. As far as new calls are concerned, a new call is admitted into the network as long as there exists an available channel in both the current cell and the first transit cell. That scheme attains zero CDP at the expense, however, of a rather high CBP.

The authors of [44] propose a handover technique similar to the *guaranteed handover service* scheme, which aims at increasing channel utilization and, thus, reducing CBP. Toward this end, the channels in that scheme are reserved only for the time intervals they are expected to be in use, hence the name *Time-based channel reservation algorithm.* Therefore, channel utilization is improved and CBP is reduced.

In [45],[46], the queuing time interval is considered to be dependent on the value of a parameter that was called *handover threshold.* This parameter should be appropriately selected in order to attain a trade-off between dropping and blocking probabilities as well as to achieve high channel utilization. In brief, a handover request is sent to the next cell at a specific time instant, which is determined by the *handover threshold* parameter. When a new call arrives, it is accepted provided that an available channel exists in the current cell. However, if the time interval until the occurrence of the first handover is shorter than the one defined by the *handover threshold* parameter, then an available channel should also exist in the succeeding cell in order for the call to be accepted. It was shown that this scheme can provide different QoS levels based on the value of the *handover threshold* parameter.

In [47],[48], CAC algorithms based on a bandwidth allocation strategy with priority queues are examined. The handover admission policy introduced distinguishes between real-time and non-real-time services. To each accepted real-time connection, bandwidth is allocated in a look-ahead horizon of 2 cells along its trajectory; while non-real-time connections reserve bandwidth only in the forthcoming cell. According to that scheme, each cell maintains four different queues, called R, S1, S2 and Q. Queue R contains those real-time connections that have reserved at least the minimum required bandwidth in the next two cells. Therefore, the handover to the next two cells is guaranteed to be successful. Queue S1 contains those real-time connections that have reserved the required bandwidth in the next cell, but not in the one after the next cell. Regarding queue S2, it contains the real-time connections that have not managed to reserve the required bandwidth in both these cells. Finally, queue Q contains the non-real-time connections that have not achieved to reserve any amount of bandwidth in the next cell. It should be noted that non-real-time connections are successfully handed over to a new cell as long as some residual bandwidth, even lower than the minimum required bandwidth for this type of calls, has been reserved in that cell. The management of the queues is such as to give priority to real-time multimedia calls over non-real-time data calls, namely the first priority is given to queue S2, the second is given to queue S1, while non-real-time connections are given the lowest priority.

The study in [49] extends the aforementioned technique and proposes a CAC algorithm that is based on the concept of multiple sliding windows. The rationale behind the proposed algorithm is to predict the amount of bandwidth that will be available at the time instant of the handover occurrence and reserve the necessary amount of bandwidth in the cells to which the call may be handed over. The highest priority is given to handover calls that are organized in a separate queue.

In [50], the authors examine the use of the knowledge of future capacity changes to trade-off some additional blocking probability, in order to meet the desired CDP. Specifically, three CAC policies based on the assumption of deterministic capacity change time instants are discussed: two for calls with exponentially distributed holding times, and one for calls whose holding time distributions have *Increasing Failure Rate* (IFR) functions. In general, the *failure rate function* $h(x)$ (also known as the *hazard rate function*) is defined as:

$$h(x) = \frac{b(x)}{1 - B(x)} \tag{6.2}$$

where $b(x)$ is the call holding time probability density function and $B(x)$ is the call holding time cumulative distribution function. Note that $h(x)\mathrm{d}x$ denotes the probability that the call will end in the next $\mathrm{d}x$ time unit given that it has been in service for x time units. A holding time distribution is said to be an IFR distribution if $h(x)$ is a non-decreasing function of x. Examples of IFR distributions are uniform, exponential, half-Gaussian distributions, and gamma-n with $n \geq 1$. Moreover, the *Admission Limit Curve* for exponentially distributed call holding times, which forms a boundary on the conditions under which a CAC policy may accept an incoming call request, has been proved to be able to serve as the basis for a CAC policy. The authors demonstrate how these CAC policies and the *Admission Limit Curve* represent progressive steps in developing optimal CAC policies for calls with exponentially distributed holding times, and they extend this process to the more general case of calls with increasing failure rate call holding times. The *Admission Limit Curve* was also investigated in [51] along with the performance of a CAC policy for increasing failure rate holding time distributions. However, in that study stochastic capacity change time instants were assumed.

6.4.2 Inter-satellite handover and CAC schemes

Although intra-satellite handovers are more frequent than inter-satellite handovers, the latter are of paramount importance to the performance of any non-GEO satellite system with partial or full *satellite diversity*. By the term satellite diversity we simply mean that a terminal has a choice of multiple visible satellites with which it can communicate. After opting for one of them, the terminal establishes a single duplex radio link with that satellite. This kind of satellite diversity is also referred to as switched diversity. Towards this end, different satellite selection criteria have been proposed and evaluated [52],[53], always with a view to minimizing CBP and CDP. The satellite selection criteria that can be found in the literature can be summarized in the following three rules:

- *Maximum capacity criterion* - The satellite with the maximum available capacity is selected. This criterion aims to attain a uniform distribution of the traffic load over the satellite constellation.
- *Maximum serving period criterion* - The satellite that offers the maximum serving time interval is selected. The aim of this criterion is to reduce the number of handovers per call.
- *Minimum distance or highest elevation angle criterion* - The closest satellite (i.e., the satellite that is seen under the highest elevation angle) is selected. This criterion aims to mitigate channel impairments.

The aforementioned satellite selection criteria can be applied to both new and handover calls. All the results that are presented from this point forward refer to Scenario 3, which is detailed in sub-Section 1.4.5 of Chapter 1.

In [52], the authors assess the *guaranteed handover* scheme as an inter-satellite handover technique for LEO constellations that require at least one satellite to be visible to both the user terminal and the Gateway Earth station.

The study in [53] extends the scheme proposed in [45],[46] for the case of inter-satellite handovers in LEO satellite diversity-based systems. The proposed scheme is evaluated for different values of the queuing time interval as well as for different constellations. Moreover, it is evaluated for nine different combinations of the satellite selection criteria.

In [54], an inter-satellite handover technique tailored for broadband LEO satellite diversity-based systems is proposed. The proposed technique constitutes a combination of the technique that is presented in [53] with the guard channels scheme. By using different parameter values for each service class, that technique aims to minimize CDP while keeping at the same time CBP at acceptable levels. Specifically, the value of the *handover threshold* parameter is different for each service class with the aim of satisfying its QoS requirements. Furthermore, the notion of the *guard class capacity* is introduced, which stands for the portion of the total capacity that is available only to calls of a specific service class. The rest of the capacity is available to calls of all service classes, and calls contend in order to reserve the capacity required for their service. Of course, the greater the mean bit-rate of the service class, the greater the *handover threshold* and the *guard class capacity* employed for this service class.

CAC and inter-satellite handover schemes geared towards multimedia LEO satellite systems are also examined in [55],[56]. In both these studies, a mobility model that takes the Earth's rotation into account was used for the assessment of the proposed schemes. In this model, satellite footprints are modeled as rectangles. The overlapping area between successive satellites in the same orbital plane is not taken into account since in that case a user should always be connected to the following satellite in order to avoid an immediate handover. However, the overlapping area between contiguous satellites in different orbital planes is taken into consideration. Moreover, terminals are uniformly distributed over the network. In addition to this, the velocity of users in fast vehicles is disregarded since it is negligible compared to the satellite's ground track speed and the Earth's rotation. The latter is considered to be equal to the velocity at the equatorial level.

Reference [55] relies on the queuing of handover requests in order to achieve low CDP. The services that the system supports are classified into two categories, namely real-time multimedia services (namely, services with stringent QoS requirements) and non-real-time data services (that is, services with loose QoS constraints). Handover requests of different service classes are stored in different queues. Priority is given to the queue where handover requests of real-time multimedia connections are stored. As soon as a call is successfully handed over to a satellite, a handover request is sent to the next candidate satellites for relaying the call. Thus, the queuing time interval can be equal to the user's sojourn time in a satellite's footprint. Moreover, the proposed scheme was examined for different combinations of the satellite selection criteria and for two different queuing policies. The first one is the well-known FIFO policy. In this scheme, the requests are served according to their arrival time. The second queuing policy that was examined is called *Last Useful Instant* (LUI) [38]. In this technique, the requests are queued according to the remaining time interval until the handover occurrence. Hence, a request is placed ahead of all the other requests in the queue that have a greater remaining queuing time.

In [55], eight different versions of the scheme are compared. Figure 6.8 illustrates the performance of the techniques for different percentages of the overlapping area. The overlapping area is defined as the percentage of the footprint's area that is overlapped by footprints of contiguous satellites. As far as the notations in the legend of Figure 6.8 are concerned, the first letter of each scheme indicates the queuing policy that was employed; namely 'F' stands for the FIFO policy, while 'L' stands for the LUI policy. The second letter denotes the satellite selection criterion that was used for new calls, whereas the third letter indicates the criterion that was employed for handover calls; in these two cases, 'C' denotes the *Maximum capacity criterion*, whereas the letter 'T' denotes the *Maximum serving period criterion*.

The schemes have been evaluated in terms of a cost function, which takes account of CBP, CDP, and the mean allocated capacity of all service classes. This cost function, which is called *General Grade of Service* (*General GoS*), in its general form can be expressed as:

$$General\ GoS = \sum_{i=1}^{N} a_i \cdot GoS_i \tag{6.3}$$

where N is the number of the service classes supported by the system and a_i is a weighting factor which is equal to

$$a_i = \frac{B_{\min_i} \lambda_i}{\mu_i} \tag{6.4}$$

where $B_{\min_i}$ denotes the minimum capacity that is required for calls of the i-th service class, whereas λ_i and μ_i are the arrival and departure rates of calls of this type of service, respectively. Concerning GoS_i, it is a function of CBP and CDP of the i-th service class and is defined as follows:

$$GoS_i = WF_1 \cdot CBP_i + WF_2 \cdot CDP \quad . \tag{6.5}$$

The terms WF_1 and WF_2 represent weighting factors, which are the same for each service class. It should be emphasized that WF_2 is much greater than WF_1 (almost tenfold greater) since the forced termination of a handover call is generally considered more irksome than the blocking of a new call. Now it is evident that a_i aims at giving an added bonus to the schemes that attain higher mean bit-rate since it reduces the effect of the corresponding GoS_i on the *General GoS*. Regarding the latter, the higher its value, the poorer the performance of the scheme and the QoS provided to the users. It becomes evident from Figure 6.8 that the FIFO policy performs similarly to the LUI policy. Notwithstanding, the FIFO policy is more appealing on account of its low complexity. Furthermore, the combination that employs the *Maximum capacity criterion* for both new and handover calls achieves the best performance. Moreover, we can note that the *General GoS* increases commensurate with the percentage of overlapping area. Nonetheless, the overlapping percentage can be beneficial for some types of services, as it is shown in [55].

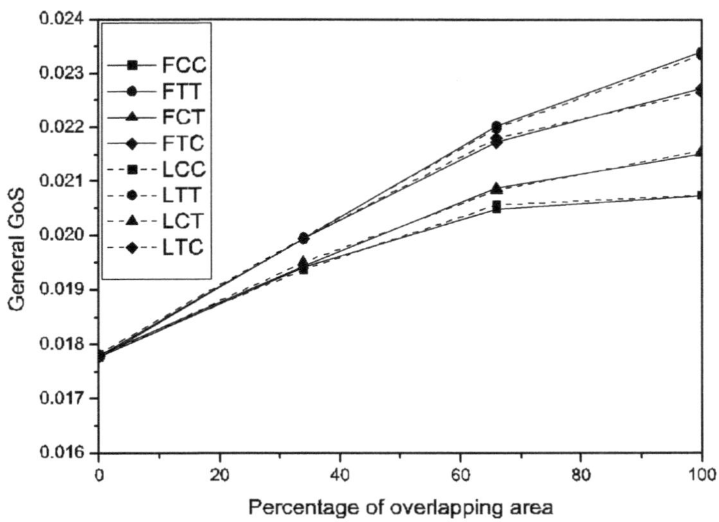

Fig. 6.8: *General GoS* of different schemes vs. overlapping percentage.

This study was extended and another CAC and inter-satellite handover scheme has been developed and assessed in [56]. The main mechanism behind this second technique is based on dynamic bandwidth de-allocation. According to the proposed mechanism, capacity reservation requests are countermanded when the capacity that they strive to reserve is unlikely to be used. In the handover schemes proposed in [52]-[55] the decision about the satellite to which the call will be handed over is taken at the time instant of the handover occurrence. This means that capacity is reserved, if possible, in all the visible satellites, and this capacity is released if the call is handed over to another satellite. On the contrary, in the scheme proposed in [56], when the capacity required for a call is reserved in one of the visible satellites, the capacity reservation requests are cast away from the queues of the other visible satellites. Hence, that scheme does not waste the limited bandwidth of the satellite channel. Simulations showed that this scheme can also capitalize upon the satellite diversity that a system may provide in order to enhance network performance. Figure 6.9 depicts *General GoS* versus overlapping percentage referring to the scheme proposed in [56].

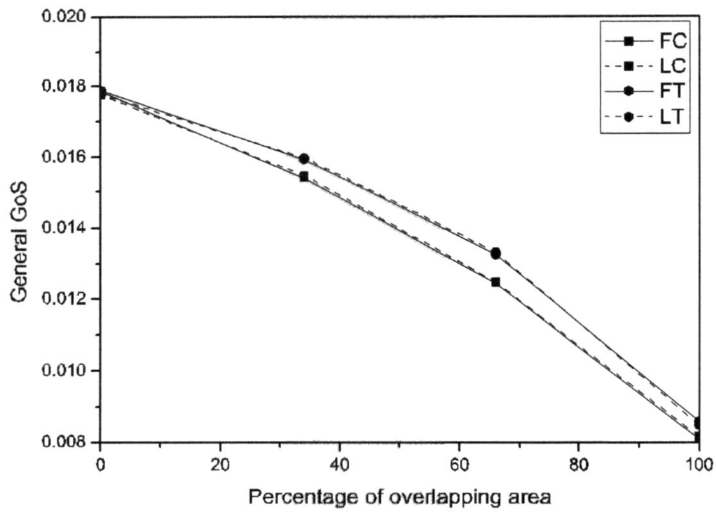

Fig. 6.9: *General GoS* of different schemes vs. overlapping percentage.

It does not make sense to use a satellite selection criterion for handover calls in this scheme, since the decision is taken before the time instant of the handover occurrence. Thus, the first letter of the acronyms in the legend of Figure 6.9 denotes the queuing policy that was employed, while the second letter indicates the satellite selection criterion that was employed for new calls. As shown in Figure 6.9, the FC and LC schemes exhibit the best performance. Recall that in Figure 6.8, the best performance was achieved by those schemes that relied upon the *Maximum capacity criterion* for new calls as well. Moreover, it can be observed that there exist significant performance disparities among the schemes that are presented in Figure 6.9 and the ones in Figure 6.8. It is apparent that the schemes presented in Figure 6.9 outperform those in Figure 6.8. The mechanism behind the schemes that were presented in [56] (i.e., those related to Figure 6.9), which allows them to attain an enhanced performance, relies on the cancellation of capacity reservation requests when the capacity that they strive to reserve is unlikely to be used. Moreover, it is evident that this scheme can capitalize upon the partial or full diversity that a LEO satellite system may provide in order to attain an improvement in system performance.

6.5 Directions for further research

This Section lists some rather interesting proposals for future research work in the field of CAC:

- Due to the costly nature of the satellite channel, integrated CAC and dynamic bandwidth allocation schemes are becoming a matter of some concern to many network operators, being these integrated schemes able to take into account both traffic pattern variations and channel conditions. In addition to this, the performance of transport layer protocols, such as TCP, is often exacerbated by intense variations in the received signal power and consequent high packet error rates. Consequently, the TCP protocol perceives an indication of congestion in the network, thus reducing the transmit information rate. In this context, a CAC algorithm able to interact with the transport layer is considerably appealing, since it allows estimating the amount of capacity that is currently in use, which is smaller than the sum of the nominal capacity of every ongoing call. In particular, the CAC algorithm should base its decisions on the goodput of the TCP connections instead of the nominal bit-rate of each connection.
- In hybrid architectures, namely integrated terrestrial-satellite networks or multi-layered satellite networks, the role of CAC is twofold: (*i*) to decide which network is the most appropriate to serve a new call; (*ii*) to decide whether or not the call can be admitted to the network. A study of a CAC algorithm able to regulate dynamically the admission of new connections in an integrated network, according to their QoS requirements, user mobility, and available resources, is of paramount importance.
- An interesting scenario involves the integration of terrestrial and satellite UMTS networks aiming at maximizing the number of connections that can be actually admitted to the network. The decision of the CAC procedure should be based on the terrestrial and satellite cell layout in the area where the connection set-up attempt occurs, the surrounding area, the mobility

and the QoS requirements of the user, and the instantaneous traffic load in the terrestrial and satellite cells. Based on the aforementioned inputs, the CAC algorithm should decide whether to admit or reject the call, the QoS guarantees that will be granted to the call, and the segment as well as the cell where it is more efficient to set-up the connection.

6.6 Conclusions

CAC constitutes an issue of paramount importance for any wireless or wired network. It is performed at the connection set-up time and determines whether or not sufficient bandwidth is available to maintain required levels of QoS. In this respect, CAC can be viewed as a preventive congestion control procedure. With the advent of ATM networks, significant research efforts have been drawn towards CAC schemes. Typically, any CAC algorithm aims at taking a decision based on two questions:

1. Does the new call impact on the QoS of ongoing calls?
2. Can the network provide the QoS requested by the new call?

Satellite systems have acquired an important role in the telecommunications arena. Over the years they have been used for a host of different services, the most important ones being television and radio broadcasts. The current trends towards the use of higher frequency bands open new opportunities to this type of systems. Future satellite networks will be able to support a wide range of multimedia applications. In this context, CAC algorithms are necessary to guarantee a fair distribution of the radio resources and to meet the QoS requirements of each service class. CAC techniques tailored for broadband GEO satellite systems have been the subject of considerable study lately.

Non-GEO satellite constellations inaugurated a new era in satellite communications in the past decade. This type of satellite systems can constitute a major asset to service providers by virtue of the appealing features that are endowed with. One of the main characteristics (and problems) of non-GEO satellite systems is the relative movement of satellites with respect to the Earth surface. Consequently, in parallel with CAC techniques, handover schemes become of great importance on account of the significant probability of service interruption. Several CAC and handover techniques have been proposed in the literature for the case of non-GEO satellite systems, aiming at providing a trade-off between call blocking probability and call dropping probability.

We are in the midst of a global revolution in information technology and satellite systems can be instrumental in the emerging network infrastructure. Nonetheless, CAC schemes for heterogeneous networks remain an issue to be addressed.

References

[1] C. C. Beard, V. S. Frost, "Prioritized Resource Allocation for Stressed Networks", *IEEE/ACM Trans. Networking*, Vol. 9, No. 5, pp. 618-633, October 2001.

[2] K. W. Ross. *Multiservice Loss Models for Broadband Telecommunication Networks*. Springer-Verlag, London, 1995.

[3] M. H. Ahmed, "Call Admission Control in Wireless Networks: a Comprehensive Survey", *IEEE Communications Surveys & Tutorials*, Vol. 7, No. 1, pp. 50-69, First Quarter 2005, [Available online: http://www.comsoc.org/livepubs/surveys].

[4] P. B. Key, "Optimal Control and Trunk Reservation in Loss Networks", *Probabil. Eng. Inform. Sci.*, Vol. 4, pp. 203-242, 1998.

[5] G. Choudhury, K. Leung, W. Whitt, "An Algorithm to Compute Blocking Probabilities in Multi-Rate Multi-Class Multi-Resource Loss Models", *Advances in Applied Probability*, Vol. 27, pp. 1104-1143, 1995.

[6] G. Choudhury, K. Leung, W. Whitt, "Efficiently Providing Multiple Grades of Service with Protection Against Overloads in Shared Resources", *AT&T Tech. J.*, pp. 50-63, July/August 1995.

[7] S. C. Borst, D. Mitra, "Virtual Partitioning for Robust Resource Sharing: Computational Techniques for Heterogeneous Traffic", *IEEE Journal on Selected Areas in Communications*, Vol. 16, No. 5, pp. 668-678, June 1998.

[8] H. G. Perros, K. M. Elsayed, "Call Admission Control Schemes: a Review", *IEEE Communications Magazine*, Vol. 34, No. 11, pp. 82-91, November 1996.

[9] K. Shiomoto, N. Yamanaka, T. Takahashi, "Overview of Measurement-Based Connection Admission Control Methods in ATM Networks", *IEEE Communications Surveys & Tutorials*, Vol. 2, No. 1, First Quarter 1999, [Available online: http://www.comsoc.org/livepubs/surveys].

[10] A. Iera, A. Molinaro, S. Marano, "Call Admission Control and Resource Management Issues for Real-Time VBR Traffic in ATM-Satellite Networks", *IEEE Journal on Selected Areas in Communications*, Vol. 18, No. 11, pp. 2393-2403, November 2000.

[11] ITU-R Recommendation S.1420, "Performance for Broadband Integrated Service Digital Network Asynchronous Transfer Mode via Satellite", 1999.

[12] S. Olariu, P. Todorova, "Resource Management in LEO Satellite Networks", *IEEE Potentials*, Vol. 22, No. 2, pp. 6-12, April/May 2003.

[13] S. Olariu, P. Todorova, "QoS on LEO Satellites: a Resource Reservation Framework", *IEEE Potentials*, Vol. 23, No. 3, pp. 11-17, August/September 2004.

[14] Y. Bernet *et al*, "Integrated Services Operation over Diffserv Networks", IETF draft, June 1999.

[15] A. Iera, A. Molinaro, "Designing the Interworking of Terrestrial and Satellite IP Networks", *IEEE Communications Magazine*, Vol. 40, No. 2, pp. 136-144, February 2002.

[16] A. Iera, A. Molinaro, S. Marano, "IP with QoS Guarantees via GEO Satellite Channels: Performance Issues", *IEEE Personal Communications*, Vol. 8, No. 3, pp. 14-19, June 2001.

[17] L. S. Ronga, T. Pecorella, E. Del Re, R. Fantacci, "A Gateway Architecture for IP Satellite Networks with Dynamic Resource Management and DiffServ QoS Provision", *International Journal of Satellite Communications and Networking*, Vol. 21, No. 4-5, pp. 351-366, July-October 2003.

[18] T. Inzerilli, S. Montozzi, "Design of an Efficient CAC for a Broadband DVB-S/DVB-RCS Satellite Access Network", in *Proc. of the First International Conference on Advanced Satellite Mobile Systems* (ASMS 2003), Frascati, Italy, July 10-11, 2003.

[19] F. Alagöz, D. Walters, A. AlRustamani, B. Vojcic, R. Pickholtz, "Adaptive Rate Control and QoS Provisioning in Direct Broadcast Satellite Networks", *Wireless Networks*, Vol. 7, No. 3, pp. 269-281, May 2001.

[20] N. Celandroni, F. Davoli, E. Ferro, "Static and Dynamic Resource Allocation in a Multiservice Satellite Network with Fading", *International Journal of Satellite Communications and Networking*, Vol. 21, No. 4-5, pp. 469-487, July-October 2003.

[21] N. Celandroni, F. Davoli, E. Ferro, A. Gotta, "Adaptive Cross-Layer Bandwidth Allocation in a Rain-Faded Satellite Environment", *International Journal of Communication Systems*, Vol. 19, No. 5, pp. 509-530, June 2006.

[22] H. Koraitim, S. Tohmé, "Resource Allocation and Connection Admission Control in Satellite Networks", *IEEE Journal on Selected Areas in Communications*, Vol. 17, No. 2, pp. 360-372, February 1999.

[23] A. Iera, A. Molinaro, G. Aloi, S. Marano, "Signalling Issues and Call Admission Control in Multimedia Satellite Networks", in *Proc. of the IEEE Wireless Communications and Networking Conference 2000*, Vol. 3, pp. 1372-1377, September 23-28, 2000.

[24] A. Iera, A. Molinaro, S. Marano, "Traffic Management Techniques to Face the Effects of Intrinsic Delays in Geostationary Satellite Networks", *IEEE Trans. on Wireless Communications*, Vol. 1, No. 1, pp. 145-155, January 2002.

[25] S. Marano, P. Race, A. Molinaro, A. Iera, "On the Performance of Connection Admission Control and traffic Management Schemes in a "DVB-RCS Suited" Satellite System", in *Proc. of the First International Conference on Advanced Satellite Mobile Systems* (ASMS 2003), Frascati, Italy, July 10-11, 2003.

[26] F. De Rango, M. Tropea, S. Marano, "Call Admission Control for Integrated Diff-Serv Terrestrial and Int-Serv Satellite Network", in *Proc. of the IEEE* 54th *Vehicular Technology Conference* (VTC2004-Spring), Milan, Italy, May 17-19, 2004.

[27] R. Q. Hu, J. Babbitt, H. Abu-Amara, C. Rosenberg, G. Lazarou, "Connectivity Planning and Call Admission Control in an On-Board Cross-Connect Based

Multimedia GEO Satellite Network", in *Proc. of the IEEE International Conference on Communications 2003* (ICC2003), Vol. 1, pp. 422-427, May 11-15, 2003.

[28] F. De Rango, M. Tropea, S. Marano, "Controlled Load Service Management in Int-Serv Satellite Access Networks", in *Proc. of the Canadian Conference on Electrical and Computer Engineering 2004*, Vol. 4, pp. 2193-2196, May 2-5, 2004.

[29] C. H. Chang, H. K. Wu, Y. O. Tseng, "Quality of Service Support for Broadband Satellite Multimedia Service", in *Proc. of the IEEE Wireless Communications and Networking Conference 1999* (WCNC 1999), Vol. 1, pp. 187-192, September 21-24, 1999.

[30] F. De Rango, M. Tropea, P. Fazio, S. Marano, "Call Admission Control with Statistical Multiplexing for Aggregate MPEG Traffic in a DVB-RCS Satellite Network", in *Proc. of IEEE Global Telecommunications Conference 2005* (GLOBECOM 2005), St. Louis, MO, USA, Vol. 6, pp. 3231-3236, November 28 - December 2, 2005.

[31] P. Pace, G. Aloi, S. Marano, "Efficient Real-Time Multimedia Connections Handling over DVB-RCS Satellite System", in *Proc. of the IEEE Global Telecommunications Conference 2004* (GLOBECOM 2004), Dallas, Texas, USA, pp. 2722-2727, November 29-December 3, 2004.

[32] Y. Qian, R. Q. Hu, C. Rosenberg, "Integrated Connection Admission Control and Bandwidth on Demand Algorithm for a Broadband Satellite Network with Heterogeneous Traffic", *IEICE Transactions on Communications*, Vol. E89-B, No. 3, pp. 895-905, March 2006.

[33] A. C. Fu, E. Modiano, J. N. Tsitsiklis, "Optimal Energy Allocation and Admission Control for Communications Satellites", *IEEE/ACM Transactions on Networking*, Vol. 11, No. 3, pp. 488-500, June 2003.

[34] J. M. Sánchez, M. Albani, C. Arbid, M. Servilio, S. H. Oh, G. Leoleis, F. Del Sorbo, G. Lombardi, A. Giralda, "Resource Management in Integrated S-T-UMTS Networks", in *Proc. of the First International Conference on Advanced Satellite Mobile Systems* (ASMS 2003), Frascati, Italy, July 10-11, 2003.

[35] IST FUTURE Project, *Deliverable D03.03: Description and Functional design of the QoS procedure for the final UMTS*, November 2002.

[36] G. Matyagina, N. Shenoy, J. Asenstorfer, "Call Admission Control for a CDMA/TDMA based ATM Satellite Access Network", in *Proc. of the International Conference on Telecommunications* (ICT'99), Cheju, Korea, June 15-18, 1999.

[37] E. Del Re, R. Fantacci, G. Giambene, "Efficient Dynamic Channel Allocation Techniques with Handover Queuing for Mobile Satellite Networks", *IEEE Journal on Selected Areas in Communications*, Vol. 13, No. 2, pp. 397-405, February 1995.

[38] E. Del Re, R. Fantacci, G. Giambene, "Handover Queuing Strategies with Dynamic and Fixed Channel Allocation Techniques in Low Earth Orbit Mobile Satellite Systems", *IEEE Transactions on Communications*, Vol. 47, No. 1, pp. 89-92, January 1999.

[39] I. Mertzanis, R. Tafazolli, B. G. Evans, "Connection Admission Control Strategy and Routing Considerations in Multimedia (Non-GEO) Satellite Networks", in *Proc. of IEEE Vehicular Technology Conference* (VTC 1997), Phoenix, AZ, USA, pp. 431-435, May 4-7, 1997.

[40] S. Cho, "Adaptive Dynamic Channel Allocation Scheme for Spotbeam Handover in LEO Satellite Networks", in *Proc. of IEEE VTC 2000*, Boston, USA, pp. 1925-1929, September 25-28, 2000.

[41] S. Cho, Ian F. Akyildiz, M. D. Bender, H. Uzunalioglu, "A New Connection Admission Control for Spotbeam Handover in LEO Satellite Networks", *Wireless Networks*, Vol. 8, No. 4, pp. 403-415, July 2002.

[42] E. Del Re, R. Fantacci, G. Giambene, "Different Queuing Policies for Handover Requests in Low Earth Orbit Mobile Satellite Systems", *IEEE Transactions on Vehicular Technology*, Vol. 48, No. 2, pp. 448-458, March 1999.

[43] G. Maral, J. Restrepo, E. Del Re, R. Fantacci, G. Giambene, "Performance Analysis for a Guaranteed Handover Service in a LEO Constellation with a Satellite-Fixed Cell System", *IEEE Transactions on Vehicular Technology*, Vol. 47, No. 4, pp. 1200-1214, November 1998.

[44] L. Boukhatem, A. L. Beylot, D. Gaïti, G. Pujolle, "TCRA: A Time-based Channel Reservation Scheme for Handover Requests in LEO Satellite Systems", *International Journal of Satellite Communications and Networking*, Vol. 21, No. 2, pp. 227-240, March/April 2003.

[45] E. Papapetrou, F. N. Pavlidou, "QoS Handover Management in LEO/MEO Satellite Systems", *Wireless Personal Communications*, Vol. 24, No. 2, pp. 189-204, January 2003.

[46] E. Papapetrou, F. N. Pavlidou, "Analytic Study of Doppler-Based Handover Management in LEO Satellite Systems", *IEEE Trans. Aerosp. Electron. Syst.*, Vol. 41, No. 3, pp. 830-839, July 2005.

[47] P. Todorova, S. Olariu, H. N. Nguyen, "A Lightweight Call Admission and Handover Management Scheme for LEO Satellite Networks", in *Proc. of the Fifth European Workshop on Mobile/Personal Satcoms* (EMPS 2002), Baveno, Italy, Sept. 25-26, 2002.

[48] P. Todorova, S. Olariu, H. N. Nguyen, "A Two-Cell-Lookahead Call Admission and Handoff Management Scheme for Multimedia LEO Satellite Networks", in *Proc. of the* 36th *Annual Hawaii International Conference on System Sciences* (HICSS-36), Big Island of Hawaii, USA, Jan. 6-9, 2003.

[49] S. Olariu, S. R. Ali Rizvi, R. Shirhatti, P. Todorova , "Q-Win - A New Admission and Handoff Management Scheme for Multimedia LEO Satellite Networks", *Telecommunications Systems*, Vol. 22, No. 1-4, pp. 151-168, January-April 2003.

[50] J. Siwko, I. Rubin, "Call Admission Control for Capacity-Varying Networks", *Telecommunication Systems*, Vol. 16, No. 1-2, pp. 15-40, January 2001.

[51] J. Siwko, I. Rubin, "Connection Admission Control for Capacity-Varying Networks with Stochastic Capacity Change Times", *IEEE/ACM Trans. on Networking*, Vol. 9, No. 3, pp. 351-360, June 2001.

[52] P. Boedhihartono, G. Maral, "Evaluation of the Guaranteed Handover Algorithm in Satellite Constellations Requiring Mutual Visibility", *International Journal of Satellite Communications and Networking*, Vol. 21, No. 2, pp. 163-182, March/April 2003.

[53] E. Papapetrou, S. Karapantazis, G. Dimitriadis, F. N. Pavlidou, "Satellite Handover Techniques for LEO Networks", *International Journal of Satellite Communications and Networking*, Vol. 22, No. 2, pp. 231-245, March/April 2004.

[54] S. Karapantazis, F. N. Pavlidou, "Design Issues and QoS Handover Management for Broadband LEO Satellite Systems", *IEE Proc. Communications*, Vol. 152, No. 6, pp. 1006-1014, December 2005.
[55] S. Karapantazis, P. Todorova, F. N. Pavlidou, "On Call Admission Control and Handover Management in Multimedia LEO Satellite Systems", in *Proc. of the* 23rd *AIAA ICSSC 2005*, Italy, Rome, September 25-28, 2005.
[56] S. Karapantazis, P. Todorova, F. N. Pavlidou, "On Bandwidth and Inter-Satellite Handover Management in Multimedia LEO Satellite Systems", in *Proc. of ASMS 2006*, Munich, Germany, May 29-31, 2006.

7

DYNAMIC BANDWIDTH ALLOCATION

Editors: Tommaso Pecorella[1], Giada Mennuti [1]

Contributors: Nedo Celandroni[2], Franco Davoli[3], Erina Ferro[2], Alberto Gotta[2], Stylianos Karapantazis[4], Giada Mennuti[1], Antoni Morell[5], Tommaso Pecorella[1], Gonzalo Seco Granados[5], Petia Todorova[6], María Ángeles Vázquez Castro[5]

[1]CNIT - University of Florence, Italy

[2]CNR-ISTI - Research Area of Pisa, Italy

[3]CNIT - University of Genoa, Italy

[4]AUTh - Aristotle University of Thessaloniki, Greece

[5]UAB - Universitat Autònoma de Barcelona, Spain

[6]FhI - Fraunhofer Institute - FOKUS, Berlin, Germany

7.1 Dynamic bandwidth allocation: problem definition

Some of the appealing advantages of satellite networks, such as the wide coverage and the configuration flexibility, make them an ideal candidate for providing multimedia services worldwide. Satellite bandwidth is, however, a commodity at a premium, and an inefficient utilization of it may negate some of the aforementioned advantages. To this end, an apportioning scheme that dynamically allocates the bandwidth among the satellite terminals, while fulfilling the QoS requirements, is of paramount importance. Moreover, the satellite scenario adds a new dimension to the treatment of bandwidth, owing to the presence of both variable physical channel operating conditions and large bandwidth-delay products. Typically, control actions in telecommunication networks need to be exerted over a wide range of time scales to cope with events that may occur with frequencies ranging from milliseconds to minutes or hours [1]-[4]. Satellite systems not only experience variable-load multimedia traffic, but also variable channel conditions and large propagation delays. The variability in operating conditions is due both to changes in the traffic loads and to the signal attenuation on the satellite links due to the degradations that result from atmospheric events, which particularly affect, for example, the transmissions in the Ka band (20-30 GHz). The variability due to changing radio channel conditions can be counteracted by means of *Adaptive Coding and Modulation* (ACM) techniques that, however, modify the available bandwidth for higher layers, thus affecting *Dynamic Bandwidth Allocation* (DBA) schemes.

Efficient bandwidth utilization and QoS provisioning are, unfortunately, two competing goals; therefore, DBA schemes seek for a trade-off between them. To address the vast majority of IP traffic, which is inherently bursty, a technique that implicitly evaluates the bandwidth requirement at each satellite terminal and manages the traffic flows is essential. The purpose of this Chapter is to present a number of solutions for assigning the satellite bandwidth to different users (Earth stations) and traffic types.

The combined action among various protocol layers (from the physical layer up to the application layer) is likely to be a good way to combat channel variability. However, this procedure could be too complex to obtain in the widest possible extent, which would imply numerous cross-layer interactions for control purposes and the related exchange of signaling information. In order to obtain optimized policies for satellite bandwidth allocation, the actions taken in a satellite network at the physical layer (where fade countermeasure techniques are applied) can be combined with actions at the data link layer (where the satellite bandwidth is allocated), thus realizing a more limited cross-layer optimization. The complexity of this procedure lies in the fast changing measurements required at the physical layer, regarding the channel state (signal power-to-noise ratio), which might produce an unstable allocation at the data link layer. Hence, the feedback information has to be properly filtered, possibly with some hysteresis to obtain a stable allocation at the data link layer.

Regarding resource allocation, another problem is the control network architecture, which can be centralized or distributed. A centralized allocation is performed by a station, which plays the role of master (or *Network Control Center*, NCC). The master station collects all information relevant to the other stations (slave stations) and performs the best bandwidth allocation. This may produce a heavy computational effort in the master station. A distributed allocation technique solves the computational problem, but requires a robust control channel and an efficient control protocol, which takes into account the large communication delay. As a consequence, the available bandwidth may be significantly reduced by the signaling protocol.

It should be observed, however, that the bandwidth allocation problem is somewhat different for different satellite network topologies. While the typical focus for GEO satellites is the efficient bandwidth assignment among terrestrial gateways, for LEO satellites handover and call prioritization procedures become crucial aspects.

In a GEO satellite system the main limit is the time delay, in a LEO satellite system this issue is mitigated, but the system complexity causes several problems. In order to achieve a continuous satellite access, a large network of LEO satellites is required with regular handovers among them. Achieving ubiquitous coverage poses a significant challenge, and the speed at which the satellites ground track moves on the Earth generates rapidly changing communication channels, subject to severe Doppler spreading. Moreover, if a constellation of LEO satellites is designed to provide global coverage, then these satellites must be able to communicate one to another, either by incorporating *Inter-Satellite Links* (ISLs) or a ground-based hub station in each footprint. All these issues contribute in making DBA an essential approach for providing the proper QoS but, at the same time, make its design very difficult.

A less treated problem, moreover, could arise from satellite-based mesh architectures. So far, the system model only considers the uplink part, relying on the assumption that downlink is not a bottleneck. In a meshed architecture with multiple, limited-bandwidth downlink spot-beams, the channel allocation will have to take into account also this aspect in order to maintain the overall QoS; this is particularly important in satellite-based switching systems [5].

7.1.1 Survey of allocation approaches

DBA schemes can be distinguished as *static* and *adaptive*.

Static algorithms

In *static* schemes, once a terminal is assigned a certain amount of capacity, this capacity remains constant for the connection's lifetime. The terminal can locally handle dynamically the bandwidth, without involving the NCC. That is, the assigned capacity can be apportioned between *High-Priority* (HP) and *Low-Priority* (LP) traffic.

Adaptive algorithms

In the case of *adaptive* schemes, each satellite terminal can send requests to the NCC in order to reserve or release channel capacity, based on its dynamic estimation of bandwidth needs.

To meet the QoS requirements of bursty and delay-sensitive traffic, the terminal can follow three approaches:

- Fixed allocation proportional to the maximum source rate, to be requested on a per-connection basis,
- Fixed allocation at a given rate using DBA for peak bursts,
- Full DBA techniques.

The first approach is inefficient for satellite systems, as bandwidth is allocated in a way that does not take into account the real needs of a station; besides, the maximum source rate is usually unknown. As regards the full DBA techniques, these can exploit the channel capacity with good efficiency, since no capacity is reserved during inactive periods. Notwithstanding, the capacity request signaling channel may become overloaded during transient changes in traffic, leading to higher delays and congestion. Consequently, a mixed approach seems to be the most flexible choice, where each terminal is assigned some fixed channels of moderate capacity, while a number of DBA channels are used during peak traffic periods.

As far as adaptive schemes are concerned, one of the challenging problems that engineers are called to grapple with is the implementation of these techniques in a GEO satellite system. The main problem stems from the high delay between the time instant that a request is sent to the NCC and the time instant at which the satellite terminal is informed about the bandwidth that has been allocated to it. This latency prevents immediate changes to the allocated capacity. Since a low latency entails better performance, a GEO satellite system represents the worst case (approximately 500 ms when the NCC is terrestrial-based or 250 ms when the majority of processing is supported by the satellite as a part of its on-board capability).

Adaptive DBA schemes are generally categorized as either *reactive* or *proactive* algorithms. *Reactive* schemes take into account the current queue length, the packet loss and the average delay in order to react to traffic fluctuations, without trying to anticipate them. Compared to *proactive* algorithms, *reactive* algorithms are easier to implement and can utilize the channel capacity more efficiently. However, QoS requirements are not easily met, since the requests are delayed by sending them to the NCC, and do not therefore necessarily represent the current bandwidth needs. In [6], the authors proposed a novel predictive bandwidth allocation and de-allocation scheme, which frees up bandwidth allocated to connections that are unlikely to be used. The look-ahead horizon of k cells is introduced, where $k = 2$. The scheme provides the lowest *Call Dropping Probability* (CDP) for real-time connections with respect to previous schemes.

Even though *reactive* schemes may perform well in LEO satellite networks, they are not well suited to GEO systems owing to the high propagation delay. *proactive* schemes aim at analyzing the traffic and predicting the required bandwidth. Usually, this is realized by providing a predictor with data up to time t (e.g., with the queue lengths, the input flows and output flows); such data are used to make a prediction at time t of the aggregated traffic in the interval $[t, t + k]$ (e.g., the traffic within the next superframe, where a superframe is the aggregation of k consecutive frames). Depending on the number of simultaneous traffic flows (i.e., TCP connections, application data streams) and the QoS model in use (i.e., DiffServ or IntServ), different traffic prediction techniques can be adopted. In a single-user per satellite terminal scenario, an IntServ-based QoS model will be assumed, whereas for a large aggregate of users per terminal, a DiffServ model seems more appropriate. When the number of data flows is very small, e.g., for a single-user per satellite terminal, traffic predictors may exploit the possibly known traffic patterns, like the TCP slow-start and the IntServ traffic information, in order to reserve the appropriate resources. If this is not viable, as in a DiffServ model approach, the traffic predictions can resort to utilizing the statistical properties of IP traffic. Hence, the required bandwidth can be estimated. In order to make adaptive predictions, i.e., capable of following changes in the traffic characteristics over time, the parameters of the predictor can be regularly updated. The performance of these schemes heavily relies upon the accurate prediction of future traffic.

7.2 DBA schemes for DVB-RCS scenarios

In a DVB-RCS return link, users are multiplexed by means of a *Multi Frequency - Time Division Multiple Access* (MF-TDMA) scheme. The DVB-RCS standard [7],[8] permits full flexibility in the way the bandwidth is divided (see a feasible example in Figure 7.1, left upper corner [9]). The adopted solution in this Section consists of an independent division of both time and frequency axis, that is, bandwidth is divided into several carriers and the time duration of the superframe is divided into timeslots. Carriers do not have necessarily the same transmission bandwidth (different types of carriers are possible) and, at the same time, the timeslot duration can be different from one carrier to another.

Return Channel Satellite Terminals (RCSTs) ask for some amount of system capacity to the NCC through capacity requests. In the DVB-RCS standard, three types of capacity request, from highest to lowest priority, are considered: CRA, RBDC, and VBDC. *Free Capacity Assignment* (FCA) usually is not taken into consideration by DBA schemes, since it may be granted by the NCC, but not requested [10]. Please refer to Chapter 1 for more details on these resource allocation methods.

Note that the requests generated by all RCSTs in a beam constitute the inputs of the bandwidth allocation problem and in principle it is not necessary to consider how RCSTs generate requests. For each bandwidth allocation update (which is done on a superframe basis) the NCC sends a *Terminal Burst Time Plan* (TBTP) to the RCSTs. This message indicates the time

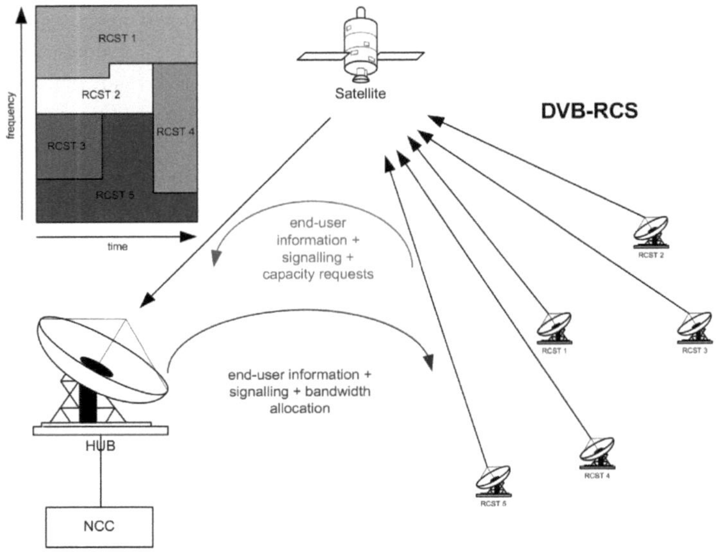

Fig. 7.1: System model. See reference [9]. Copyright ©2006 IEEE.

and the frequency that each RCST should use to transmit (see Figure 7.1 [9]).

In any case, the standard does not give strict constraints on the algorithms to be used in the resource allocation process; hence, it is possible to develop advanced techniques by using the standard request types. The only weakness of the standard is related to the lack of information contained in the requests; hence, two requests of the same type will have to be considered as equal, even if the requesting RCSTs have to deliver different kinds of traffic (e.g., volume-based requests for high priority and low priority traffic).

The next improvements in DVB-RCS-based allocation strategies will be focused on two topics, both related to a cross-layer approach. The first one will be to consider the effects of fading countermeasures; the second one will be to define a simple interface for upper layers, in order to develop a cross-layer QoS manager, able to tune the allocation process to the actual QoS requirements, possibly considering a pricing system, i.e., taking into account the user willingness to pay. A possible protocol architecture to support cross-layer interactions is proposed in sub-Section 1.6.2 referring to the BSM standard.

7.3 Recent developments on DBA techniques

7.3.1 DVB-RCS dynamic channel allocation using control-theoretic approaches

One of the main issues with *proactive* DBA is the accurate prediction of future traffic. Traffic predictors are usually affected by errors due to unexpected network behaviors (e.g., packet loss, network congestion, etc.), TCP behavior and, more generally, uncertainty in the user interactions. Coupling the traffic predictors with appropriate control-theoretic techniques, however, allows maintaining the required QoS with an acceptable computational effort.

In a DVB-RCS GEO satellite system, the NCC receives the bandwidth requests of each RCST and decides whether to satisfy or not these requests on the basis of a fair policy of resource sharing among all the RCSTs. In order to meet the desired QoS, both the request algorithm and the NCC allocation strategy are of paramount importance.

In [11]-[13] the authors compared some different allocation strategies based on traffic prediction, assuming that each RCST is used to transmit a heavy aggregate of traffic. Figure 7.2 shows the proposed system model. It can be observed that the bandwidth controller must take into account the traffic predictions, the actual queue sizes and the packet scheduler behavior, to satisfy the bandwidth requests. In the figure, the NCC is depicted as a simple delay with a "disturb", due to the possibility of denying a bandwidth request.

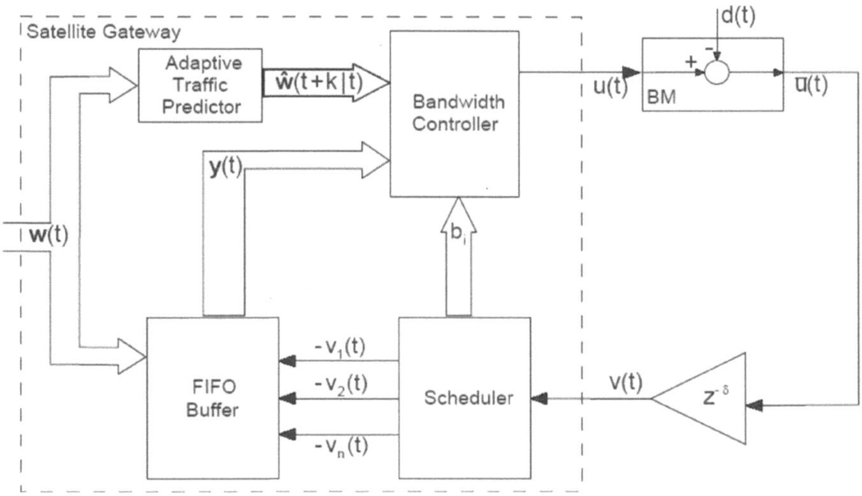

Fig. 7.2: RCST system model.

Simulations have evaluated the bandwidth loss, measured as the number of slots per frame allocated to an RCST but not used to transmit data, and the QoS performance of two DBA techniques based on a *Receding Horizon Controller* and a *Smith Predictor Controller* (RHC and SPC, respectively) versus three fixed (*Fix n* in Figure 7.3, where n is the number of slots/frame assigned to a single RCST) allocation schemes with different bit-rate values [13]. A self-similar traffic with Hurst parameter $H = 0.8$ has been used to feed each RCST. Figure 7.3 reports the bandwidth loss distribution (a) and the delay for the *Expedited Forwarding* (EF), *Assured Forwarding* (AF), and *Best Effort* (BE) DiffServ [14]-[17] traffic classes [(b), (c) and (d), respectively]. It can be observed that the bandwidth loss is greatly reduced by using DBA techniques, whereas the overall QoS of the traffic classes is acceptable, in particular by using the RHC scheme.

However, despite the great advantages of DBA techniques, both in terms of QoS satisfaction and efficient resource utilization, some new problems arise when they are applied. In particular:

- "Greedy" traffic flows can compromise the whole satellite system's QoS.
- Compatibility between different control techniques and bandwidth request methods should be validated.
- Security issues on the signaling channel should be analyzed in order to prevent denial of service attacks based on fake bandwidth reservations.

Those issues should be investigated and carefully addressed before using DBA techniques in any actual system.

7.3.2 Dynamic bandwidth de-allocation

Several approaches for bandwidth and handover management have been studied in the recent literature in the case of mobile satellite systems. Publications in this area investigate only bandwidth allocation and the intra-satellite handover management. In reference [18], an advanced bandwidth management strategy is proposed and evaluated, allowing for bandwidth allocation/deallocation and a novel inter-satellite handover management scheme, tailored for multimedia LEO satellite networks with satellite diversity. The main mechanism is based on bandwidth de-allocation. According to the proposed scheme, capacity reservation requests for handover calls are removed from the queues when the capacity that they strive to reserve is unlikely to be used. Simulations confirmed the usefulness of bandwidth de-allocation mechanism. Other details of this scheme have been already discussed in sub-Section 6.4.

7.3.3 Dynamic bandwidth allocation with cross-layer issues

Some examples of cross-layer DBA schemes are here briefly discussed, limiting the description to recent works [19]-[25]. An overview of cross-layer approaches

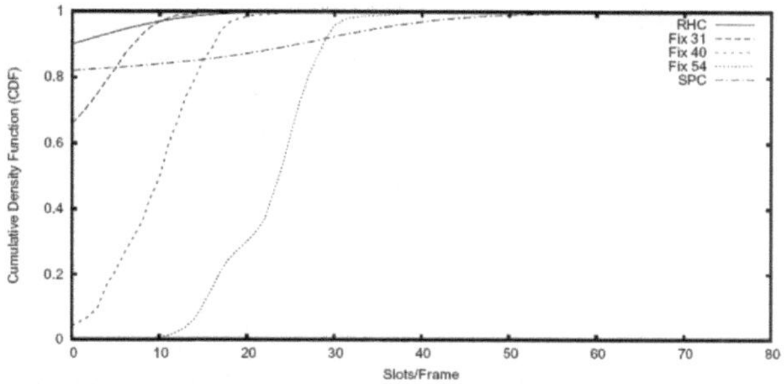

(*a*) Bandwidth loss distribution for different DVB-RCS allocation schemes

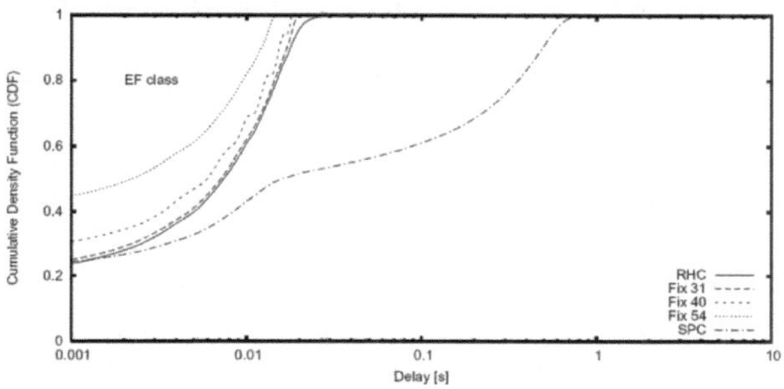

(*b*) EF packet delay distribution for different DVB-RCS allocation schemes

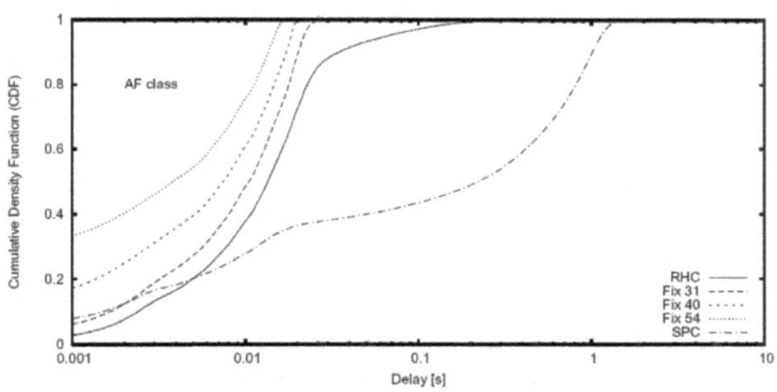

(*c*) AF packet delay distribution for different DVB-RCS allocation schemes

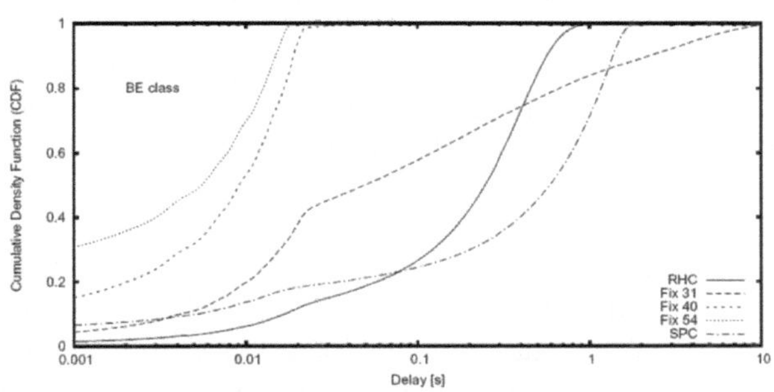

(*d*) BE packet delay distribution for different DVB-RCS allocation schemes

Fig. 7.3: DBA in GEO satellite systems.

in this context can be found in [26],[27].

The first example considered concerns the presence of mixed *Guaranteed Bandwidth* (GB) real-time traffic and *Best Effort* (BE) traffic. GB is subject to CAC, which is exerted independently by local controllers, situated at the Earth stations (i.e., RCSTs), within an amount of bandwidth that is assigned to these stations over a certain allocation time interval. The bandwidth allocation is the responsibility of a master station and can be done periodically or on-demand. Best-effort traffic is represented by an inelastic model (i.e., the congestion control mechanism of TCP is not explicitly taken into account), used to compute the loss probability of cells (*Asynchronous Transfer Mode*, ATM or DVB), stored in the Earth stations' buffers; this traffic utilizes the bandwidth that remains available after serving the GB class. The cross-layer interaction stems from the fact that a fade countermeasure, based on bit and coding rate adaptation, is used at the physical layer, whose influence on the bandwidth allocation is accounted for by "redundancy coefficients" [representing the inverse of the ratios of the *Information Bit Rate* (IBR) in the specific channel condition to the one in clear sky]. Various methods have been considered for bandwidth allocation, and the overall structure has been evaluated in the presence of real fading traces [19],[20]. Figure 7.4 [20] represents call blocking, call dropping (due to a temporary lack of bandwidth) and cell loss probabilities for three different allocation strategies: (*i*) cross-layer *Optimized Centralized* (OC, where the bandwidth is allocated on demand by the master station, which solves a centralized optimization problem); (*ii*) cross-layer *Optimized Proportional* (OP, where optimal allocation requests are computed locally by the Earth stations and then passed to the master, which re-scales them and distributes the bandwidth proportionally); (*iii*) *Simple Proportional* (SP, based on offered load, with no cross-layer dynamic allocation). The reported results refer to a 10,000 s simulation, with 10 Earth stations, 5 of which experience different fading conditions, whereas 5 operate in clear sky. In these graphs, the probabilities for each point in time are computed by averaging over all stations in the system, and over a time window of 1,000 s. The fading is dynamically variable, according to real traces.

The advantage of the cross-layer allocations lies in maintaining blocking probability values below a given threshold (5% in the specific case), while minimizing the call dropping and the BE traffic cell loss probabilities in the stations' buffers.

The second example deals with DBA in the presence of only inelastic packet traffic with two stations, whose traffic loads periodically alternate between a lower and a higher value. Figure 7.5 [22] illustrates the convergence properties of a gradient descent technique, based on *Infinitesimal Perturbation Analysis* (IPA) [21]-[23]. Station 2 is in clear sky, whereas station 1 also experiences fading variations, besides those in traffic load. The bandwidth allocation provided by the IPA gradient estimation, based only on on-line measurements, is capable to face both dynamic effects in order to minimize

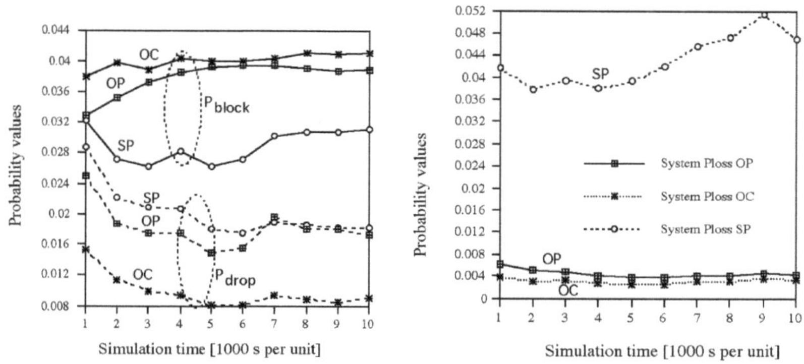

Fig. 7.4: Call blocking and dropping (left) and cell loss (right) probabilities.

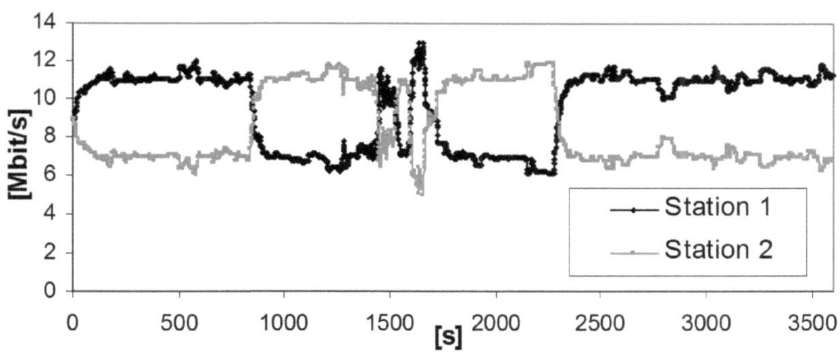

Fig. 7.5: IPA gradient descent allocation, under traffic load and fade changes. See reference [22]. Copyright ©2006 IEEE.

the overall loss volume.

The problem considered in [21],[22] is a pure parametric optimization. In order to avoid transient periods in the convergence of the on-line gradient descent technique, a different point of view can be adopted [23] where *open-loop feedback* control strategies (i.e., stemming from a *functional* optimization approach) are approximated by means of neural networks.

Finally, a DBA cross-layer optimization, aiming at achieving the "best" compromise between the TCP goodput maximization and fairness, has been treated in [24],[25], in a GEO bent-pipe satellite scenario. The numerical details of the example shown here are the same as in [25], with a combination of long-lived TCP NewReno connections, sharing various bottleneck links, determined by 10 different fading classes (stemming from different source-destination pairs), under the Hotbird 6 link budget [28] and real fading traces. The "instantaneous" goodput is determined by the dynamic bandwidth and redundancy allocation, which aims at counteracting fading effects and achieving a compromise between maximizing the total goodput and maintaining fairness among connections.

Figure 7.6 [25] shows the behavior as a function of time of the overall goodput (the points are the results of a moving average over a 10 s window) for two classes operating under different fading conditions (note that the carrier power-to-noise spectral density ratio, C/N_0, for the source-destination pairs is also shown). The used strategies attempt to maximize the total goodput and to maintain fairness in different ways (see [25] for a description of these strategies):

- The "merge" strategy is the best choice between two alternative methods ("tradeoff" and "range", respectively) that establish a balance between goodput and fairness;
- The "proportionally fair" technique maximizes the sum of the logarithms of the individual goodputs, so as to attain a *Nash Bargaining Solution* (NBS);
- The "BER threshold" strategy simply adjusts the redundancy to keep always BER below a given limit, and assigns the bandwidths proportionally to the redundancy and the number of connections of each class (no cross-layer action).

The advantages of the cross-layer strategies, shown in detail in [25], are not only in terms of goodput, but also in terms of fairness.

7.3.4 Joint timeslot optimization and fair dynamic bandwidth allocation in a system employing adaptive coding

In [29], an enhanced and multi-beam DVB-RCS system is addressed, considering both *Adaptive Coding* (AC) and dynamic framing. AC arises when the transmission is severely affected by channel conditions (as in the Ka band). In order to keep the link active, framing design must be flexible enough to adapt in time and frequency, to allow for the use of different carriers (this technique is also known as *Dynamic Resource Allocation*, DRA) and/or different protection-levels of channel coding (AC).

The problem of optimal framing has been already addressed in the literature. For example, in [30] a method is presented for optimal super-frame pattern design for the DVB-RCS MF-TDMA return link, so that the system data throughput is maximized. The authors formulate the design problem as a non-linear combinatorial optimization problem. However, the developed method considers *static framing* and, therefore, it is not extensible to Ka band

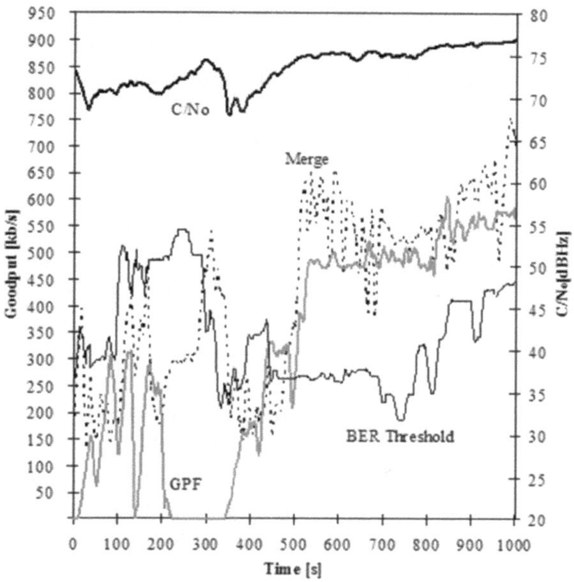

(a)

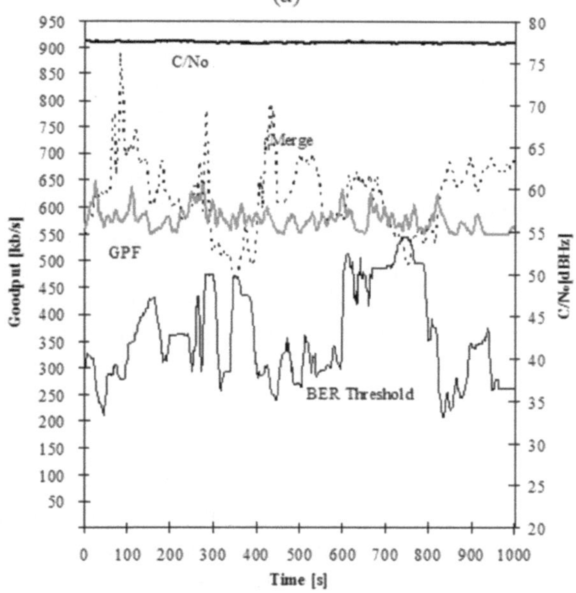

(b)

Fig. 7.6: Merge, Proportionally Fair and BER Threshold (thr = 10^{-6}) strategies. A class in fading (*a*); a class in clear sky (*b*). See reference [25]. Copyright ©2006 IEEE.

transmissions, where adaptive physical layer is used and adaptive framing is more appropriate. In [31], bandwidth segmentation on a super-frame basis is also presented, but slots are of fixed duration within the different types of carriers.

Current systems typically make use of fixed framing for its simplicity. Instead, the DVB-RCS terminal considered in [29] is assumed to have transmission capabilities sufficiently agile to realize multimedia communications under adverse channel conditions. This is achieved by using the dynamic-slot MF-TDMA feature of the standard, integrated with AC that is adapted on a super-frame basis. This strategy segments the total bandwidth into different types of carriers, according to user traffic demands and weather conditions. It is assumed that slots can be assigned to users with different coding rate on the same carrier, which leads to coexisting slots of different size, which is called *adaptive framing*, due to AC. The total bandwidth is segmented into several carriers that can be of different bandwidths and the slots contained in the frames can be of various durations, according to the chosen coding rate; users can be granted slots of different durations on different carriers (sequentially). Differently from other studies, bandwidth is segmented not only in the presence of different traffic types, but also assuming realistic dynamic weather conditions, to which coding rate is adapted.

Capacity is allocated giving priority to heavy rain-affected users, then considering less affected ones, and ending with clear sky users, while there is still bandwidth available. The major issue to keep into account concerns the limits of capacity that can be allocated, due to adaptive framing.

In this study, the time dimension is partitioned into super-frames, a super-frame into frames and frames into slots. The super-frame length is 26.5 ms and seven different coding rates (1/3, 2/5, 1/2, 2/3, 3/4, 4/5, 6/7) are considered; the modulation is QPSK. Regarding the frequency dimension, it is assumed that the total bandwidth can be dynamically segmented, from super-frame to super-frame, and that up to four different carrier types can be used in a super-frame: 540 kHz (carrier type I), 270 kHz (carrier type II), 135 kHz (carrier type III) and 67.5 kHz (carrier type IV). The roll-off factor is 0.35, providing symbol rates of 400, 200, 100 and 50 kbaud, respectively. The number and type of active carriers is adapted to the traffic requests and the needs of the users, which vary according to channel conditions. The transmitted packet can be an ATM cell or an MPEG packet (for numerical evaluations we will only refer to ATM cells). With AC, the length of the slots transmitting such fixed-length packet becomes variable and, therefore, the number of slots contained by a given type of carrier becomes variable, as well.

Not all the users are necessarily always active. Active users are divided into categories, according to both their symbol energy to noise-plus-interference spectral density ratio, $E_s/N_{o,tot}$ and traffic characteristics. Traffic is assumed to be uniform and the considered classes are: *Constant Bit Rate* (CBR), *Variable Bit Rate* (VBR), and BE. For simplicity and without loss of generality, one user is assumed to ask only for one of these traffic classes, so that each user will have allocated slots of a given fixed length on the carrier type corresponding to its $E_s/N_{o,tot}$ [29]. Traffic demands are queued according to the type of DVB-RCS capacity request, which can be CRA, RBDC, and VBDC. Capacity requests are prioritized: CRA has the highest priority and VBDC the lowest. CBR traffic is assigned to CRA as a whole, whereas VBR traffic is assigned to CRA and RBDC. Similarly, BE traffic is also divided between RBDC and VBDC.

The number of carriers of each type is computed at every super-frame, given priority to the users affected by rain. Assuming a given $E_s/N_{o,tot}$ for the user and some given requests for the current super-frame, a closed-form estimation of the number of carriers required per *carrier type* is computed in terms of an estimation of the number of slots as follows:

$$n_{n_i}^{C}(s) = n_{n_i}^{C,CBR}(s) + n_{n_i}^{C,VBR}(s) = N_{n_i}^{C}(s)\left[\frac{(r_{CBR}^{C}+r_{VBR}^{C})T_s}{\eta_i L(\eta_i)}\right], \quad i = 1, 2, \ldots, N^{AC} \tag{7.1}$$

$$n_{n_i}^{R}(s) = n_{n_i}^{R,VBR}(s) + n_{n_i}^{R,BE}(s) = N_{n_i}^{R}(s)\left[\frac{(r_{VBR}^{R}+r_{BE}^{R})T_s}{\eta_i L(\eta_i)}\right], \quad i = 1, 2, \ldots, N^{AC} \tag{7.2}$$

$$n_{n_i}^{V}(s) = n_{n_i}^{Vol}(s) + n_{n_i}^{V,BE}(s) = N_{n_i}^{V}(s)\left[\frac{V+r_{BE}^{V}T_s}{\eta_i L(\eta_i)}\right], \quad i = 1, 2, \ldots, N^{AC} \tag{7.3}$$

where s is an index making reference to the super-frame, which consists of 10 frames and lasts 265 ms, $n_{n_i}^{X,Y}(s)$ is the number of requested slots corresponding to capacity request type X (X = C, R or V, which correspond to CRA, RBDC or VBDC requests, respectively) of traffic class type Y (Y = CBR, VBR or BE) requiring spectral efficiency η_i, $n_{n_i}^{X}(s)$ is the total number of requested slots corresponding to capacity request type X, $N_{n_i}^{X}(s)$ is the number of users requesting capacity type X, T_s is the duration of a super-frame in seconds. N^{AC} is the number of possible coding rates, and r_Y^X is the bit-rate requested by traffic class Y that is mapped to request type X. V is a possibly additional amount of bits requested as volume (instead of bit-rate), which results in $n_{n_i}^{Vol}(s)$ slots, and $L(\eta_i)$ is the length in bits of the packet.

The number of carriers of each type is estimated from the total number of slots needed according to (7.1)-(7.3). The fragmentation of the bandwidth into carriers is performed, starting from the heavy rain-affected users down to the clear sky ones, while there is still bandwidth available. With all these assumptions, a key result has been obtained in [29] by applying cross-layer design for DVB-RCS with AC. The user satisfaction strongly depends on the distribution of users relative to the spatial distribution of channel conditions. As a conclusion, smarter scheduling policies should be designed, taking into account this effect in order to design fairer bandwidth allocation schemes. In what follows, a smarter scheduling policy is proposed, based on a joint optimization accounting for both a fair bandwidth distribution among users and channel conditions and timeslot duration.

Proposed framework

Recall that in DVB-RCS, the TBTP (composed of several *frames* (F) of duration T_F) is updated and transmitted every super-frame. If BW is the total system bandwidth, then the scheduler is in charge of solving an allocation problem for each $BW \times T_F$ block (note that it can also be applied to the whole super-frame). BW is divided into different carrier types to serve different users, accounting for different *Service Level Agreements* (SLAs), location and terminal equipment. The problem to be faced consists in multiplexing N users into C carriers of BW_i bandwidth that transmit into a frame of T_F seconds.

An ETSI specification [32] imposes a number of constraints to the problem, namely:

- The total transmission capacity (i.e., carriers) in the satellite beam is divided in *areas*.
- The symbol rate and slot timing must be the same for all carriers in one area. Coding rates are not necessarily the same.
- A given RCST belongs to one (and only one) area and can use only one carrier at a given time.

Hence, it is possible to simplify the problem creating sub-problems, one for each group of carriers of the same type (see Figure 7.7, on the left [9]). It is meaningful to consider that the RCSTs in one area, while transmitting in a common carrier type, use the same transmission rate. Note that the DVB-RCS standard defines an adaptive-coding physical (PHY) layer with several possible coding rates, so the mapping of users to areas is basically defined by the quality of the link (channel conditions). As before, the minimum transmission unit (a layer-2, MAC, packet) can be an ATM cell (53 bytes) or a *Moving Picture Experts Group* (MPEG) container (188 bytes). The following analysis is related to the case of ATM cells.

Following the previous discussion, the aim here is to obtain TBTP reduced signaling for frame description (excessive signaling in the *Frame Composition Table*, FCT, entails a reduction in bandwidth efficiency). A timeslot with common duration for all areas is imposed ([1]), allowing a very simple assignment

[1] Note that fixing a timeslot duration common to all areas introduces some unused bandwidth that depends on both the timeslot duration and the packet length (ATM cell in our case). However, once a given RCST has been assigned to a certain timeslot, it can change its transmission rate inside the timeslot without affecting the transmission timing of the other RCSTs. This argumentation validates the robustness of the solution proposed.

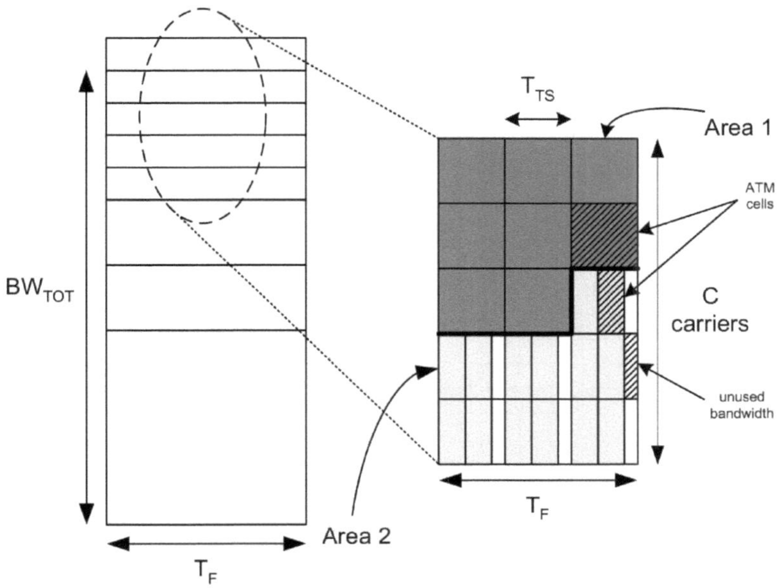

Fig. 7.7: Scheduling (bandwidth allocation) problem. See reference [9]. Copyright ©2006 IEEE.

procedure (after having known the number of timeslots per area): from left to right and from top to bottom (according to the reading order). Regarding signaling issues, this is translated into a simple FCT, since it indicates the common timeslot type (which is described in the *Time Composition Table*, TCT) and how many times it is repeated in the carrier. On the basis of the area rate, one or more ATM cells can be transmitted in a single timeslot.

A possible timeslot and ATM cell assignment is shown in Figure 7.7, on the right [9]. The problem of how to assign timeslots to areas and ATM cells to RCSTs is discussed later, after introducing the scheduling hierarchy concept [32].

Scheduling hierarchy

The general scheduling problem (which may involve thousands or more RCSTs) may be complex to solve. Therefore, it seems reasonable to reduce it to some smaller problems by imposing some known structure (that can also facilitate signaling). This is an idea similar to that proposed in [33] (particularly in centralized optimization algorithms). According to [32], some minimum resources are guaranteed to the *service providers*. Since the relative RCSTs for each *service provider* can be distributed over different areas, in [32] the scheduling hierarchy presents the *segment concept*, i.e., a grouping of

RCSTs in a given area with a minimum predefined amount of resources that must be guaranteed to them (see Figure 7.8 [9]).

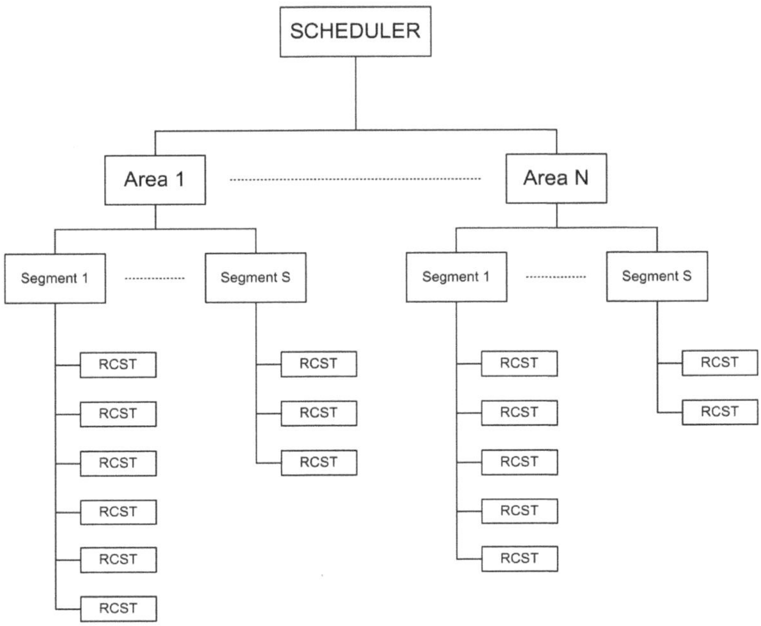

Fig. 7.8: Scheduling (bandwidth allocation) hierarchy in DVB-RCS. See reference [9]. Copyright ©2006 IEEE.

Hence, the scheduling strategy proposed is (resources can be imagined as ATM cells, but the solution can be applied identically to MPEG packets):

- Compute per-area aggregated user requests and guaranteed minimum.
- Assign the guaranteed minimum to the areas. Allocate the remaining resources with a "fair" algorithm (see the next paragraph).

For each area, the problem of distributing the assigned resources among the area terminals is solved similarly to the problem of distributing all the satellite resources among areas:

- Collect the user requests, but now per segment.
- Assign the minimum amounts to the segments.
- Allocate the remaining area resources among the segments, with a "fair" algorithm.

Finally, the resources allocated to each segment are distributed among the RCSTs associated to that segment taking into account the priorities established by the capacity requests:

- Allocate the resources to the users, depending only on their CRA requests (highest priority).
- Assign the remaining resources on the basis of the RBDC requests.
- Assign the remaining resources on the basis of the VBDC requests.

In the next paragraph the problem of fair assignment of resources is stated and solved.

Fair resource allocation

The following maximization problem is presented in [33], where a fair allocation of P resources among N *entities* (areas, segments or terminals) is achieved:

$$\begin{aligned} &\max_{x_1, \ldots, x_N} \prod_{i=1}^{N} x_i \\ &\textit{subject to} \sum_{i=1}^{N} x_i \leq P \\ &d_{\min_i} \leq x_i \leq d_{\max_i} \end{aligned} \tag{7.4}$$

where x_i is the amount of resources allocated to entity i, d_{min_i} is the part of resources guaranteed to i, whereas d_{max_i} (also indicated in what follows as d_i) is the request of i.

Compared to [33], the main difference is that now a minimum resource allocation must be guaranteed to each entity. As consequence, the solution is slightly changed. Note that it is simple to convert (7.4) in a convex optimization problem [34] with the application of the logarithm function to the objective. Moreover, the resulting problem is analytically solvable by means of the *Karush-Kuhn-Tucker* (KKT) conditions [34], that force the following solution:

$$x_i = \begin{cases} \frac{1}{\lambda}, & d_{\min_i} \leq \frac{1}{\lambda} \leq d_{\max_i} \\ d_{\min_i}, & \frac{1}{\lambda} \leq d_{\min_i} \\ d_{\max_i}, & \frac{1}{\lambda} \geq d_{\max_i} \end{cases} \tag{7.5}$$

where λ is a positive value that implies $\sum_{i=1}^{N} x_i \leq P$.

It is possible to achieve the solution in a graphic way, by simply filling a container (shaped accordingly with guaranteed resources and demands) with an amount P of water (Figure 7.9 [9]).

Since (7.4) is solvable, the solution firstly assigns the minimum amounts (namely, "pale water") and then "fairly" distributes the rest (namely, "strong water"). In this case, the solution is generally computed for a real-valued

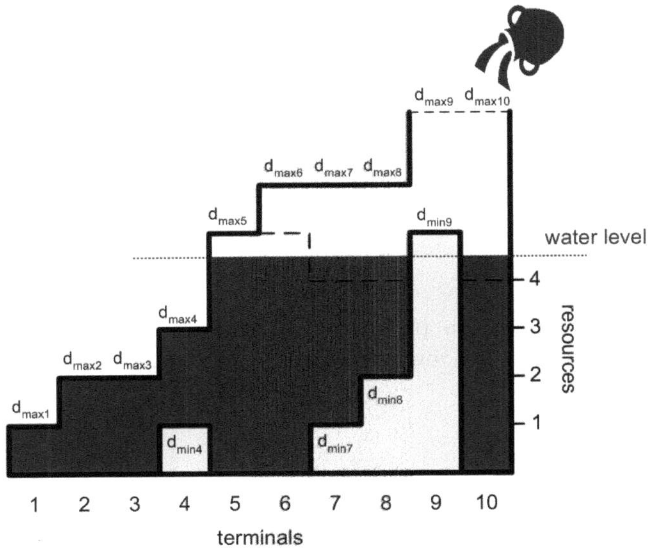

Fig. 7.9: Fair resource distribution solution. See reference [9]. Copyright ©2006 IEEE.

problem, but it is possible to obtain the particularization to the integer case (i.e., with integer variables x_i), simply assuming one extra resource (round up) to a subgroup of users, sharing the same number of resources, and round down the remaining ones (dotted line in Figure 7.9).

Focusing now on the highest level in the scheduling hierarchy of Figure 7.8, i.e., the timeslot distribution among areas, a possible design option consists in setting the timeslot duration T_{TS} to fit exactly an ATM cell of the area that transmits with the lowest rate. Higher rates allow transmitting more than one ATM cell per timeslot (see Figure 7.7, on the right). Simply analyzing this example, one can realize that some bandwidth remains unused. One obvious question is if it is possible to set up T_{TS} in order to reduce this inefficiency (this aspect will be addressed in the next paragraph). Note that T_{TS} is forced to be in the interval $[T_{min}, T_{max}]$, where these values are left as design parameters (the reader can find an example later on in the text). However, the lower limit must fulfil $T_{min} \geq t_1$, being t_1 the largest possible ATM cell duration in the system.

Area size selection and timeslot optimization

The problem in (7.4), already providing fair resource distribution, is extended here to include the timeslot optimization. For the sake of simplicity, it is not essential to consider now the minimum guaranteed resources. If N_{ATM_i} is the number of ATM cells allocated to area i, N_i is the number of timeslots allocated to area i, and T_{TS} the timeslot duration, the analyzed problem maximizes the product $\prod_{i=1}^{N} N_{ATM_i}$:

$$\begin{aligned} &\max_{T_{TS},N_1,\ \dots\ ,N_N} \prod_{i=1}^{N} N_i \cdot K_i\left(T_{TS}, t_i\right) \\ &\textit{subject to} \sum_{i=1}^{N} N_i \leq N_{TOT}\left(C, T_F, T_{TS}\right) \\ &0 \leq N_i \cdot K_i\left(T_{TS}, t_i\right) \leq d_i \\ &T_{\min} \leq T_{TS} \leq T_{\max} \end{aligned} \tag{7.6}$$

where t_i is the time duration of an ATM cell transmitted at rate r_i for the i-th area, K_i is the number of ATM cells that fit in a timeslot (depending on both the ATM cell duration, t_i, and T_{TS}) and N_{TOT} is the total number of timeslots (depending on the number of carriers, the frame duration and the timeslot duration).

The input data d_i must be in principle considered as MAC layer information. However, it may be interesting to think about cross-layer mechanisms to enable some network influence in d_i (this is the case, for example, when the RCSTs send network layer information to the NCC or the requests made at the RCSTs take into account that information). Moreover, the proposed technique requires PHY cross-layer information (the area rates r_i) and it influences both the MAC and PHY layer of the RCSTs (the latter being done through T_{TS} adjustment).

It is possible to solve the problem fixing T_{TS} and then optimizing over the N_i's. Let $\left\{N_{i_{opt}}^1\right\}$ be the solution to this problem. Fixing these values, optimization over T_{TS} is a one-variable optimization problem. Imagine the solution is $T_{TS_{opt}}^1$. Iterations of this mechanism would drive into the optimal joint solution if the problems were jointly convex, so it is mandatory to fix both problems.

Fixing T_{TS}, the problem:

$$\begin{aligned} &\max_{N_1,\ \dots\ ,N_N} \prod_{i=1}^{N} N_i \cdot \prod_{i=1}^{N} K_i\left(T_{TS}, t_i\right) \\ &\textit{subject to} \sum_{i=1}^{N} N_i \leq N_{TOT}(C, T_F, T_{TS}) \\ &0 \leq N_i \leq \left\lceil \frac{d_i}{K\left(T_{TS}, t_i\right)} \right\rceil \end{aligned} \tag{7.7}$$

is convex, where the ceiling function ($\lceil\cdot\rceil$) is necessary in the integer case in order to avoid the situation of one area that requests some ATM cells, but does not receive any timeslot. In this case, the problem in (7.7) is equivalent to the integer version of (7.4) and, thus, the solution is known.

The following problem for the timeslot optimization (developing expressions for the K_i's and N_{TOT}) is achieved fixing the N_i's:

$$\begin{aligned} &\max_{T_{TS}} \prod_{i=1}^{N} N_i \cdot \left\lfloor \frac{T_{TS}}{t_i} \right\rfloor \\ &\textit{subject to} \sum_{i=1}^{N} N_i \leq C \cdot \left\lfloor \frac{T_F}{T_{TS}} \right\rfloor \\ &0 \leq N_i \leq \left\lceil \frac{d_i}{\left\lfloor T_{TS}/t_i \right\rfloor} \right\rceil \\ &T_{\min} \leq T_{TS} \leq T_{\max}. \end{aligned} \tag{7.8}$$

The floor function ($\lfloor \cdot \rfloor$) is obviously necessary to convert this problem into non-convex and, hence, the joint problem too. However, the problem can be easily solved if the "integrality" that the floor function introduces is exploited. Look at the following observation:

"*Departing from a feasible value of T_{TS} and increasing it, it can only reduce the objective value unless a multiple value of some of the t_i's is reached*".

It is possible to note that the only meaningful values of T_{TS} are the multiples of the t_i values, into the interval $[T_{min}, T_{max}]$. The values comprised between any of these special values do not allow to place an extra ATM cell inside any timeslot at the expenses of a potential decrease in N_{TOT}. In the case under consideration, where there are few areas and the same amount of t_i's, the possible T_{TS} values are not so many and (7.8) can be simply solved via exhaustive (but small) search.

The optimization procedure for the joint problem consists of:

- Identify the possible values of T_{TS}.
- Suppress equal values coming from multiples of different t_i's.
- Optimize the N_i's for each possible value.
- Select $\{T_{TS}, N_i\}$ with best objective value in (7.6).

In order to guarantee the optimal solution, the joint convexity is not necessary, as there are only a few valid values of T_{TS} and it is sufficient to explore the optimality of each of them. Next, some results showing the importance of taking a good choice of T_{TS} are given.

Let us consider a scenario with $C = 111$ carriers of 540 kHz bandwidth and $T_F = 26.5$ ms. Imagine a DVB-RCS situation, with the RCSTs transmitting via 7 different coding rates and, hence, 7 different ATM cell durations are possible (namely, one area per coding rate is defined). The relation among areas, coding rates and ATM cells duration is presented in Table 7.1 [9]. A *Quadrature Phase Shift Keying* (QPSK) modulation is assumed, transmitted through a raised cosine pulse with roll-off factor equal to 0.35. Consider that the timeslot duration can be adjusted between $T_{min} = t_1$ and $T_{max} = 3t_1$.

Area identifier	Coding rate	ATM cell duration
1	$r_1 = 1/3$	$t_1 = 1.59$ ms
2	$r_2 = 2/5$	$t_2 = 1.325$ ms
3	$r_3 = 1/2$	$t_3 = 1.06$ ms
4	$r_4 = 2/3$	$t_4 = 0.795$ ms
5	$r_5 = 3/4$	$t_5 = 0.706$ ms
6	$r_6 = 4/5$	$t_6 = 0.6625$ ms
7	$r_7 = 6/7$	$t_7 = 0.6183$ ms

Table 7.1: Definition of areas. See reference [9]. Copyright ©2006 IEEE.

In what follows, it is explained how to compute the aggregated demand of RCSTs (number of requested ATM cells) per area. The *Aggregated System Demand* (ASD) is defined as the mean of the sum of all demands in all areas. Such demand is distributed among the areas according to a given distribution p. Note that areas with higher rates accumulate more requests since it is expected that most of the RCSTs can be found in areas with good weather conditions. Low rate areas are designed to satisfy the transmission requirements of RCSTs affected by rain and a small part of RCSTs can be assumed in that situation (considering that RCSTs are uniformly distributed in space). As an example, take into account the distribution $p = [1/15, 1/15, 2/15, 3/15, 3/15, 3/15, 2/15]$. After obtained ASD per area (it is a mean value), a realization of demand in each area using a uniform *probability density function* (pdf) with the given mean ASD is computed.

For convenience, let us define a reference ASD value, which corresponds to the capacity transported by the system when only the highest rate transmits and $T_{TS} = t_7$ (i.e., the maximum possible transported capacity). In the particular case studied, $ASD_{ref} = 4662 ATM_{cell}/frame$, which corresponds to 74.6 Mbit/s, and ASD can be greater than the ASD_{ref} value. Note that it is possible to consider a scenario where the highest rate area asks the reference ASD while the other areas ask their own "maximum" transport capacity (depending on the area rate and, obviously, less than the reference ASD).

In the results, computed via the Monte Carlo method, the fair allocation algorithm that solved (7.6) is compared to an opportunistic design, analyzing in both cases the effect of timeslot optimization. For the sake of completeness, assume that the opportunistic design finds the optimal values of the following problem:

$$\max_{T_{TS},N_1,\,\ldots\,,N_N} \sum_{i=1}^{N} N_i \cdot K_i\left(T_{TS}, t_i\right)$$
$$subject\ to \sum_{i=1}^{N} N_i \leq N_{TOT}\left(C, T_F, T_{TS}\right) \tag{7.9}$$
$$0 \leq N_i \cdot K_i\left(T_{TS}, t_i\right) \leq d_i$$
$$T_{\min} \leq T_{TS} \leq T_{\max}.$$

In order to obtain the relative solution, it is necessary to assign all the demand (until there are resources left) of the highest rate area and iterating this procedure for each area (ordered by transmission rate) until the lowest rate area is reached (if possible depending on the available resources). Note that this design assures maximum transported capacity at the expenses of offering poor QoS to users with degraded channel conditions, in general.

The first analysis in Figure 7.10 [9] studies the *Bandwidth Occupation* (BO), defined as:

$$BO = \frac{\sum_{i=1}^{7} N_i \cdot t_i}{C \cdot T_F}. \tag{7.10}$$

With the optimization of T_{TS}, the occupation for both fair and opportunistic strategies is significantly improved.

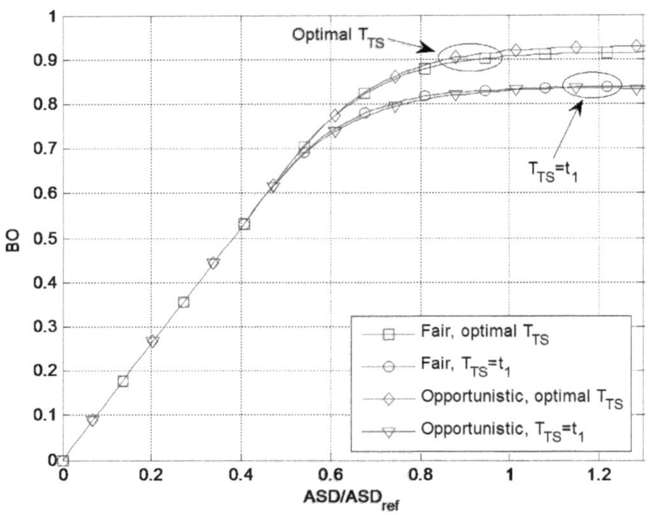

Fig. 7.10: Bandwidth occupation. See reference [9]. Copyright ©2006 IEEE.

The *Transported Capacity* (TC) is another performance index studied and it is defined as:

$$TC = \frac{\sum_{i=1}^{7} N_i \cdot K_i}{ASD_{ref}}. \tag{7.11}$$

In Figure 7.11 [9], the sum of the assigned ATM cells in all areas normalized by the reference ASD value (in fact it is a maximum transport capacity value) is shown. With the optimization over T_{TS}, the transported capacity is significantly improved: over 6% more capacity in the fair case and near 8% increase in the opportunistic design. This result shows that the increase in BO, due to T_{TS} optimization (Figure 7.10) effectively implies a TC increase. The opportunistic design could reach the maximum TC value as ASD increases (independently of the requests distribution), whereas the fair algorithm will generally saturate at a lower value (between 0.62 and 0.69 in the studied case).

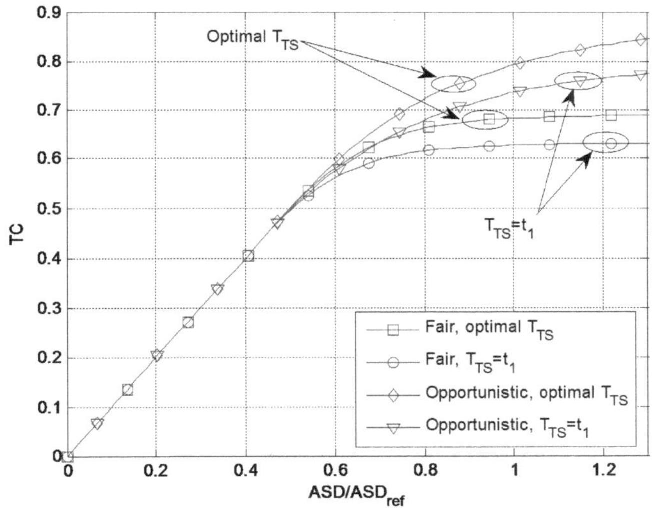

Fig. 7.11: Transported capacity. See reference [9]. Copyright ©2006 IEEE.

In what follows, the fairness issue is addressed by measuring the fairness differences between the solutions. This is done by using the fairness index definition in [35]. For a given solution $N_{ATM_1}, \ldots, N_{ATM_7}$, new variables can be defined as:

$$y_1 = \frac{N_{ATM_1}}{N^*_{ATM_1}}, \ \ldots, y_7 = \frac{N_{ATM_7}}{N^*_{ATM_7}} \tag{7.12}$$

and the computation of the *Fairness Index* (FI) is as follows:

$$FI = \frac{\left(\sum_{i=1}^{7} y_i\right)^2}{7 \cdot \sum_{i=1}^{7} y_i^2} \tag{7.13}$$

where $N^*_{ATM_i}$ is the most "fair" solution obtained with the fair algorithm with optimal T_{TS} (the "fair" solution is defined in this way).

FI is obtained for the following 2 solutions (see the results in Figure 7.12 [9]):

- Solution 1: the fair solution with $T_{TS} = t_1$.
- Solution 2: the opportunistic solution with optimal T_{TS}.

It is important to note that whereas solution 1 exhibits good fairness performance, solution 2 reduces it significantly.

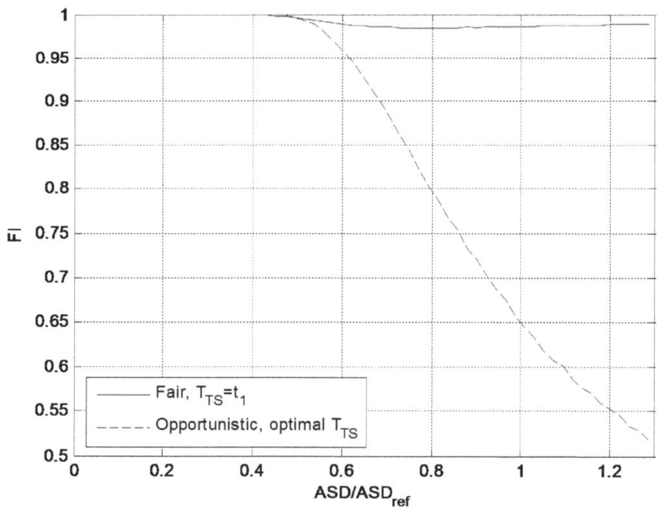

Fig. 7.12: Fairness study. See reference [9]. Copyright ©2006 IEEE.

At the end of this sub-Section, the study of the occupation efficiency for the different significant values of T_{TS} (when only one area is requesting resources) is addressed. Let us assume a very high demand to transmit, thus using the maximum possible bandwidth in the proposed framework (see the results in Table 7.2 [9]). Some T_{TS} values achieve a better occupation efficiency than others, depending on which areas are considered as "active". In particular, the configuration $T_{TS} = 4t_4$ is the one that gives better results in the general case (when all areas are active and the distribution p is totally unknown). This configuration is the most robust choice in the max-min sense: knowing nothing about the mapping of users to areas, the max-min robust design corresponds to the one that gives the best (max) performance for the worst (min) possible user distribution.

Bandwidth occupation efficiency																	
area/T_{TS}	t_1	$3t_7$	$3t_6$	$3t_5$	$3t_4$	$4t_7$	$4t_6$	$4t_5$	$5t_7$	$4t_4$	$5t_6$	$5t_5$	$6t_7$	$6t_6$	$6t_5$	$7t_7$	$7t_6$
1	0.98	0.80	0.80	0.74	0.61	0.61	0.55	0.55	0.49	0.98	0.86	0.86	0.74	0.74	0.74	0.61	0.61
2	0.82	0.66	0.66	0.61	0.51	0.51	0.92	0.92	0.82	0.82	0.72	0.72	0.61	0.92	0.92	0.77	0.77
3	0.65	0.53	0.53	0.98	0.82	0.82	0.74	0.74	0.65	0.98	0.86	0.86	0.74	0.74	0.98	0.82	0.82
4	0.98	0.80	0.80	0.74	0.92	0.92	0.83	0.83	0.74	0.98	0.86	0.86	0.74	0.92	0.92	0.77	0.77
5	0.87	0.71	0.71	0.98	0.82	0.82	0.74	0.98	0.87	0.87	0.76	0.95	0.82	0.82	0.98	0.82	0.82
6	0.82	0.66	1.00	0.92	0.77	0.77	0.92	0.92	0.82	0.82	0.89	0.89	0.77	0.92	0.92	0.77	0.89
7	0.76	0.93	0.93	0.86	0.72	0.95	0.86	0.86	0.95	0.95	0.83	0.83	0.86	0.86	0.86	0.83	0.83
mean	0.84	0.73	0.78	0.83	0.74	0.77	0.79	0.83	0.76	0.91	0.83	0.85	0.75	0.84	0.90	0.77	0.79

Table 7.2: Bandwidth occupation study. See reference [9]. Copyright ©2006 IEEE.

This Section has presented an alternative framework for bandwidth allocation in the DVB-RCS scenario, which is compliant with the latest ETSI technical specifications. In the hierarchical bandwidth allocation procedure deduced, the general fair allocation algorithm takes into account minimum resource guaranteed. The timeslot selection has been optimized and its implications analyzed, obtaining that the timeslot optimization is reasonably independent of the scheduling policy (either implementing fair, opportunistic or other strategies).

7.3.5 Dynamic bandwidth allocation for handover calls

In LEO and MEO satellite constellations, the handover problems can affect the QoS of the connections. In [36], bandwidth for handover is dynamically allocated, by calculating the possible handovers from neighboring beams, on the basis of users' location information. The reservation mechanism provides a low handover blocking probability with respect to a fixed guard channel strategy. However, employing user location information seems not reasonable, because updating locations would cause high processing load to the on-board handover controller and increase the complexity of terminals. This method seems only suitable for fixed users.

In [37], the authors have introduced two different mobility models for satellite networks. In the first model, only the motion of satellites is taken into account, whereas, in the second one, other motion components, like Earth rotation and user movements, are considered. The key idea of the algorithm is that, in order to prevent handover failure during a call, bandwidth will be reserved in a particular number S of spot-beams that the call would handover into.

In [38], a probabilistic resource reservation strategy for real-time services was proposed. The sliding window concept is adopted to predict the necessary amount of reserved bandwidth for a new call in its future handover spot-beams. As for real-time services, a new call request is accepted if the originated spot-beam has available bandwidth and resource reservation is successful in future handover spot-beams. As for non real-time service, new call requests are accepted if the originated spot-beam satisfies its maximum required bandwidth.

In [6],[39], a selective look-ahead strategy is proposed where real-time and non-real time service classes are differently treated. Bandwidth allocation only pertains to real-time connection handovers. To each accepted connection, bandwidth allocation is performed in a look-ahead horizon of k cells along its trajectory. This algorithm offers low call dropping probability, i.e., a reliable management of call handovers of and acceptable call blocking probability for new calls.

7.4 Conclusions

This Chapter has presented a set of dynamic bandwidth allocation techniques and identified associated research topics. We can conclude this Chapter by highlighting these two types of DBA problems and related techniques:

- *Handover-constrained techniques*, mainly used for LEO satellites, where the main problem is to acquire a resource among a number of different satellites, since the communication lifetime is long enough to require a number of handovers;
- *Bandwidth-constrained techniques*, affecting mainly GEO systems, where the main issue is to cope with the high delay-bandwidth product that makes the reactive approaches unfeasible for delay-constrained traffic types.

The problem of multi-tier satellite systems, i.e., satellite systems using a combination of multiple orbital systems, like GEO+LEO, has not been considered, but it could be challenging, due to the multiple use of the different techniques among the various tiers. This problem requires further investigations as it involves also intra-tier and inter-tier routing schemes.

Most of the described DBA techniques are inherently satellite-dependent; each satellite system should adapt or implement its own techniques in order to maximize system efficiency. A common theme is that optimizing 'efficiency' does not always means maximizing the bandwidth occupancy, but it is a concept more related to fulfilling the system goals in terms of QoS, user satisfaction and, ultimately, system capacity to maximize the network operator's revenue. Hence, one of the possible approaches to further study DBA techniques is to embed a *cost-function* into the DBA decision process, in order to introduce an abstraction layer between the *raw* user bandwidth requests and the actual bandwidth allocation decision algorithms.

Another topic that needs further investigation is represented by the fairness of the proposed techniques. Most techniques that involve terminal-based decisions (like in most DVB-RCS systems) can be heavily affected by fairness issues in a multi-vendor and multi-algorithm environment, thus creating serious issues in real-world deployments. At present, this problem is still an open point and should be addressed either by allowing the centralized decision process to take into account the different behaviors, or by defining some fairness threshold that every user equipment implementation must comply with. We must observe that the first option is not viable in the long-term, as it requires extra-work in the bandwidth allocation decision unit, along with the knowledge of every implementation, and this is not always possible. The second option requires the definition of precise fairness metrics and test suites to certify the user terminal fairness.

The DBA implementation is therefore a key element for the efficient operation of many satellite systems. Design choices in DBA techniques can greatly impact the overall system performance, and the evolution of appropriate techniques and analysis methods will remain important research topics for future generations of systems.

References

[1] K. W. Ross. *Multiservice Loss Models for Broadband Telecommunication Networks*. Springer-Verlag, London, UK, 1995.

[2] H. J. Chao, X. Guo. *Quality of Service Control in High-Speed Networks*. Wiley, New York, NY, 2002.

[3] J. Walrand, P. Varaiya. *High-Performance Telecommunication Networks*, 2nd Ed., Morgan Kaufmann. San Francisco, CA, 2000.

[4] M. H. Ahmed, "Call Admission Control in Wireless Networks: a Comprehensive Survey", *IEEE Communications Surveys and Tutorials*, Vol. 7, No. 1, pp. 50-69, 1st Quarter 2005.

[5] F. Chiti, R. Fantacci, D. Tarchi, S. Kota, T. Pecorella, "QoS Provisioning in GEO Satellite with Onboard Processing Using Prediction Algorithms", *IEEE Wireless Communications Magazine*, Vol. 12, No. 5, pp. 21-27, October 2005.

[6] P. Todorova, S. Olariu, H. N. Nguyen, "A Two-Cell-Lookahead Call Admission and Handoff Management Scheme for Multimedia LEO Satellite Networks", in *Proc. of the* 36th *Hawaii International Conference on System Sciences* (HICSS-36), Big Island, Hawaii, 2003.

[7] ETSI, "Digital Video Broadcasting (DVB); Interaction channel for satellite distribution system", EN 301 790, 2005.

[8] ETSI, "Digital Video Broadcasting (DVB); Interaction channel for satellite distribution system; Guidelines for the use of EN 301 790", TR 101 790, 2006.

[9] A. Morell, G. Seco-Granados, M. A. Vázquez-Castro, "Joint Time Slot Optimization and Fair Bandwidth Allocation for DVB-RCS Systems", in *Proc. of the IEEE GLOBECOM 2006*, San Francisco, California, USA, November 27 - December 1, 2006.

[10] J. Neale, A. K. Mohsen, "Impact of CF-DAMA on TCP via Satellite Performance", in *Proc. of the Global Telecommunications Conference 2001* (GLOBECOM '01), Vol. 4, pp. 2687-2691, November 2001.

[11] L. Chisci, R. Fantacci, T. Pecorella, "Predictive Bandwidth Control for GEO Satellite Networks", in *Proc. of IEEE International Conference on Communications* (ICC 2004), Paris, France, pp. 3958-3962, June 2004.

[12] L. Chisci, R. Fantacci, T. Pecorella, "Strategies for Distributed Bandwidth Control in Communication Networks with High Bandwidth Delay Product", in *Proc. of the* 43rd *IEEE Conference on Decision and Control*, Atlantis, Paradise Island, Bahamas, Vol. 4, pp. 3732-3737, December 2004.

[13] L. Chisci, T. Pecorella, R. Fantacci, "Dynamic Bandwidth Allocation in GEO Satellite Networks: a Predictive Control Approach", *Control Engineering Practice*, Vol. 14, No. 9, pp. 1057-1067, September 2006.
[14] K. Nichols, S. Blake, F. Baker, D. Black, "Definition of the Differentiated Services Field (DS Field) in the IPv4 and IPv6 Headers", IETF RFC 2474, Dec. 1998.
[15] S. Blake, D. Black, M. Carlson, E. Davies, Z. Wang, W. Weiss, "An Architecture for Differentiated Services", IETF RFC 2475, December 1998.
[16] B. Davie, A. Charny, J. C. R. Bennett, K. Benson, J. Y. Le Boudec, W. Courtney, S. Davari, V. Firoiu, D. Stiliadis, "An Expedited Forwarding PHB (Per-Hop Behavior)", IETF RFC 3246, March 2002.
[17] D. Grossman, "New Terminology and Clarifications for Diffserv", IETF RFC 3260, April 2002.
[18] S. Karapantazis, P. Todorova, F. N. Pavlidou, "On Bandwidth and Inter-Satellite Handover Management in Multimedia LEO Satellite Systems", *Advanced Satellite Mobile Systems Conference 2006* (ASMS 2006), Herrsching am Ammersee, Germany, May 29-31, 2006.
[19] N. Celandroni, F. Davoli, E. Ferro, "Static and Dynamic Resource Allocation in a Multiservice Satellite Network with Fading", *International Journal of Satellite Communications and Networking, Special Issue on Satellite IP Quality of Service*, Vol. 21, No. 4-5, pp. 469-487, July-October 2003.
[20] N. Celandroni, F. Davoli, E. Ferro, A. Gotta, "Adaptive Cross-layer Bandwidth Allocation in a Rain-faded Satellite Environment", *International Journal of Communication Systems*, Vol. 19. No. 5, pp. 509-530, June 2006.
[21] F. Davoli, M. Marchese, M. Mongelli, "Optimal Resource Allocation in Satellite Networks: Certainty Equivalent Approach Versus Sensitivity Estimation Algorithms", *International Journal of Communication Systems*, Vol. 18, No. 1, pp. 3-36, February 2005.
[22] F. Davoli, M. Marchese, M. Mongelli, "Discrete Stochastic Programming by Infinitesimal Perturbation Analysis: the case of Resource Allocation in Satellite Networks with Fading", *IEEE Transactions on Wireless Communications*, Vol. 5, No. 9, pp. 2312-2316, September 2006.
[23] M. Baglietto, F. Davoli, M. Marchese, M. Mongelli, "Neural Approximation of Open-Loop Feedback Rate Control in Satellite Networks", *IEEE Transactions on Neural Networks*, Vol. 16, No. 5, pp. 1195-1211, September 2005.
[24] N. Celandroni, F. Davoli, E. Ferro, A. Gotta, "Adaptive Bandwidth Partitioning Among TCP Elephant Connections Over Multiple Rain-Faded Satellite Channels", in *Proc. of the 3rd Internat. Workshop on QoS in Multiservice IP Networks*, Catania, Italy, Feb. 2005; in *Lecture Notes in Computer Science*, 3375, Springer-Verlag, Berlin, pp. 559-573, 2005.
[25] N. Celandroni, F. Davoli, E. Ferro, A. Gotta, "Long-Lived TCP Connections via Satellite: Cross-Layer Bandwidth Allocation, Pricing and Adaptive Control", *IEEE/ACM Transactions on Networking*, Vol. 14, No. 5, pp. 1019-1030, October 2006.
[26] N. Celandroni, F. Davoli, E. Ferro, A. Gotta, "An Overview of Some Techniques for Cross-Layer Bandwidth Management in Multi-Service Satellite IP Networks", in *Proc. of Workshop on "Advances in Satellite Communications: New Services and Systems"*, IEEE GLOBECOM '05, St. Louis, MO, pp. WO4:4.1-WO4:4.6, November/December 2005.

[27] N. Celandroni, F. Davoli, E. Ferro, A. Gotta, "Networking with Multi-Service GEO Satellites: Cross-Layer Approaches for Bandwidth Allocation", *International Journal of Satellite Communications and Networking*, Vol. 24. No. 5, pp. 387-403, September/October 2006.

[28] Web site with URL: http://www.eutelsat.com/satellites/13ehb6.html.

[29] M. A. Vázquez Castro, M. Ruggiano, L. S. Ronga, M. Werner, "Uplink Capacity Limits for DVB-RCS Systems with Dynamic Framing and Adaptive Coding", in *Proc. of AIAA&Ka Band Joint Conference*, Rome 2005.

[30] K. D. Lee, Y. H. Cho, S. J. Lee, H. J. Lee, "Optimal Design of Superframe Pattern for DVB-RCS Return Link", *ETRI Journal*, Vol. 24, No. 3, pp. 251-254, June 2002.

[31] K. D. Lee, K. N. Chang, "A Real-Time Algorithm for Timeslot Assignment in Multirate Return Channels of Interactive Satellite Multimedia Networks", *International J. Select. Areas Communication*, Vol. 22, No. 3, pp. 518-528, April 2004.

[32] ETSI, "Satellite Earth Stations and Systems (SES); Broadband Satellite Multimedia (BSM) Services and Architectures: QoS Functional Architecture", TS 102 462, December 2005.

[33] A. Girard, C. Rosenberg, M. Khemiri, "Fairness and Aggregation: A Primal Decomposition Study", *Networking 2000, Lecture Notes in Computer Science 1815*, Springer-Verlag, pp. 667-678, May 2000.

[34] L. Boyd and S. Vandenberghe. *Convex optimization.* Cambridge University Press, 2003.

[35] R. Jain, D. Chiu, W. Hawe, "A Quantitative Measure of Fairness and Discrimination for Resource Allocation in Shared Systems", *Tech. Rep. DEC TR-301, Digital Equipment Corp.*, September 1984.

[36] S. Cho, "Adaptive Dynamic Channel Allocation Scheme for Spotbeam Handover in LEO Satellite Networks", in *Proc. of IEEE Vehicular Technology Conference 2000* (Fall VTC 2000), Boston, MA, pp. 1925-1929, September 2000.

[37] I. Mertzanis, R. Tafazolli, B. G. Evans, "Connection Admission Control Strategy and Routing Considerations in Multimedia (Non-GEO) Satellite Networks", in *Proc. of IEEE Vehicular Technology Conference 1997* (VTC Spring 1997), Phoenix, AZ, pp. 431-435, May 1997.

[38] M. El-Kadi, S. Olariu, P. Todorova, "Predictive Resource Allocation in Multimedia Satellite Networks", in *Proc. of IEEE GLOBECOM 2001*, San Antonio, TX, Vol. 4, pp. 2735-2739, November 2001.

[39] P. Todorova, S. Olariu, H. N. Nguyen, "A Selective Look-Ahead Bandwidth Allocation Scheme for Reliable Handoff in Multimedia LEO Satellite Networks", in *Proc. of ECUMN'2002*, Colmar, France, pp. 36-43, April 2002.

Part III

Cross-Layer Techniques for Satellite-Independent Layers

8

RESOURCE MANAGEMENT AND NETWORK LAYER

Editors: Ulla Birnbacher[1], Wei Koong Chai[2]

Contributors: Paolo Barsocchi[3], Ulla Birnbacher[1], Wei Koong Chai[2], Antonio Cuevas[4], Franco Davoli[5], Alberto Gotta[3], Vincenzo Mancuso[6], Mario Marchese[5], Maurizio Mongelli[5], José Ignacio Moreno Novella[4], Francesco Potortì[3], Orestis Tsigkas[7]

[1]TUG - Graz University of Technology, Austria

[2]UniS - Centre for Communication Systems Research, University of Surrey, UK

[3]CNR-ISTI - Research Area of Pisa, Italy

[4]UC3M - Universidad Carlos III de Madrid, Spain

[5]CNIT - University of Genoa, Italy

[6]UToV - University of Rome "Tor Vergata", Italy

[7]AUTh - Aristotle University of Thessaloniki, Greece

8.1 Introduction

The Internet protocols have become the worldwide standard for network and transport protocols and are increasingly used in satellite communication networks. Also traditional telecommunication and broadcast applications like VoIP and video streaming are transported over the Internet, although it does not support natively applications with tight QoS requirements. In satellite communication networks, further challenges arise, as bandwidth resources are limited and physical transmission time adds some more pressure on delay constraints. Since resources are limited, the efficient assignment of bandwidth to different data streams has always been an issue for satellite communications. However, supporting QoS for IP-based applications results in additional requirements for resource allocation. In order to provide QoS for applications, several layers of the protocol stack of a satellite communication system will need to be adapted or have to interact with each other in some way. This Chapter will concentrate on different resource management schemes at the MAC layer (layer 2) for supporting IP QoS (layer 3).

This Chapter begins with an overview of the current IP QoS frameworks in Section 8.2. In Section 8.3, the discussion is focused on the interaction of layer 2 and layer 3 in satellite environments for the support of IP QoS. This Section ends with an example of implementation for a variant of one of the most popular IP QoS frameworks. The following Section 8.4 provides an in-depth work on achieving QoS requirements by a cross-layer approach over SI-SAP. Section 8.5 looks into another aspect of resource management: the QoS provisioning for terminals supporting dual network access (WiFi and satellite). Implicit cross-layer design methodology is used in Section 8.6 for switched Ethernet over LEO satellite networks. Finally, this Chapter is concluded in Section 8.7. In the studies carried out in this Chapter, Scenario 2 (i.e., GEO-based DVB-S/-RCS systems; see Chapter 1, Section 1.4) has been adopted, except for the considerations made in Section 8.6, where Scenario 3 (i.e., LEO satellite) has been considered.

8.2 Overview IP QoS framework

In order to support the emerging Internet QoS, some QoS frameworks have been proposed. These service models and mechanisms evolve the IP architecture to support new service definitions that allow preferential or differentiated treatment to be provided to certain traffic types. *Integrated Services* and *Differentiated Services* have already been introduced in Section 3.3, but are discussed below in more detail with satellite networks in mind, including *Multiprotocol Label Switching* (MPLS).

8.2.1 Integrated services

The *Integrated Services* (IntServ) model [1] requires resources, such as bandwidth and buffers, to be reserved *a priori* for a given traffic flow to ensure that the QoS requested by this traffic flow is fulfilled. The IntServ model includes additional components beyond those used in the best-effort model such as packet classifiers, packet schedulers, admission control and signaling. A packet classifier is used to identify flows that have to receive a certain level of service. A packet scheduler manages the service provided to different packet flows to ensure that QoS commitments are met. Admission control is used to determine whether a router has the necessary resources to accept a new flow.

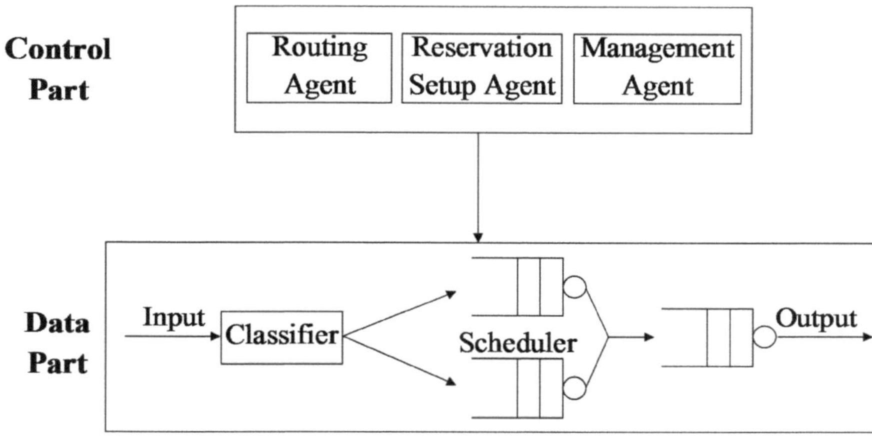

Fig. 8.1: Implementation reference model for routers with IntServ [2].

A notable feature of the IntServ model is that it requires explicit signaling of QoS requirements from end-systems to routers. The *Resource Reservation Protocol* (RSVP) [3] performs this signaling function and is a critical component of IntServ. RSVP is a soft state signaling protocol. It supports receiver-initiated establishment of resource reservations for both multicast and unicast flows. Recently, RSVP has been modified and extended in several ways to reserve resources for aggregation of flows, to set up MPLS explicit label switched paths with QoS requirements, and to perform other signaling functions within the Internet.

Two services have been defined under the IntServ model: *guaranteed service* [4] and *controlled-load service* [5]. The guaranteed service provides a firm quantitative bound on the end-to-end packet delay for a flow. This is accomplished by controlling the queuing delay on network elements along the data flow path. The guaranteed service model does not, however, provide bounds on jitter (inter-arrival times between consecutive packets). The controlled-load service can be used for adaptive applications that can tolerate some delay, but are sensitive to traffic overload conditions. This type of application typically operates satisfactorily when the network is lightly loaded, but its performance degrades significantly when the network is heavily loaded. Controlled-load service, therefore, has been designed to provide approximately the same service as best-effort service in a lightly loaded network regardless of actual network conditions. Controlled-load service is described qualitatively in that no target values of delay or loss are specified.

The IntServ architecture represents a fundamental change to the current Internet architecture, which is based on the concept that all flow-related state information should be in the end-systems. The main problem of the IntServ model is scalability, especially in large public IP networks, which may potentially have millions of active micro-flows concurrently in transit, since the amount of state information maintained by network elements tends to increase linearly with the number of micro-flows.

8.2.2 Differentiated services

One of the primary motivations for *Differentiated Services* (DiffServ) [6] was to devise alternative mechanisms for service differentiation in the Internet that mitigate the scalability issues encountered with the IntServ model. Scalable mechanisms are deployed within the DiffServ framework for the categorization of traffic flows into behavior aggregates, allowing each behavior aggregate to be treated differently, especially when there is shortage of resources such as link bandwidth and buffer space.

A DiffServ field in the IPv4 header has been defined. Such field consists of six bits of the part of the IP header, formerly known as TOS octet, and it is used to indicate the forwarding treatment that a packet should receive at a node. Within the DiffServ framework, a number of *Per-Hop Behavior* (PHB) groups have been also standardized. Using the PHBs, several classes of services can be defined using different classification, policing, shaping, and scheduling rules.

Conceptually, a DiffServ domain consists of two types of routers, namely *core router* and *edge router*. Core router resides within the domain and is generally in charge of forwarding packets based on their respective *DiffServ Code Point* (DSCP). The edge router is located at the boundary of the network domain which will either further connect to another domain (inter-domain) or to end-users. It can be further categorized as ingress router which operates on traffic flowing into the domain and egress router which operates on traffic exiting the domain.

In order for an end-user to receive DiffServ from its *Internet Service Provider* (ISP), it may be necessary for the user to have a *Service Level Agreement* (SLA) with the ISP. An SLA may explicitly or implicitly specify a *Traffic Conditioning Agreement* (TCA), which defines classifier rules, as well as metering, marking, discarding, and shaping rules. Packets are classified, and possibly policed and shaped at the ingress routers of a DiffServ network according to SLAs.

When a packet traverses the boundary between different DiffServ domains, the DiffServ field of the packet may be re-marked according to existing agreements between the domains. DiffServ allows only a finite number of

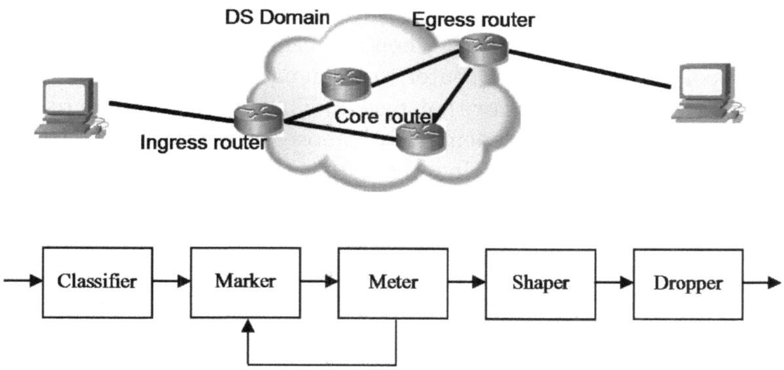

Fig. 8.2: DiffServ network; (top) DiffServ domain illustration; (bottom) logical view of DiffServ packet classifier and traffic conditioner.

service classes to be indicated by the DiffServ field.

The main advantage of the DiffServ approach relative to the IntServ model is scalability. Resources are allocated on a per-class basis and the amount of state information in the routers is proportional to the number of classes rather than to the number of application flows. A second advantage of the DiffServ approach is that sophisticated classification, marking, policing, and shaping operations are only needed at the boundary of the network.

The DiffServ control model essentially deals with traffic management issues on a per-hop basis and consists of a collection of micro-control mechanisms. Other *traffic engineering* capabilities, such as capacity management (including routing control), are also required in order to deliver acceptable QoS in DiffServ networks.

At the current stage, the DiffServ approach is still evolving. Two directions of its development can be categorized: namely, *absolute* DiffServ and *relative* DiffServ [7]. The absolute DiffServ approach is the more traditional approach detailed above. The newer and simpler approach is the relative DiffServ, whereby QoS assurances are provided relative to the ordering between several traffic or service classes rather than specifying the actual service level or quality of each class. This approach is lightweight in nature, since it minimizes computational cost as it does not require sophisticated mechanisms such as admission control and resource reservation. As such, recently, it has gain in popularity not only in terrestrial networks, but also in wireless [8],[9] and satellite systems [10].

8.2.3 Multiprotocol Label Switching (MPLS)

MPLS is an advanced forwarding scheme, which extends the Internet routing model and enhances packet forwarding and path control [11]. MPLS stands for *Multiprotocol Label Switching*; note that the word 'multiprotocol' is used, because these techniques are applicable to any network layer protocol.

In conventional IP forwarding, when a packet of a connectionless network layer protocol travels from one router to the next, each router makes an independent forwarding decision for that packet. That is, each router re-examines the packet's header and independently chooses a next hop for the packet, based on the results of the routing algorithm. Choosing the next hop can be thought of as the composition of two functions. The first function partitions the entire set of possible packets into a set of *Forwarding Equivalence Classes*; the second function maps each class to a next hop.

In the MPLS forwarding paradigm, the assignment of a particular packet to a given class is done just once, at the ingress to an MPLS domain, by *Label Switching Routers* (LSRs). As the forwarding decision is concerned, different packets that get mapped into the same class are indistinguishable and will follow the same path. The class to which the packet is assigned is encoded as a short fixed-length value known as a *label*. When a packet is forwarded to its next hop, the label is sent along with it; that is, packets are labeled before they are forwarded. At subsequent hops, there is no further analysis of the packet network layer header. Rather, the label is used as an index into a table, which specifies the next hop, and a new label. The old label is replaced by the new label, and the packet is forwarded to its next hop. Most commonly, a packet is assigned to a class based (completely or partially) on its destination IP address. However, the label never is an encoding of that address.

A *Label Switched Path* (LSP) is the path between an ingress LSR and an egress LSR through which a labeled packet traverses. The path of an explicit LSP is defined at the originating (ingress) node of the LSP. MPLS can use a signaling protocol such as RSVP or *Label Distribution Protocol* (LDP) to set up LSPs. MPLS is a very powerful technology for Internet traffic engineering because it supports explicit LSPs, which allow constraint-based routing to be implemented efficiently in IP networks.

8.3 Resource management for IP QoS

Resource management schemes at MAC layer (layer 2) are essential in supporting IP QoS (layer 3). The current IP QoS frameworks (i.e., IntServ and DiffServ) define several service classes to cater for users with different QoS requirements. The resource management scheme must be able to allocate dynamically the available resources in an IP-based satellite network to achieve the requirements of the defined service classes. This includes a mapping scheme between layer 3 and layer 2, dynamic bandwidth allocation and scheduling mechanisms.

In this Section, the specific scenario under consideration is a DiffServ satellite domain with DVB-RCS architecture for multimedia fixed unicast users. The choice of DiffServ is mainly in view of the problems of the IntServ framework, such as scalability and deployment. For DiffServ, in general, there are three types of PHBs being used: namely *Expedited Forwarding* (EF), *Assured Forwarding* (AF) and *Best Effort* (BE). EF PHB caters for low loss, low delay and low jitter services. The AF PHB consists of four AF classes, where each class is allocated with different amounts of buffer and bandwidth. Hence, each subscriber with a specific *Subscribed Information Rate* will receive assured performance for traffic within such rate while excess traffic may be lost depending on the current load of the AF class. Finally, the BE PHB is the same as the original best effort IP paradigm.

For the DVB-RCS architecture, there are four transmission capacity allocation schemes; namely *Continuous Rate Assignment* (CRA), *Rate Based Dynamic Capacity* (RBDC), *Volume Based Dynamic Capacity* (VBDC) and *Free Capacity Assignment* (FCA). For the description of these different resource allocation schemes, please refer to Chapter 1, sub-Section 1.4.3.

Before mapping the DiffServ PHBs to DVB-RCS resource allocation schemes, it is vital to note that the entire DiffServ domain is assumed to be properly dimensioned. This is because there is no one mapping scheme that can achieve high efficiency in all types of traffic mixture. A particular scheme, which performs well in one scenario, may perform poorly in another. The network management and dimensioning problem is not within the scope of this study.

Usually, EF PHB is used to transport non-delay tolerant application traffics such as VoIP and video conferencing. To achieve the stringent QoS requirements of this class of applications, the use of CRA in the MAC layer is a must. However, considering system efficiency, a minimal use of RBDC combined with CRA is plausible. The entire DiffServ domain has to be properly dimensioned as noted above. For example, if a very high traffic percentage is of the EF type, then the satellite bandwidth will be quickly consumed with all the slots being reserved with CRA, thus causing high blocking and drop rate. As for AF PHB, the combined use of RBDC and VBDC is proposed with RBDC as the main resource provider. Under low load, packets belonging to each class of AF will receive similar treatment. However, to differentiate between the AF classes, a different maximum RBDC value (i.e., maximum bit-rate that can be allocated with RBDC) can be defined so that the higher AF class will receive better treatment. If the request is higher than the maximum RBDC, the users can still request for VBDC resources. For BE traffic, the use of VBDC and FCA is proposed.

8.3.1 Relative DiffServ by MAC Scheduling

An alternative scenario on resource management schemes at MAC layer (layer 2) to support IP QoS (layer 3) is the work on attempting to realize *relative* service differentiation in a *Bandwidth on Demand* (BoD) satellite IP network. The *Proportional Differentiated Service* (PDS) [7] model is one of the most recent developments of DiffServ in the direction of *relative* service differentiation. It strives to strike a balance between the strict QoS guarantee of IntServ and the scalability of DiffServ. Similarly to DiffServ, PDS segregates traffics into a finite number of service classes. However, it does not provide them with absolute QoS guarantees. Instead, it controls the performance gap between each pair of service classes, i.e., *quantitative* relative differentiation amongst the supported classes.

Formally, the PDS model requires

$$\frac{\sigma_i}{\sigma_j} = \frac{r_i}{r_j}; \quad \forall i, j \in \{1 \ldots N\}. \tag{8.1}$$

where each class is associated with a *Differentiation Parameter* (DP), r_i, and σ_i is the performance metric of interest for class i, e.g., throughput, packet loss or queuing delay. N is the total number of supported service classes in the network.

In this Section, classes are numbered in decreasing priority order, i.e., the lower the class index, the better the service provided to it. All DPs are normalized with reference to the highest priority class (= 1):

$$0 < r_N < r_{N-1} < ... < r_2 < r_1 = 1$$

This Section demonstrates how such a model can be realized in an IP-based broadband multimedia BoD GEO satellite network with resource allocation mechanisms analogous to the DVB-RCS system standard [12].

Figure 8.3 [10] illustrates the main nodes of the network architecture:

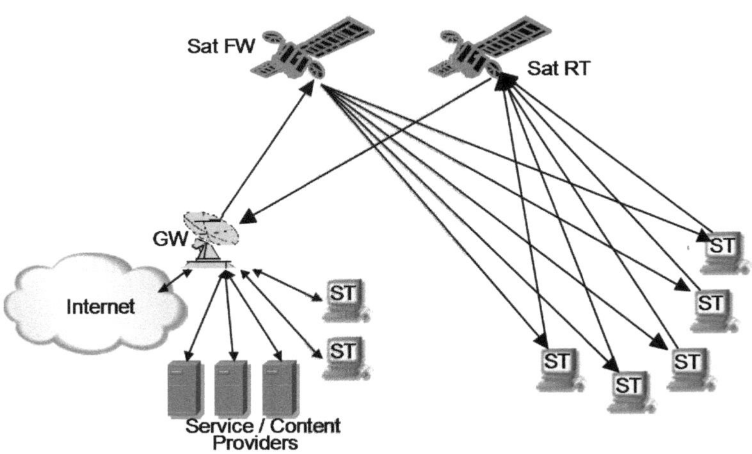

Fig. 8.3: Reference satellite system, resembling the DVB-RCS architecture. See reference [10]. Copyright ©2005 IEEE.

- Satellite(s): The satellite is assumed to be equipped with *On-Board Processor* (OBP) and the scheduler is located on-board.
- *Traffic Gateway* (GW): In line with the DVB-RCS definition, GWs are included to provide interactive services to networks (e.g., Internet) and service providers (e.g., databases, interactive games, etc.).
- *Satellite Terminal* (ST): STs represent the users. They may represent one (residential) or more users (collective).

Time Division Multiple Access (TDMA) is used for the forward path whereas on the return path, *Multi Frequency - TDMA* (MF-TDMA) is assumed. In an MF-TDMA frame, the basic unit of the link capacity is the *Time Slot* (TS) with multiple TSs grouped in TDMA frames along several frequency carriers. In this Section, fixed MF-TDMA frame is considered whereby the bandwidth and duration of successive TSs is static. For more details on MF-TDMA characteristics, please refer to Chapter 1.

The BoD scheme used in this Section is derived from [13]. It is a cyclic procedure between two stages: the *resource request estimation* stage and the *resource allocation* stage. It involves the BoD entity located at the ST and BoD scheduler located onboard the satellite. The BoD entity handles all packets of the same class which are stored in the same queue, i.e., there will be x BoD entities in an ST if this ST supports x classes. In the *resource request estimation* stage, the BoD entities (i.e., STs) periodically compute and send *Slot Requests* (SRs) to the BoD scheduler, when there are new packet arrivals at their queues. In the *resource allocation* stage, upon reception of SRs, the BoD scheduler allocates TSs to each requesting BoD entity based on a certain scheduling discipline and policies defined by the network operator. It then constructs and broadcasts the *Terminal Burst Time Plan* (TBTP) that contains all the resource allocation information to the BoD entities. Figure 8.4 [10] gives the BoD timing diagram, which also describes the basic tasks involved.

Due to the unique characteristics of satellite networks, the realization of such framework is very different from those solutions provided for terrestrial and wireless systems. For terrestrial wired networks, the scheduler only needs to schedule the departure of each contending packet *locally* within a router. In wireless and satellite domain, the access to the transmission medium is often controlled in a distributed manner by a MAC protocol. Hence, packets from one node may contend with packets from other nodes. This leads to the consideration of using layer 2 scheduling to realize the model instead of purely depending on layer 3. Based on the layer 3 QoS classes, the MAC layer scheduler will decide how best to schedule the packets in order to achieve the QoS required.

Moreover, there are several fundamental architectural and environmental differences between terrestrial wireless networks and satellite networks supporting dynamic bandwidth allocation mechanisms. Firstly, for a BoD-based satellite architecture, resource has to be requested by the STs before they can

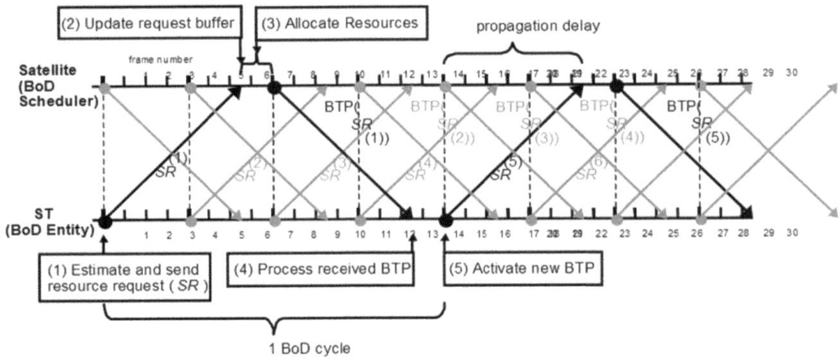

Fig. 8.4: BoD timing diagram. See reference [10]. Copyright ©2005 IEEE.

make use of it, so that the scheduler ends up scheduling requests for resource rather than packets. Secondly, there is a non-negligible propagation delay between the STs and the scheduler that may, depending on the access control algorithm, inflate the waiting time of a packet in the ST queue. The impact of this semi-constant delay has to be taken into account by the scheduler in providing relative service differentiation.

The *Satellite Waiting Time Priority* (SWTP) scheduler is a satellite adaptation of the *Waiting Time Priority* scheduler [7], proposed by Kleinrock in [14]. SWTP schedules SRs from BoD entities rather than individual packets. SWTP has been shown to be able to provide proportional queuing delay to several classes of MAC frames in the context of BoD environment. Its main elements are as follow.

1. *Resource request.* Formally, if Q_i^m is the set of newly arrived packets at the i-th queue of BoD entity m, i.e., packets that came within the last resource allocation period, q the set cardinality, and τ_j the arrival time of packet j, $1 \leq j \leq q$, indexed in increasing order of arrival times, then the BoD entity m computes at time t the SR timestamp ts_i^m, according to the arrival time of the last packet that arrived in the queue during the last resource allocation period, namely: $ts_i^m = t - \tau_q$.
2. *Resource allocation*: the BoD scheduler computes the priority of each SR. The priority $P_i^m(k)$, assigned to SR_i^m in the k-th resource allocation period is

$$P_i^m(k) = r_i \cdot \left(w_i^{SR}(k) + \alpha\right) \tag{8.2}$$

where α accounts for the propagation delay of TBTP and the processing delay of BoD entities, while $w_i^{SR}(k) = t - ts_i^m$ and ts_i^m is the timestamp information encoded in each SR. Finally, r_i denotes here the *Delay*

Differentiation Parameter (DDP): each one of the N MAC classes is attached with a specific r_i, $1 \leq i \leq N$. At each allocation period, the SWTP allocates TSs by considering requests in decreasing priority order: requests are fully satisfied as long as they do not exceed the available capacity. All unsatisfied requests will be buffered for the next allocation period. At the next allocation period, the priorities of the buffered SRs will be recalculated to account for the additional waiting time of SRs at the scheduler.

The setup of the simulations is as follow. The capacities for all links are configured to be 2048 kbit/s. Unless explicitly stated otherwise, the network is set to have DDPs: 1, 1/2, 1/4, 1/8. By this setting where the differentiation is exactly half of the next adjacent class, the ideal *performance ratio* according to the PDS model will be 0.5. The IP packet size used is 500 bytes, while MAC frames are of 48 bytes with additional 5 bytes due to header (ATM case).

Figure 8.5 shows the queuing delay for each service class, while Figure 8.6 presents the corresponding delay ratios under constant bit-rate traffic [10]. The ideal value for the ratios is 0.5 for all cases. From the plotted results, it is clear that the SWTP scheduler can indeed emulate closely the PDS model. Since the PDS model requires that the 'spacing' between any two service classes strictly adheres to the ratio of the DDPs for the specified service classes, the scheduler should not be dependent on the traffic distribution between service classes. Figure 8.7 [10] shows the result of this test at a utilization of 95%: the achieved ratios are very near to the ideal value of 0.5.

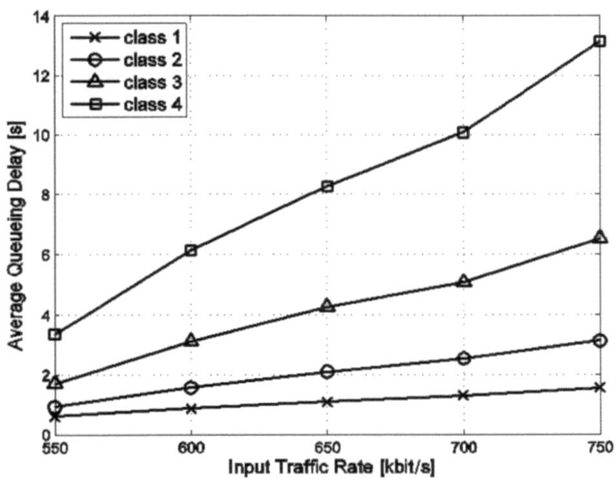

Fig. 8.5: Queuing delay of different service classes following the specified spacing of the model. See reference [10]. Copyright ©2005 IEEE.

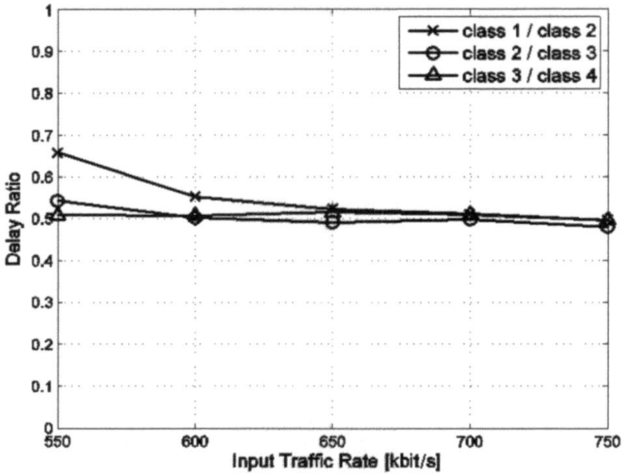

Fig. 8.6: Delay ratios achieved that are close to the ideal delay ratios. See reference [10]. Copyright ©2005 IEEE.

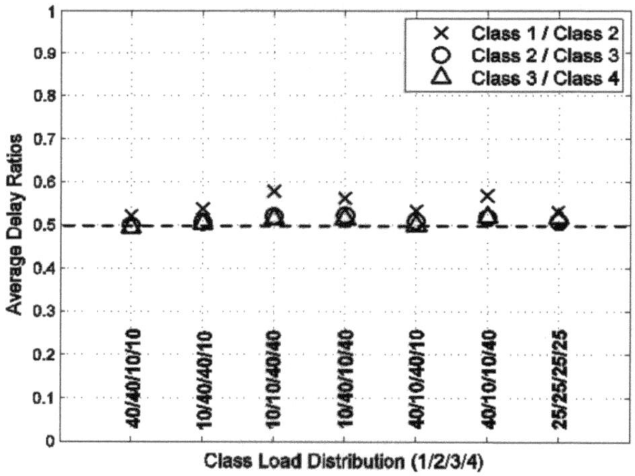

Fig. 8.7: SWTP emulating the PDS in different load distributions with all values achieved close to the ideal value. See reference [10]. Copyright ©2005 IEEE.

To illustrate the capability of SWTP in accurately controlling the spacing between different service classes, three sets of DDPs have been defined below.

- Set A: [1, 1/2, 1/4, 1/8]
- Set B: [1, 1/2, 1/3, 1/4]
- Set C: [1, 1/4, 1/5, 1/6].

Simulations with utilization of 95% have been conducted based on these DDP sets and the results given in Figure 8.8 [10] show the *normalized ratios* of all the three cases, where the normalized ratios are defined as the achieved performance ratios divided by the respective ideal ratios. With the ideal value as 1.0, it can be concluded that SWTP is indeed able to control the class spacing. However, due to the long propagation delay, the spacing between the highest and lowest DDP should not be too large to ensure reasonable delay for the lowest class.

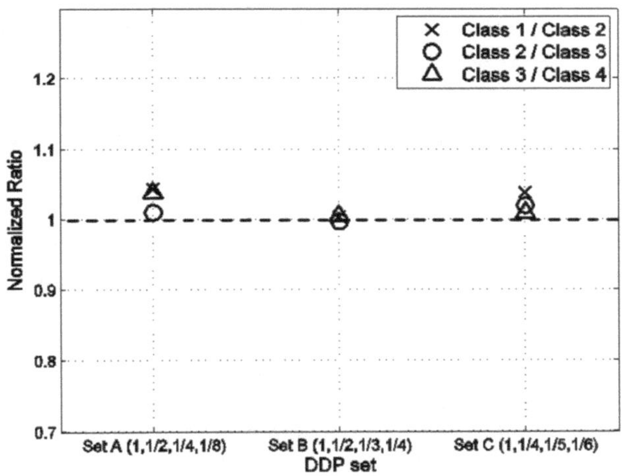

Fig. 8.8: SWTP with 3 sets of DDPs: all normalized delay ratios are close to the ideal value. See reference [10]. Copyright ©2005 IEEE.

The behavior of SWTP in short timescale is investigated to ensure that the predictability requirement of the PDS model is satisfied. Figure 8.9 [10] shows the individual packet delays upon departure in a four-class scenario for a period of 100 ms. The graph shows that SWTP can consistently provide the appropriate spacing for the service classes.

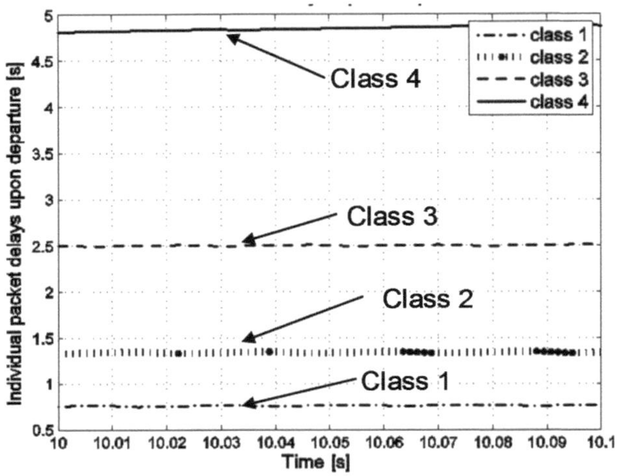

Fig. 8.9: Short time scale behavior of SWTP showing its predictability property. See reference [10]. Copyright ©2005 IEEE.

8.4 QoS mapping over satellite-independent service access point

In what follows, we are specifically concerned with the cross-layer interaction between the network and the MAC layer, in order to preserve QoS requirements, or, in more precise terms, to operate a mapping between the QoS mechanisms operating at the two layers. Within a more general view, with reference to the ETSI *Broadband Satellite Multimedia* (BSM) protocol architecture [15],[16], we might refer to the inter-working between the *Satellite-Independent* (SI) and the *Satellite-Dependent* (SD) architectural components at the SI-SAP (*Satellite-Independent - Service Access Point*), by taking into account both the change in encapsulation format and the traffic aggregation in the passage from SI to SD queues. Note that the ETSI BSM architecture has been described in Chapter 1, Section 1.5.

Cross-layer RRM problems, involving network and MAC layers, have been extensively considered in [17]-[19]. Reference [20] also provides guidelines and architectural details. In particular, in [17]-[19] *Dynamic Bandwidth Allocation* (DBA) is applied by computing bandwidth requests for each Earth station's DiffServ queue, which are passed to a centralized scheduler, typically residing in a *Master Control Station* (MCS). The latter assigns the bandwidth proportionally to the requests received; the remaining capacity is assigned on a free basis. Such scheme has been called *Combined Free/Demand Assignment Multiple Access* (CF/DAMA).

In a similar context, the problem of QoS mapping between adjacent layers has been recently treated in [21]-[23]. Rather than considering specifically the network and the MAC layers, the problem is posed in the more general ETSI BSM scenario mentioned above. In the presence of IP DiffServ queues at the higher layer, the problem consists in dynamically assigning the bandwidth (service rate) to each SD queue, so that the performance required at the IP layer is guaranteed. By considering a fluid model and the loss volume as the performance indicator of interest, the *Infinitesimal Perturbation Analysis* (IPA) technique of Cassandras *et al.* [24] (already mentioned in Chapter 7 in a different scenario) is applied in order to maintain on-line the equalization between the loss volumes at the two different layers (by assuming that the resource allocation at the SI layer is capable of satisfying the requirements). In doing so, both traffic and fading variations are taken into account. Further details on the application of the IPA technique are provided in sub-Section 8.4.2.

8.4.1 Model-based techniques for QoS mapping and support

Earth stations use reservation mechanisms (bandwidth requests) to transmit their traffic flows (voice or MPEG video, bandwidth reserved for DiffServ aggregates, MPLS pipes, etc.), which may be carried with priority at the satellite link level within some specific DVB service classes. The control process works upon requests for bandwidth allocation, which can be satisfied within a *Round Trip Time* (RTT) for the request to reach the scheduler and the response to be received (referred to as *DBA cycle time* in [17]). Hence, whenever traffic flows are characterized by a relatively low burstiness (e.g., the peak-to-average ratio of their rates is close to 1), simple DAMA schemes (e.g., VBDC) can be employed to manage the traffic of Earth stations [19]. The bandwidth allocation can be controlled in this case by means of CAC functions. When burstiness is higher, DBA is applied by computing bandwidth requests (on the basis of a model) for each Earth station's DiffServ queue, which are passed to a centralized scheduler that assigns the bandwidth proportionally to the requests received; the remaining capacity is assigned on a free basis, according to CF/DAMA. Various traffic models have been used to represent the burst-level behavior of real-time *Variable Bit Rate* (VBR) traffic; among them, we can consider voice with silence detection and VBR-encoded MPEG video. In this case, two control functionalities at different time scales should be employed, namely, CAC at the call level and DBA at the burst level, to guarantee at the same time both a specified degree of QoS and an efficient bandwidth utilization.

In [17], models capturing both *Short Range Dependent* (SRD) and *Long Range Dependent* (LRD) behaviors have been used to represent the arrival processes of traffic aggregates to the *User Terminal* (UT) IP queues in a DiffServ scenario. They are based on *Markov-Modulated Poisson Processes* (MMPP) and *Pareto-Modulated Poisson Processes* (PMPP), giving rise to MMPP/G/1 and PMPP/G/1 queuing systems, respectively. The adopted service-dependent QoS metric is the probability that the length of each service queue exceeds a given threshold; we consider the constraint that this probability must be kept below a specified value, beyond which the station is considered in outage. The scheduling of the MAC queues must be such that this constraint is fulfilled for the IP-level queues (i.e., those corresponding to EF, AF and BE services within a given Earth station). No fading variations are taken into account, but, as noted in [17], the effect of fade countermeasures might be included as a reduction in the available uplink bandwidth. Note that if the state of the sources can be assumed to change more slowly than the DBA cycle time, within which the allocated bandwidth remains constant, the queuing behavior in these intervals can be approximated by a much simpler M/D/1 system.

8.4.2 A measurement-based approach for QoS mapping and support

The work done in [21]-[23] takes a different look at the QoS mapping and support problem, by disregarding the use of models, but rather relying on measurement-based optimization techniques. This framework is that of ETSI-BSM [15],[16] (let us consider for example the RBDC scheme). In such a context, two basic facts are taken into account: the change of *information unit* (e.g., from IP to IP-over-DVB) and the heterogeneous traffic aggregation, since, for hardware implementation constraints, the number of available SD queues can be lower than that of SI queues (see also Chapter 1, sub-Section 1.4.3). Figure 8.10, taken from [21], reports and example.

The problem is then how much bandwidth must be assigned to each SD queue, so that the SI IP-based SLA (i.e., the performance expected) is guaranteed. In doing this, the effect of fading on the satellite channel is also taken into account. As in other works (see, e.g., [25]), when the fade countermeasure in use is modulation and coding rate adaptation, the effect of fading is modeled as a reduction in the bandwidth (i.e., the service rate) effectively 'seen' by a layer 2 traffic buffer.

IP *Packet Loss Probability* (PLP) is one of the SLA performance metrics considered in [23] (the other being IP *Packet Average Delay*). However, we concentrate here on PLP. The mathematical framework is based on *Stochastic Fluid Models* (SFM) of the SI-SAP traffic buffers [24],[26]. N SI queues and, without loss of generality, one single SD queue are considered for the analytical formulation (Figure 8.11).

Let $\alpha_i^{SI}(t)$ be the input process entering the i-th traffic buffer at the SI layer at time t, $i = 1, \ldots, N$. After entering one single buffer [with service rate $\theta_i^{SI}(t)$] at the SI layer, each $\alpha_i^{SI}(t)$ process is conveyed to a single SD buffer [whose service rate is $\theta^{SD}(t)$] at the SD layer after a format change. ${}^{i}L_V^{SI}\left[\alpha_i^{SI}(t), \theta_i^{SI}(t)\right]$ denotes the loss volume of the i-th IP buffer according to the bandwidth allocation $\theta_i^{SI}(t)$.

Let $\alpha^{SD}(t)$ be the input process of the buffer at the SD layer at time t. The $\alpha^{SD}(t)$ process derives from the output processes of the SI buffers.

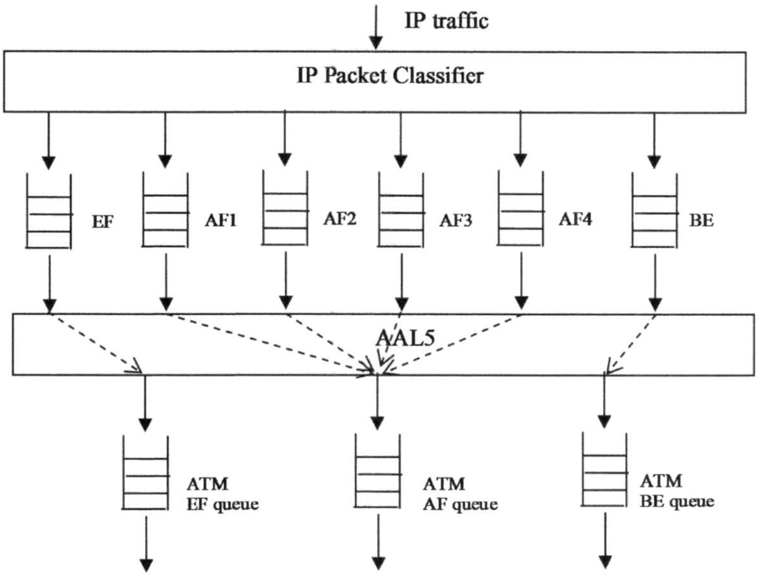

Fig. 8.10: Queuing at the SI-SAP interface: satellite-independent (DiffServ) over satellite-dependent layer (ATM). See reference [21]. Copyright ©2005 IEEE.

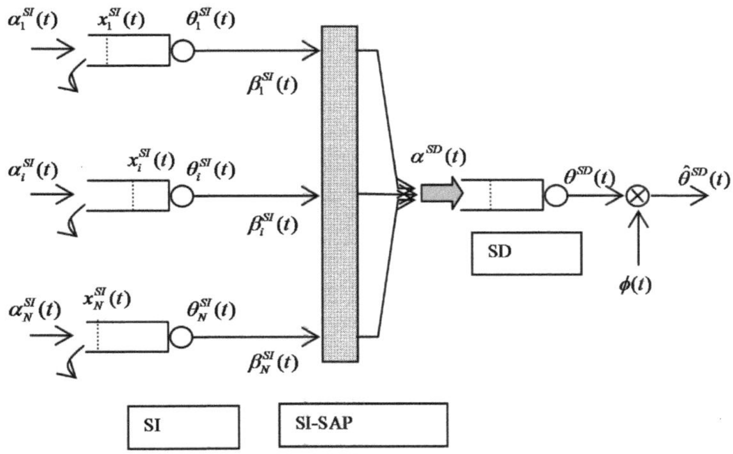

Fig. 8.11: Stochastic processes and buffer set for the envisaged SI-SAP queuing model.

The loss volume of the i-th traffic class within the SD buffer is indicated by ${}^{i}L_{V}^{SD}\left[\alpha^{SD}(t), \theta^{SD}(t)\cdot\phi(t)\right]$. It is a function of the following elements: the SD input process $\alpha^{SD}(t)$, the fading process $\phi(t)$ and the SD bandwidth allocation $\theta^{SD}(t)$. It is remarkable that ${}^{i}L_{V}^{SD}(\cdot)$ cannot be obtained in closed-from.

The problem reveals to be the equalization of the QoS measured at the different layers of the protocol stack (i.e., SI and SD):

***QoS Mapping Optimization* (QoSMO) Problem**: find the optimal bandwidth allocation, ${}^{Opt}\theta^{SD}(t)$, so that the cost function $J(\cdot, \theta^{SD}(t))$ is minimized:

$$ {}^{Opt}\theta^{SD}(t) = \arg\min_{\theta^{SD}(t)} J(\cdot, \theta^{SD}(t)); J(\cdot, \theta^{SD}(t)) = \mathop{E}_{\omega\in\Theta} L_{\Delta V}(\cdot, \theta^{SD}(t)) \tag{8.3} $$

$$ L_{\Delta V}(\cdot, \theta^{SD}(t)) = \sum_{i=1}^{N}\left[{}^{i}L_{V}^{SI}(\alpha_i^{SI}(t), \theta_i^{SI}(t)) - {}^{i}L_{V}^{SD}(\alpha^{SD}(t), \theta^{SD}(t)\cdot\phi(t))\right]^2 . $$

In (8.3), ω denotes a sample path of the system, i.e., a realization of the stochastic processes involved in the problem (coming from quantities $\phi(t)$, $\alpha_i^{SI}(t)$, $i = 1, \dots, N$, $\alpha^{SD}(t)$). Note that the cost function [see the second row in (8.3)] weighs the sum of the quadratic deviations of the loss volumes at the two layers, over all traffic classes associated with SI queues.

This QoSMO problem is very complex to be solved. Two approaches are considered below; one is based on the *equivalent bandwidth* concept and the other is based on IPA.

Traditionally, *equivalent bandwidth* techniques are based on the statistical characterization of the traffic generated by users' applications. The only simply applicable statistics, useful for the SD rate provision, are the mean (m) and the standard deviation (σ) of the α^{SD} process. Hence, a popular equivalent bandwidth technique, actually applicable in this context, is ruled by (8.4) below [27]. Let us consider the following notations: $k = 1, 2, \dots$ the time instants of the SD rate reallocations, $m_{\alpha^{SD}}(k)$ and $\sigma_{\alpha^{SD}}(k)$ the mean and the standard deviation, respectively, of the SD input process measured over the time interval $[k, k+1]$. Therefore, the bandwidth provision $\theta^{SD}(k+1)$ at the SD layer, assigned for the time interval $[k+1, k+2]$, may be computed as:

$$ \theta^{SD}(k+1) = m_{\alpha^{SD}}(k) + a\cdot\sigma_{\alpha^{SD}}(k) \tag{8.4} $$

where $a = \sqrt{-2\ln(\varepsilon) - \ln(2\pi)}$ and ε represents the upper bound on the allowed PLP. Such allocation method is called *Equivalent Bandwidth* approach (EqB) in what follows.

In [23], another measurement-based equivalent bandwidth algorithm is proposed that can face:

- Heterogeneity of the QoS requests in the aggregated trunk;
- Change of encapsulation format;
- Fading counteraction;
- No knowledge of SD input process's statistical properties;
- No knowledge of SD buffer size.

To match these requirements, the derivative of the cost function $L_{\Delta V}(\cdot)$ is used:

$$\frac{\partial L_{\Delta V}(\cdot,\theta^{SD})}{\partial\theta^{SD}} = 2\cdot\sum_{i=1}^{N}\frac{\partial^i L_V^{SD}(\theta^{SD})}{\partial\theta^{SD}}\left[{}^iL_V^{SD}(\theta^{SD}) - {}^iL_V^{SI}(\theta_i^{SI})\right]. \quad (8.5)$$

Using IPA (see, e.g., [24],[26] and references therein), each $\frac{\partial^i L_V^{SD}(\theta^{SD})}{\partial\theta^{SD}}$ component can be obtained in real-time only on the basis of some traffic samples acquired during the system evolution. Let $[k,\, k{+}1]$ be the time interval between two consecutive SD bandwidth reallocations. The interval of time in which the buffer is not empty are defined as busy periods. The derivative estimation is computed at the end of the decision epoch $[k,\, k{+}1]$ as follows:

$$\left.\frac{\partial^i L_V^{SD}\left(\theta^{SD}\right)}{\partial\theta^{SD}}\right|_{\hat\theta^{SD}(k)} = \phi\left(k\right)\cdot\sum_{\varsigma=1}^{N_k^i}\left.\frac{\partial^i L_{k,\varsigma}^{SD}\left(\theta^{SD}\right)}{\partial\theta^{SD}}\right|_{\hat\theta^{SD}(k)} \quad (8.6)$$

$$\left.\frac{\partial^i L_{k,\varsigma}^{SD}(\theta^{SD})}{\partial\theta^{SD}}\right|_{\hat\theta^{SD}(k)} = -\left({}^i\nu_\varsigma^k\left(\hat\theta^{SD}\left(k\right)\right) - {}^i\xi_\varsigma^k\left(\hat\theta^{SD}\left(k\right)\right)\right) \quad (8.7)$$

where ${}^iL_{k,\varsigma}^{SD}(\theta^{SD})$ is the ς-th contribution to the SD loss volume of the i-th traffic class for each busy period B_k^ς within the decision interval $[k,\, k{+}1]$; ξ_ς^k is the starting point of B_k^ς; ν_ς^k is the instant of time when the last loss occurs during B_k^ς; N_k^i is the number of busy periods within the interval $[k,\, k{+}1]$ for service class i. It must be noted that $\hat\theta^{SD}(k)$ represents the SD bandwidth reduction due to fading within the time interval $[k,\, k{+}1]$ (i.e., $\hat\theta^{SD}(k) = \theta^{SD}(k)\cdot\phi(k)$, where $\phi(k)$ represents the bandwidth reduction seen at the SD layer, due to redundancy applied at the physical layer to counteract channel degradation).

The proposed optimization algorithm is based on the gradient method, whose descent step is ruled by (8.8):

$$\theta^{SD}(k+1) = \theta^{SD}(k) - \eta_k\cdot\left.\frac{\partial L_{\Delta V}\left(\cdot,\theta^{SD}\right)}{\partial\theta^{SD}}\right|_{\hat\theta^{SD}(k)} ;\quad k = 1,\ 2,\ \ldots \quad (8.8)$$

In (8.8), η_k denotes the gradient step size and k the reallocation time instant. This method is called *Reference Chaser Bandwidth Controller* (RCBC).

8.4.3 Performance evaluation and discussion

These rate control mechanisms (i.e., RCBC and EqB) have been investigated through simulations [21],[23]. An *ad-hoc* C++ simulator has been developed for the SI-SAP environment described above, considering a *general* satellite system. In what follows, for the sake of simplicity, only the traffic aggregation problem is faced by assuming no channel degradation over the satellite channel.

The case considered is that of two SI traffic buffers. The first one conveys the traffic of 30 VoIP sources. Each VoIP source is modeled as an exponentially-modulated on-off process, with mean "on" and "off" times equal to 1.008 s and 1.587 s, respectively. All VoIP connections have peak rate of 64 kbit/s. The IP packet size is 80 bytes. The SI service rate for VoIP assures an SLA target PLP below 10^{-2} (SI VoIP buffer size is 30 IP packets). The second buffer is dedicated to a video service. "Jurassic Park I" video trace, taken from the Web site referenced in [28], is used. The SI rate allocation for video (also measured through simulations), is 350 kbit/s. It assures a PLP $= 10^{-3}$, which is the target SLA for video (the SI video buffer size is 10,500 bytes). Both outputs of the SI buffers are conveyed towards a single queue at the SD layer. DVB encapsulation (header 4 bytes, payload 184 bytes) of the IP packets through the LLC/SNAP (overhead 8 bytes) is implemented in this case. The SD buffer size is 300 DVB cells.

In Figure 8.12 (firstly presented in [21]), the SD bandwidth provision produced by RCBC is compared with EqB. The loss probability bound ε for EqB is set to 10^{-3}, being the most stringent PLP constraint imposed at the SI level. The time interval between two consecutive SD bandwidth reallocations is denoted by T_{RCBC} and T_{EqB}, for RCBC and EqB respectively. Note that in the following graphs, for the sake of simplicity, T denotes T_{RCBC} (T_{EqB}) in the RCBC (EqB) case.

T_{RCBC} is fixed to 7 minutes, while T_{EqB} is set to the following values:

$$\{T_{RCBC} \cdot 1/3, T_{RCBC} \cdot 1/2, T_{RCBC}, T_{RCBC} \cdot 2, T_{RCBC} \cdot 4\}$$

in different tests in order to highlight the possible inaccuracy introduced by the real-time computation of the EqB statistics using different time scales.

According to Figure 8.12, RCBC captures the bandwidth needs of the SD layer in a single reallocation step. Whereas, EqB produces strong oscillations in the SD rate assignment. It is also clear from Figure 8.12 that the IPA-based estimation (8.5) is more robust than the on-line estimation of $m_{\alpha^{SD}}$ and $\sigma_{\alpha^{SD}}$. The IPA sensitivity estimation drives RCBC toward the optimal solution of the QoSMO problem.

The SD buffer's video PLP, averaged over the entire simulation horizon, is shown in Figure 8.13 (taken from [21]). The performance of RCBC, referenced to as "SD layer RCBC" is very satisfying: actually, the RCBC video PLP is $7.56 \cdot 10^{-4}$. A result "below threshold" has been measured for

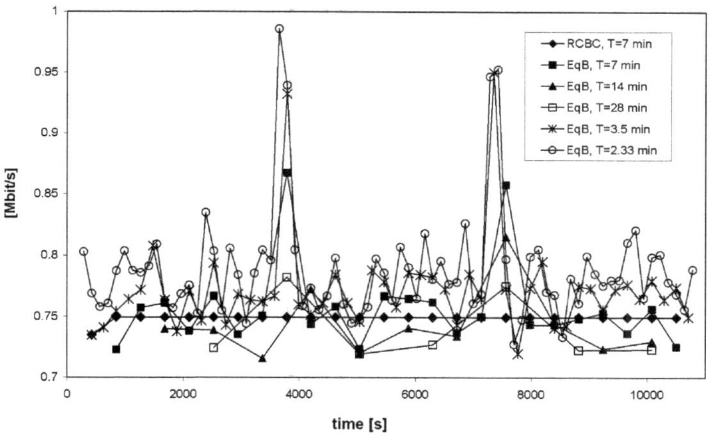

Fig. 8.12: Aggregation of VoIP and Video. SD allocations. RCBC versus EqB. See reference [21]. Copyright ©2005 IEEE.

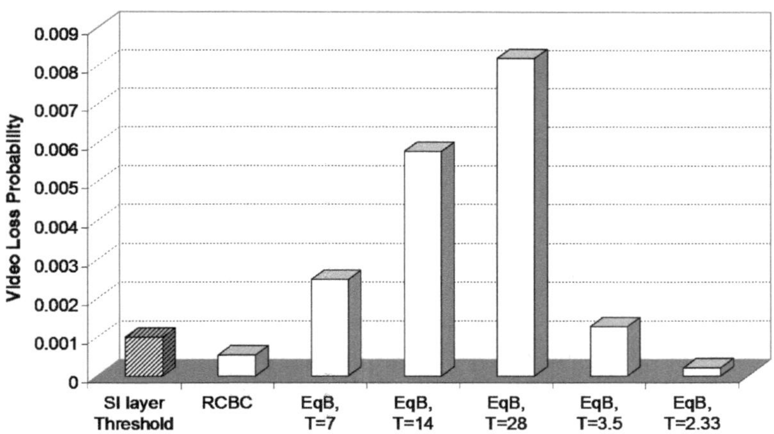

Fig. 8.13: Aggregation of VoIP and Video. Video PLP. See reference [21]. Copyright ©2005 IEEE.

EqB only for frequent reallocations ($T_{\text{EqB}} = T_{RCBC} \cdot 1/3 = 2.33$ minutes). The corresponding bandwidth allocations, averaged over the simulation duration, are shown in Figure 8.14 (taken from [21]). RCBC not only allows saving bandwidth compared to the "SD layer EqB T = 2.33 min" strategy, but offers a performance comparable to the other EqB cases, whose offered PLP is far from the SI threshold. In brief, RCBC finds the optimal operation point of the system, namely, the minimum SD bandwidth provision needed to track the SI QoS thresholds.

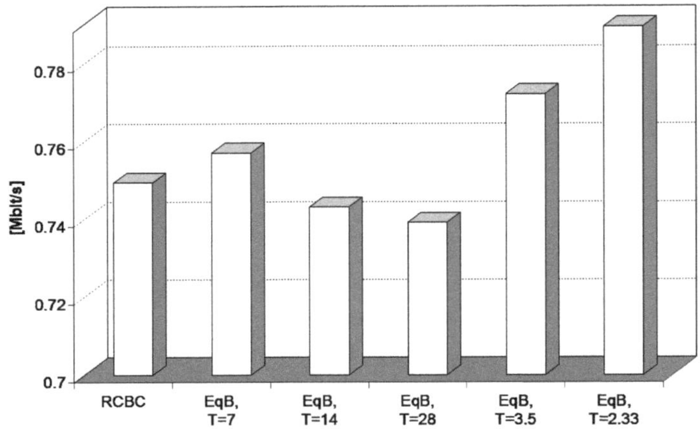

Fig. 8.14: Aggregation of VoIP and Video. Average SD bandwidth provision. See reference [21]. Copyright ©2005 IEEE.

8.5 QoS provisioning for terminals supporting dual network access - satellite and terrestrial

When terminals support dual network access -satellite and terrestrial (WLAN, UMTS, etc.) links- it is quite critical to select the appropriate network for each application, depending on both the resources available and the kind of application involved. In some instances (such as real-time tele-operation), it is not only a matter of user satisfaction, but also of satisfying critical service goals. For example, the QoS provision may be related to the deadline fulfillment: violating a deadline may cause a sea farm hitting the sea bottom or a remote probe bump into a rock.

This Section provides an analysis on relevant technologies in this context and focuses on QoS frameworks to support terminal mobility between satellite, wireless, and terrestrial networks. In particular, we analyze the problem of the multiple access to different networks (which includes satellite, wireless, and terrestrial networks) in order to support more than one access network at the same time. In such a context, the focus is on network selection based on QoS parameters. We work on QoS parameter identification at layer 2 for selected applications as well as IP-oriented solutions for network mobility and network selection. Let us consider two specific topics:

- Redundant codes in hybrid networks and
- Mechanisms for error recovery in WiFi access points.

Redundant codes in hybrid networks

Hybrid networks consisting of satellite links and mobile *ad hoc* networks present a series of challenges due to different packet-loss patterns, delay, and, usually, scarce available bandwidth. In this scenario, redundant encoding, in the form of *Forward Erasure Correction* (FZC) codes [29],[30], can provide an effective protection against losses in multicast videoconferencing and video streaming applications. The use of efficient encoding techniques can provide further reduction on bandwidth requirements.

A real test-bed based on a remote video streaming server interconnected via a GEO-satellite pipe to a local WLAN (both 11 Mbit/s and 5 Mbit/s cases have been considered, according to IEEE 802.11b) is presented in [31], by adopting the multicast network protocol. The satellite pipe is based on the commercial Skyplex network [32] that operates in the Ka band with the Hotbird 6 transponder. The developed platform, described in [33], is shown in Figure 8.15. The purpose is to provide users with a low-cost, high-availability platform for performing experiments with IP packets over the Skyplex platform. Such devices have been also used to experiment the FZC encoding.

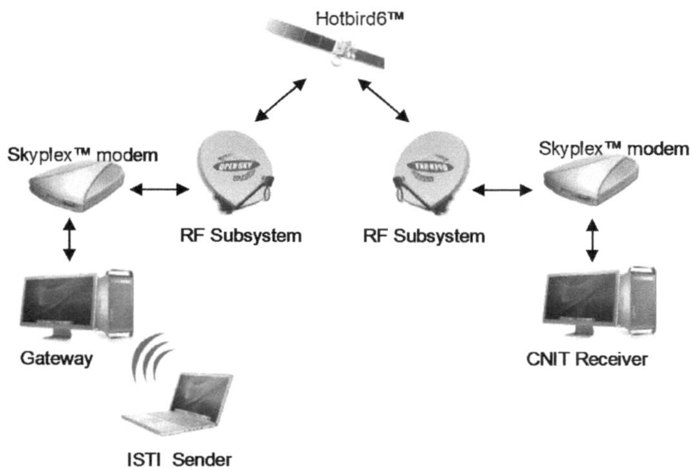

Fig. 8.15: Test-bed platform architecture.

The obtained experimental measurements show the performance of FZC codes based on Vandermonde matrix [34], for multicast video streaming applications. Basically, k blocks of source data are encoded to produce n blocks of encoded data (with $n > k$), such that any subset of k-encoded blocks suffices to reconstruct the k-block source data. Considering the real implementation, the encoder works at the transport layer by fetching k information packets from the video stream and then transmitting $k+l$ UDP packets (k of information + l of redundancy) towards the receiving host. The encoder adds a preamble of 4 bytes for the sequence number, which is then cancelled by the decoder. The decoder fetches k of the $k+l$ packets per block and recovers the information, provided that no more than l packets are lost in a single block of packets. The receiver then feeds the MPEG-4 decoder with the received stream. Different MPEG-4 movies have been broadcast towards the terrestrial WLAN through the satellite channel by using a standard video decoder and home-made FZC encoding/decoding software. The authors in [31] used this software to work between the application layer and the UDP transport layer; however, this software can also be used between layer 2 and 3.

The performance evaluation has been based both on subjective perception of QoS and on objective parameters of QoS. The used subjective assessment has been the perceptual quality, called the *Mean Opinion Score* (MOS), which requires several observers and many tests in order to provide a reasonable statistical spread of results ([1]). The reference measure used for an objective video quality assessment has been the *Peak Signal to Noise Ratio* (PSNR), calculated on the difference signal between the original error-free sequence and the received decoded sequence. Other general objective parameters of QoS, such as packet delivery delay and packet loss, have been considered in [31]. In what follows the numerical results are presented.

Packet Loss - The experiment in un-coded mode (no FZC) shows that packet loss can be reduced from 13% to about 6% by changing the transmission rate from 11 to 5.5 Mbit/s (see Table 8.1). In this case, the channel occupancy increases by 56%. Interestingly, we can note that with FZC and 200/100 coding ratio the packet loss is almost zero.

Coding ratio	Packet Loss	Residual Bandwidth [Mbit/s]
Uncoded 11 Mbit/s	12.98%	4.77
110/100	8.17%	4.67
Uncoded 5.5 Mbit/s	5.39%	4.24
120/100	3.25%	4.58
130/100	0.83%	4.48
200/100	0%	3.83

Table 8.1: Packet loss and residual bandwidth after FZC encoding.

[1] According to ITU-R Recommendation 500-5, MOS values are: imperceptible (5), perceptible but not annoying (4), slightly annoying (3), annoying (2), very annoying (1).

Packet delivery delay - The maximum packet delivery delay is evaluated as the time necessary to recover all the information when a number of packets equal to the redundancy packets are lost. Table 8.2 shows the packet delivery delay versus the coding ratio.

Coding ratio	Max [ms]	Mean [ms]	Var.[ms^2]
110/100	105.6	52.149	0.769
120/100	115.2	54.99	0.792
130/100	124.8	52.637	0.805
200/100	192	53.805	1.675

Table 8.2: Maximum, mean and variance of delivery delay. See reference [31]. Copyright ©2005 IEEE.

Mean Opinion Score (MOS) - Thirty persons have answered to three quality questions (overall, video, and audio quality) for each considered coding ratio; MOSs have been calculated. The percentages of people, who have considered the video acceptable, are shown in Table 8.3.

Uncoded 11 Mbit/s	110/100 11 Mbit/s	Uncoded 5.5 Mbit/s	120/100 11 Mbit/s	130/100 11 Mbit/s	200/100 11 Mbit/s
7.7%	15.4%	26.9%	34.6%	100%	100%

Table 8.3: Acceptability of received video. See reference [31]. Copyright ©2005 IEEE.

PSNR - The PSNR-based video quality metric (henceforth denoted as VQM_P - *Peak Video Quality Measurement*) uses a form of the logistics function that is recommended in references [35],[36] and evaluates the mean of how much transmitted frames differ from the original ones. Sixty seconds of the transmitted video have been compared with the received video to evaluate the VQM_P parameter. Results are presented in Figure 8.16, where VQM_P is compared for different coding ratio values.

Error recovery in WiFi access points

The second topic of our study on the interconnection of WLAN and satellite networks deal with some mechanisms for error recovery when a *Fast HandOver* (FHO) occurs between different IEEE 802.11b *Access Points* (APs). Fast handover techniques using paradigms like make-before-break or bi-casting reduce the L3 handover time to extremely short delays that are acceptable for all the applications [37]. However, in *layer 2* (L2), the handover time is

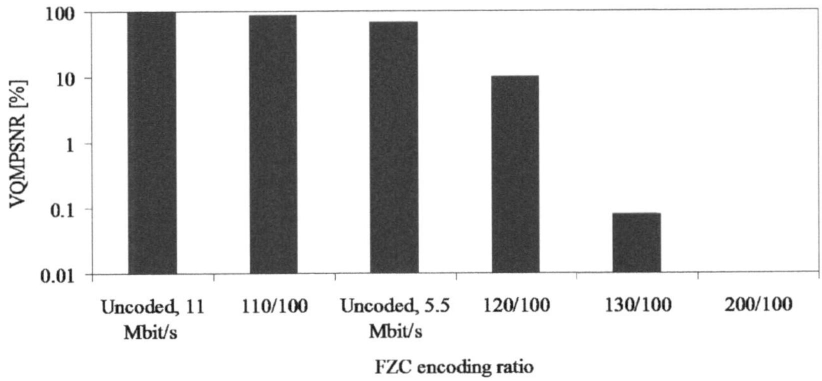

Fig. 8.16: Peak Video Quality Measurement.

very high for some technologies. So, when doing an intra-technology handover (e.g., the interface changes its AP and/or channel), an unacceptable disruption may occur. For instance, in IEEE 802.11b (WiFi) an FHO procedure can take from about 500 ms in a standard mode, to less than 10 ms in an optimized mode, where the number of frequencies, scanned in order to establish the new communication, is sensibly reduced [38]. During such a time, all transmitted information can be lost. In such a context, the adoption of robust FZC codes can permit to recover totally the lost information. This procedure may cost in terms of bandwidth utilization and computational complexity, due to the generation of the redundant information. To minimize these facts we propose to use FZC in the last hop, i.e., between the *Mobile Node* (MN) and its *Access Router* (AR). In what follows, AP and AR terms are used interchangeably. The *Core Network* (CN) will then not need extra computational power and use more bandwidth in its access link.

During the L2 disruption, both the ARs and the MN can stop sending packets and buffer them and send them when the connectivity is re-established. Indeed, during an FHO, the MNs have means to predict that there is going to be an L2 disruption. For instance, they may know the instant in which the FHO takes place or finishes or its duration, but, perhaps, not with accuracy or some parameters may be unknown. Buffering by its own is not a perfect solution; hence, we propose to complement it with FEC techniques of the FZC type. Our solution consists in adding (or increasing) the FEC used between MN and ARs during the predicted FHO duration. Table 8.4 permits to understand the advantage of FEC techniques with respect to pure buffering. This table depicts an FHO (beginning and end instants of the L2 disruption are indicated) and also shows when the disruption actually happens. If we use buffering, the communication is cut between the disruption indications and then the buffered packets are sent. But using FEC, MN and

ARs continue sending packets and the communication is cut only during the actual disruption. When this disruption ends, lost packets may be recovered by exploiting FEC capabilities. When "L2 disruption end" is indicated, FEC can be stopped. Of course the advantage of FEC versus buffering depends on how big is the shift between disruption indication and actual disruption.

	TIME →					
Communication status (mobile node)	Normal Tx	L2 disruption indication	L2 disruption	L2 disruption	L2 disruption end indication	Normal Tx
Mobile node behavior		Buffering	Buffering	Buffering	Buffering	
Correspondent node behavior	pkt Rx					pkt Rx
Mobile node behavior using FEC techniques		FEC	FEC	FEC	FEC	
Correspondent node behavior when the MN uses FEC during handover	pkt Rx	pkt Rx			pkt Rx	pkt Rx

Table 8.4: Buffering versus FEC techniques while the transmitter node undergoes an FHO.

A first problem to solve is how the MN will indicate its own software and the AR that it is going to perform an FHO. That may depend on the actual L2 technology and even, in some technologies, the AR will be the one telling the MN that it has to move. In WiFi, when the MN detects that the signal level of its current AR decreases bellow a threshold, it initiates the FHO procedure (scanning new channels in new ARs and then doing the FHO to the selected channels). This issue can trigger two actions in the MN: it starts doing FEC and tells the AR to do so as well.

The second issue to solve is to determine the ideal amount of redundancy in the FEC technique. Hence, we must calculate the maximum number of packets lost during L2 disruption (we must estimate the total disruption time and the packet rate). This number of packets is the redundancy that must be included in the FEC. In Table 8.5 this disruption time (shadowed parts) corresponds to 3 packets and thus 3-packet redundancy is added (packets 3, 4 and 5). Note that information packets are labeled with a letter and redundancy packets with a number.

Also the buffering needed at the receiver (both MN side and the AR one, depending on the direction of transmission) must be determined. For doing so, we suppose the worst case scenario, when the disruption occurs while sending the information packets and not when sending redundancy packets. In such a case, to recover the information, the whole block composed of information and redundancy packets (for instance block formed by packets D to 5), must be received before being able to extract the information packets. This determines the minimum buffering time of 8 packets in Table 8.5.

Time	5	6	7	8	9	0	1	2	3	4	5	6	7	8	9	0	1	2	3	4	5	6	7	8	9	0	1	2	3
Informat. frames	A		B		C			D			E		F			G			H			I			J				
Tx frames						A	B	C	1	2					D	E	F	G	3	4	5			H	I	J	6	7	
Rx frames						A	B	C	1	2							F	G	3	4	5			H	I	J	6	7	
Buffer at Rx														A			B			C			D			E			F

Table 8.5: FEC technique example. Shadowed cells correspond to disruption.

Note that these aspects can also apply if the FEC technique is employed between the CN and the MN, being the ARs transparent to that. However, there are many advantages of employing the FEC technique in the MN-AR link and at link level. First, in our proposed FHO scheme, the AR already must have special functionality like bicasting, thus adding this FEC functionality will not complicate too much the AR. Second, redundancy is only present in the last hop, besides in this last hop, FEC techniques may already be employed because the link (e.g., air link) may be prone to errors, so, perhaps our solution can be implemented just modifying the parameters of the existing FEC. Finally, in a multicast scenario, the source will send the data and the ARs would be the ones to add the appropriate FEC redundancy.

8.6 Switched Ethernet over LEO satellite: implicit cross-layer design exploiting VLANs

The aim of this Section is to introduce the possibility of exploiting Switched Ethernet facilities in a LEO network using *Inter-Satellite Links* (ISLs). We refer here to Scenario 3 (see Chapter 1, sub-Section 1.4.5) for the study carried out in this part of Chapter 8. The interest is to support QoS-aware resource management procedures by tacking into consideration the mapping between logical and physical network topologies (that are both well known), like in the case of terrestrial switched-networks. In fact, the optimization of connection paths can be performed, in a large network scenario, by means of routing algorithms or switching mechanisms. The former solution is particularly suited in case of networks whose topology is not completely known or predictable, while the switching approach has been adopted in the scope of single domain networks with known topology, or single-provider-managed area networks (LAN, WAN, MAN). LEO satellite can be classified as an extension of LAN/MAN, in which the network topology changes over time in a regular and predictable way, thus the switching approach can be a natural candidate for network management issues. Furthermore, an Ethernet-like switching solution (i.e., the one proposed by IEEE 802.1) addresses the problem of the interoperation between network layer and the satellite-specific MAC. The mechanism is known as *Logical Link Control* (LLC), and provides a useful framework to deploy an IP-MAC mapping with QoS control.

IEEE 802.1 standards offer a set of facilities meant to build, operate and manage a network comprising one or more transmission technologies and one or more communication methods (i.e., physical and MAC layers). Bridging of different technologies is obtained by using special devices (i.e., bridges) that are able to translate, when needed, frames from a given MAC format to a different one. In particular, IEEE 802 standards propose a special LLC layer [39], located just above the MAC layer [40], that is common to all network segments and whose frames can be transported in any kind of MAC frames. LLC is the glue that allows the system to interconnect different physical and MAC solutions of the IEEE 802 family.

Switches can be used in order to reduce the number of competitors to the same shared medium in a network, by segmenting the overall network and bounding the frame transmission in a limited range. In a switched-network, users can receive and transmit at the same time by means of two separate channels, the so-called *Full Duplex Switched Ethernet* facilities that can be suitably used in Gigabit LANs.

Now, it is worth noting that:

- LLC functionalities are analogous to protocol adaptation functions that are used for the transportation of IP traffic over satellite devices;
- LEO payload with bridging/switching modules is envisioned for future satellite networking;
- Full-duplex techniques are consistent with satellite communication systems, where different channels are commonly adopted for uplink and downlink.

Thus, we firstly questioned how much existing Ethernet-like solutions could be reused to obtain a protocol harmonization for satellite and terrestrial devices; secondly, we investigated how existing mechanisms should be enhanced to mach with satellite-specific issues, and in particular with LEO mobility problems.

Our research turns in the exploitation of a cross-layer design of layers 2 and 3. In fact, layer 2 switching is performed on the basis of end-to-end connections to be established (known from the IP demand, which does not involve all the LEO network at once, but only a set of sub-networks, possibly separated), and by taking into account the knowledge of logical path changes due to the turnover of LEO satellite positions. In other words, we deploy virtual sub-networks (*Virtual Local Area Networks*, VLANs) that span that portion of the LEO network needed to accommodate the IP demand, while continuously adjusting the logical VLAN topology on the basis of the predicted satellites motion. As a result, it is possible to choose data-link connections in order to balance the network traffic, to optimize the usage of MAC resources and to differentiate users and services (i.e., groups to be handled in VLANs). In turn, an enhanced degree of connectivity, robustness and QoS is provided to the IP level.

8.6.1 Protocol harmonization and implicit cross-layer design via IEEE VLAN

Protocol harmonization is meant for QoS-aware internetworking of communication segments, avoiding intricacies due to satellite gateways needed to access a satellite sub-network, and also due to routing and path discovery to be managed for time-varying topologies, typical of non-GEO satellite fleets. As a consequence, a LEO satellite constellation could be merged at layer 2 with legacy terrestrial networks, and end-to-end communications could seamlessly use satellite and terrestrial links.

Cross-layer design of layers 2 and 3 is meant to avoid the negative impact of topological changes, deep signal shadowing, data loss, and reconfiguration of layer 2 links needed to span the entire network and to avoid loops. However, note that we deal here with *implicit* cross-layer design, since we aim at optimizing the resource management in the MAC layer to support IP QoS and robustness. The optimization of resource management is performed by joining the effectiveness of *spanning tree* algorithms and the possibility to configure remotely and proactively LEO switching devices on-the-fly. To this aim, a centralized path/VLAN *manager* is required that selects ISLs to be activated at any time instant, provides path redundancy by means of multiple disjointed VLANs, and avoids service discontinuity by switching the IP traffic from a VLAN to another.

The rationale of our proposal is based on the consideration that most of topology changes in the network can be foreseen from the knowledge of the satellite motion and from a statistical analysis of signal strength at receivers, so that the switched-network can be proactively managed.

In order to manage efficiently a switched-network, it is necessary to maintain only a sub-set of inter-node links to form a connected loop-free graph (i.e., a tree-like logical topology). This permits to confine broadcast traffic, eliminates looping frames, and, mostly important, makes easier the routing of data frames by exploiting a tree that connects all nodes without redundancy. However, the extraction of a tree from the original network graph has to be performed in accordance with traffic management criteria. Indeed, multiple trees could be adopted to segment the network traffic, to create virtual sub-networks, to perform load balancing, or also to provide some redundancy if a link abruptly fails. These advanced features are offered by IEEE standards for VLANs [41] that include LLC for switching and bridging, *Spanning Tree Protocol* (STP) and its variants *Rapid STP* (RSTP) [42] and *Multiple STP* (MSTP) [43], and VLAN tagging methods. IEEE VLAN and MSTP can be suitably adopted in order to simplify the management of a huge number of connections, even though adopting satellite switching implies the constitution of very large WANs or MANs where IP routing is unnecessary. Important advantages can be obtained, such as the possibility to exploit a particularly broad connectivity, or the possibility to eliminate IP route discovery latency, frequent inconsistencies in IP routing tables due to LEO topology changes, and path elaboration delays.

Although spanning tree protocols are able to rearrange their configuration after a link or a node fails, the satellite VLAN approach only works if spanning trees and VLANs are proactively adjusted when the LEO logical topology changes. Note that legacy reconfiguration procedures could take several seconds (due to the huge network diameter) during which the network graph results unconnected. However, the adoption of proactive mechanisms is reasonable since: (*i*) the LEO satellite fleet is known *a priori*; (*ii*) the LEO fleet can be designed so that at lest one satellite is always visible in the target coverage area, and at least two satellites are visible during a handover event. Thus, the proactive management simply consists of setting up a new VLAN with a new spanning tree *before* the handover is performed, thus forcing a *VLAN handover* before the *physical handover*. Note that, as for the VLAN handover procedure, it simply requires to change the tagging of frames at the edge of the satellite path from the old VLAN tag to the new one.

Considering the service offered to end-users, the adoption of VLANs with proactively managed multiple spanning trees allows avoiding: (*i*) IP data-flow discontinuities due to physical topology changes, (*ii*) waste of large time intervals in spanning tree reconfigurations, triggered by the failure of a link or a node, and (*iii*) waste of bandwidth due to possible flooding effects after reconfigurations.

8.6.2 Performance evaluation

Multiple VLANs allow the network provider to hide network topology changes. In fact, during a topology transition, a new path will be available *before* the old path goes down. We include these different paths within different VLANs (i.e., addressed by different VLAN tags in the frame header) and switch to the new path *during* the topology transition. The VLAN manager knows the topology transition and enforces a VLAN tag change (i.e., a VLAN handover) at the edge of the network, so that each frame will follow a path in the VLAN identified by its new tag. In practice, we use multiple VLANs as redundant sub-networks.

Table 8.6 [44] reports what happens to frames generated in a message exchange between two terrestrial users at the edge of a Teledesic-like LEO network, when UDP is used, comparing the case in which a VLAN handover is adopted to a scenario without such a feature. We consider the case where only two users are in the network and only one connection is active ([2]). Summarizing, we can say that the absence of VLANs implies service discontinuities and long flooding phases after reconfigurations (due to the unidirectional nature of UDP exchange considered). Using VLANs, discontinuities of the service are avoided: packets are not filtered and end-to-end connectivity is not broken. The flooding after reconfiguration is bounded to the VLAN used after the handover; whereas, an STP-based approach would flood the entire network (STP uses a single tree for the entire network). Similar considerations can be made when TCP is used, with the exception of the occurrence of short flooding phases, due to frequent TCP ACKs.

UDP		
Event	***Action (No VLAN)***	***Action (two VLANs)***
Service request from Client	Flooding	Flooding in VLAN#1
	Switches learn Client address	VLAN#1 switches learn Client address
Service response from Server	Switching/no flooding	Switching/no flooding
	Switches learn Server address	VLAN#1 switches learn Server address
Service data sent	Switched/no flooding	Switching/no flooding
Topology change	Re-compute Spanning Tree	Handover to VLAN#2
	LAN temporarily disconnected	Re-compute CIST and VLAN#1 Spanning Tree
	All frames discarded	VLAN#1 temporarily disconnected
		VLAN#2 flooded, filtering databases are empty
Downstream (Upstream) frames after reconfiguration	Network flooded by "new" switches until an up- (down-) stream frame is sent by Client (Server)	VLAN#2 flooded by all switches until an up- (down-) stream frame is sent by Client (Server)

Table 8.6: How topology changes affect UDP connections. Note that each network region that is managed by MSTP needs a *Common Internal Spanning Tree* (CIST) to interconnect all nodes in that region. See reference [44]. Copyright ©2005 IEEE.

Table 8.7 [44] collects a set of actions performed by network entities at the occurrence of specific TCP-related events. The advantage of using proactive VLANs is clear from the comparison with the "legacy" behavior of switches,

[2] This is the worst-case scenario, since switching devices need bidirectional flows in order to learn the route towards a remote user, otherwise frames are flooded in the VLAN.

as described in the central column of the table, and the behavior of the VLAN with support for off-line reconfiguration, as described in the rightmost column: VLANs reduce flooding effects and off-line reconfiguration eliminates service discontinuities. However, a flooding phase is still needed in order to fill the *filtering database* (a layer-2 routing table built by bridges by monitoring incoming frames - source address and incoming port - to discover the outgoing port to be used to forward frames without the need of flooding) of the new VLAN.

TCP		
Event	***Action (No VLAN)***	***Action (two VLANs)***
Request from Client (TCP SYN)	Flooding	Flooding in VLAN#1
	Switches learn Client address	VLAN#1 switches learn Client address
Response from Server (TCP ACK)	Switching/no flooding	Switching/no flooding
	Switches learn Server address	VLAN#1 switches learn Server address
TCP SYN-ACK	Switched/no flooding	Switched/no flooding
Service data flow	Switched/no flooding	Switched/no flooding
Client's ACK	Switched/no flooding	Switched/no flooding
Topology changes	Re-compute Spanning Tree	Switch to VLAN#2
	LAN temporarily disconnected	Re-compute CIST and VLAN#1 Spanning Tree
	All frames discarded	VLAN#1 temporarily disconnected
Downstream (Upstream) frame after reconfiguration	Flooded by "new" switches until an up- (down-) stream frame is sent by Client (Server)	Flooded in VLAN#2 switches until an up- (down-) stream frame is sent by Client (Server)

Table 8.7: How topology changes affect TCP connections. See reference [44]. Copyright ©2005 IEEE.

In what follows, we show some results obtained in a scenario similar to the one depicted in Figure 8.17 [44], by means of the OPNET [45] simulator with modified bridging/switching devices. Satellite orbits are designed following the design principles of the Teledesic system, where LEO orbits have a 1375 km altitude and the satellite capacity is set to 32 Mbit/s for both uplink and downlink with terrestrial users, even though channels can be allotted to users as multiple of the 16 kbit/s basic channel. Two terrestrial bridges/switches are considered (T1 and T2 in the figure). Users are located close to the terrestrial bridges; the longest path between two users in the simulation has a delay bounded to 200 ms. We tested both UDP and TCP-based applications with and without VLAN supports.

Let us describe the generation of both UDP traffic and the TCP one. In

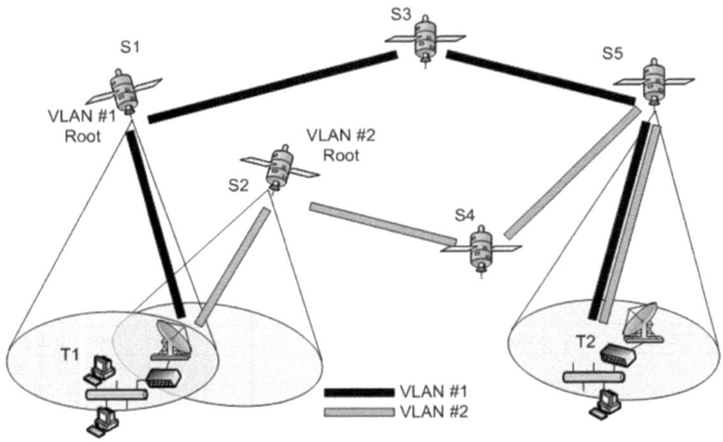

Fig. 8.17: Possible network topology with two VLANs available. See reference [44]. Copyright ©2005 IEEE.

particular, UDP traffic is obtained via simulation of unidirectional constant-rate video connections; also packet distribution is constant. Three UDP groups of users have been considered that differ in terms of both bit-rate (i.e., 256, 128 and 64 kbit/s, respectively for Class I, II and III users), and the average request inter-arrival time (which is exponentially distributed, so that the arrival process of connection requests is Poisson). Mean inter-arrival times are, respectively: 45 s for Class I, 22.5 s for Class II, and 11.25 s for Class III. Each connection has a duration exponentially distributed with mean value of 180 s. The number of users in a group is selected so that each UDP group offers 1 Mbit/s traffic in average, directed from nodes located in T1 to T2.

As for TCP-based traffic, the following results have been obtained by considering the separate contribution of three groups of FTP users. Every user, located in T1, requests files of B bytes, where B is exponentially distributed with a mean of 5,000,000 bytes, while the file request inter-arrival time is exponentially distributed with a mean of 5 s. User groups are differentiated based on the available resources allotted in the access link: Class I (*High Rate*) has an aggregate guaranteed rate of 512 kbit/s for the downstream and 128 kbit/s for the upstream; Class II (*Medium Rate*) has an aggregate guaranteed rate of 128 kbit/s for the downstream and 32 kbit/s for the upstream; eventually, Class III (*Low Rate*) has an aggregate guaranteed rate of 64 kbit/s for the downstream and 16 kbit/s for the upstream. Each group saturates its link capacity due the high file request rate (5 files per second are requested, i.e., about 25 MByte per second, which requires at least 120 Mbit/s plus the protocol overhead: the system is overloaded and the number of FTP requests overwhelms the number of FTP sessions that reach the end of transmission). As a matter of fact, simulations confirm the behavior described in Table 8.6 for UDP and the considerations about TCP in Table 8.7. Details are provided below.

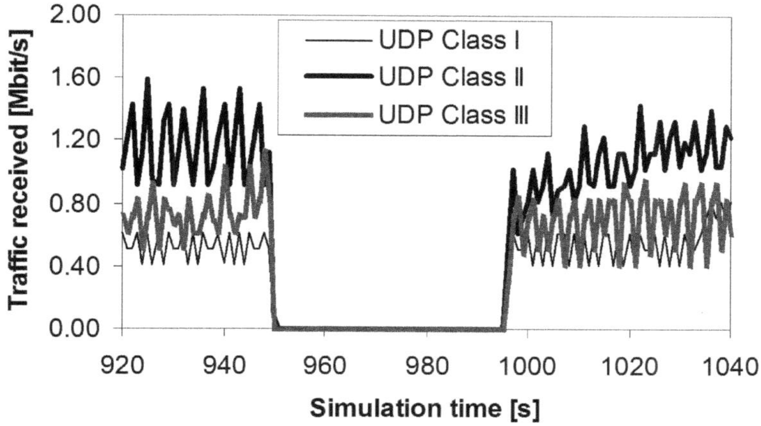

Fig. 8.18: UDP throughput with STP, no VLANs.

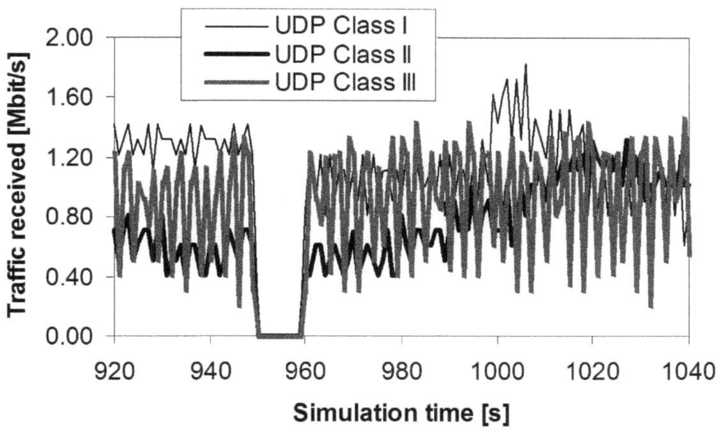

Fig. 8.19: UDP throughput with RTSP, no VLANs.

Figures 8.18 to 8.20 show the throughput of a unidirectional UDP connection between two remote hosts. In the simulations, a physical topology change occurred at $t = 950$ s, and one can notice that a traditional STP approach requires up to 45 s to recover the path; using RSTP this time is shortened, but several seconds, about 10 s, are still needed to reconfigure the large switched-network. On the contrary, preconfigured VLANs allow a seamless handover, without service discontinuities. In Figure 8.20, a VLAN handover is enforced at $t = 940$ s, just a few seconds before the physical topology change. Similar considerations could be made by considering bidirectional UDP flows, where the traffic is generated in each direction as in the unidirectional case.

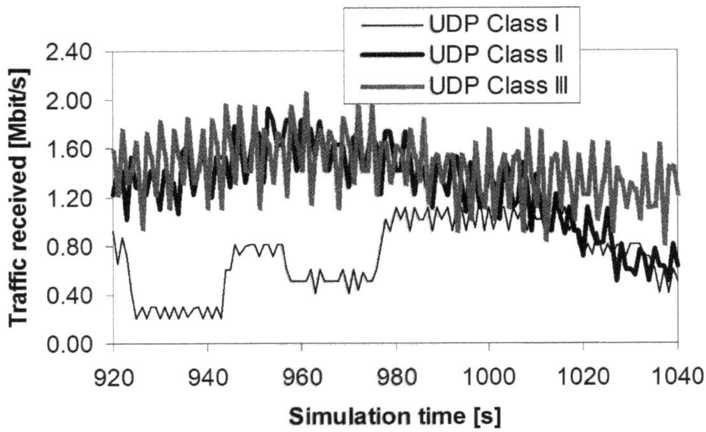

Fig. 8.20: UDP throughput with VLAN handover.

Figures 8.21 and 8.22 depict the throughput of a TCP connection for hosts requesting FTP files from a network server. In this case, traffic flows are bidirectional, due to the presence of ACK packets in the return channel, even though the connection is strongly asymmetric. In these simulations, a topology change occurred at $t = 800$ s. By using STP (Figure 8.21) or RSTP, we can notice a service interruption with a duration similar to that experienced in UDP simulations, but the effect is partially masked by the build up of long queues at the last satellite-to-ground station link, especially for the TCP Class III, which is allotted the minimum resources. It is worth noting that after the network reconfiguration, each traffic group aggregate suffers from high fluctuation due to the synchronization of TCP flows after the outage period. In particular, Class I experiences a very drastic fluctuation, while lower rate traffic classes grow very slowly. Eventually, if we consider the adoption of VLAN (Figure 8.22), with a handover operated at $t = 790$ s, no significant variation can be noted in the traffic aggregate of each class. Again, off-line configured VLANs allow ground stations to switch seamlessly between VLANs, and avoid service discontinuities.

As for the flooding effects due to topology changes, first we consider unidirectional UDP flows in the network, from site T1 to site T2. Figures

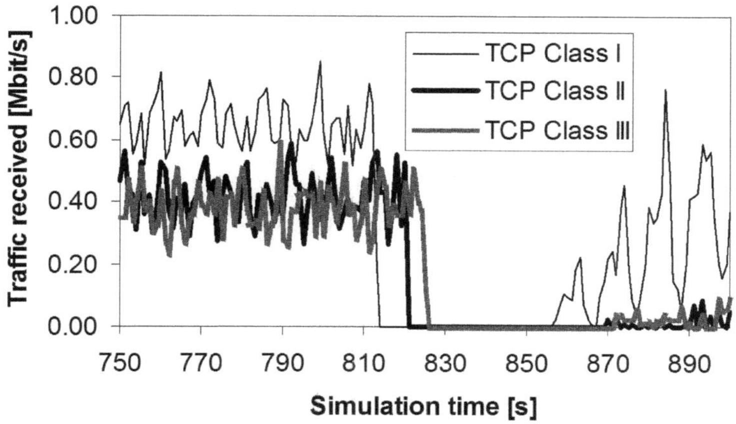

Fig. 8.21: TCP throughput with STP, no VLANs.

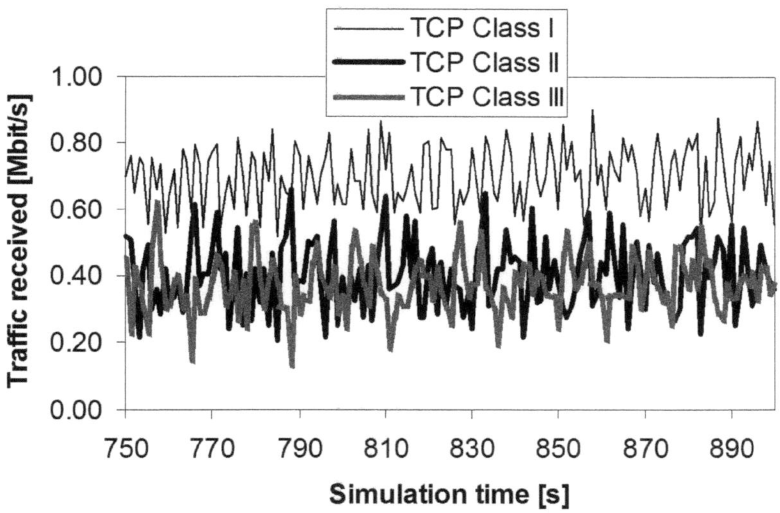

Fig. 8.22: TCP throughput with VLAN handover.

8.23 and 8.24 represent data flooded by switches when no appropriate entries are found in the filtering database. Each flooded data frame is accounted for only once, no matter if multiple switches will flow again the same frame. In practice, a flooding phase starts after an automatic route change, performed by RSTP (or STP, not showed here). This is the reason why Figure 8.23 shows flooded packets for multiple sources after the first disrupted path is recovered, which is not mandatory for the data path we are interested to.

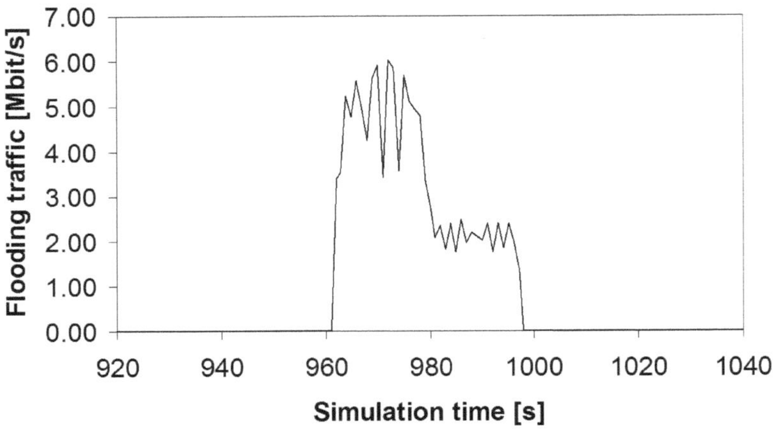

Fig. 8.23: Unidirectional UDP connections: normalized aggregated flooding (RSTP).

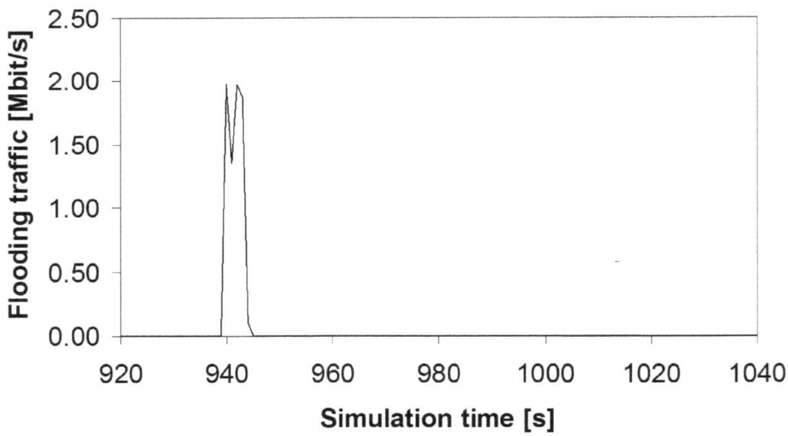

Fig. 8.24: Unidirectional UDP connections: normalized aggregated flooding (VLAN).

Thus, the flooding phase ends only after the network is fully reconfigured and a new frame is sent in the reverse path for each user (i.e., after a new request is sent per each UDP traffic class, which is represented, in these simulations, by a single user). Figure 8.24 shows that by adopting VLAN-based network management, a simple VLAN handover is required a few seconds before the original path goes down. However, VLAN handover requires a brief flooding phase just after the handover, since the filtering database learning phase has to be performed as well.

As for the flooding effects in case of bidirectional UDP connections, in the same network and traffic conditions as before, it can be found a very limited flooding due to the fact that an intense traffic is used in both directions, so that database learning phases are very short.

Figure 8.25 shows the burden of flooding data for TCP connections when RSTP is adopted. Data are related to frames carrying TCP segments (data and ACKs) that experience at least one duplication in a generic network node. Due to the asymmetric nature of TCP upstream and downstream (i.e., the different size and bandwidth occupied by data and ACKs), it is appropriate to distinguish between the amount of flooded bits and the number of flooded frames: we represent the flooding in terms of flooded packets, which give a normalized estimate of how much dangerous the flooding can be. Note that flooding occurs while the network is reconfiguring itself. However, RSTP operation allows data to be flooded immediately after the link failure. In fact, when using spanning trees, each link is represented as an arch of an oriented graph. The orientation of each arch is from the root to the leaves of the tree. Thus, a link connects an up-node to a down-node. Using RSTP, the down-node is in charge of sensing the link failure and starting the recovery phase; the down-node is also allowed to use alternative links to reach the root of the tree. The most important cause of flooding in RSTP is given by frames that reach a down-node of a broken link. Eventually, flooding is almost completely avoided by using VLANs, as stated by Figure 8.26.

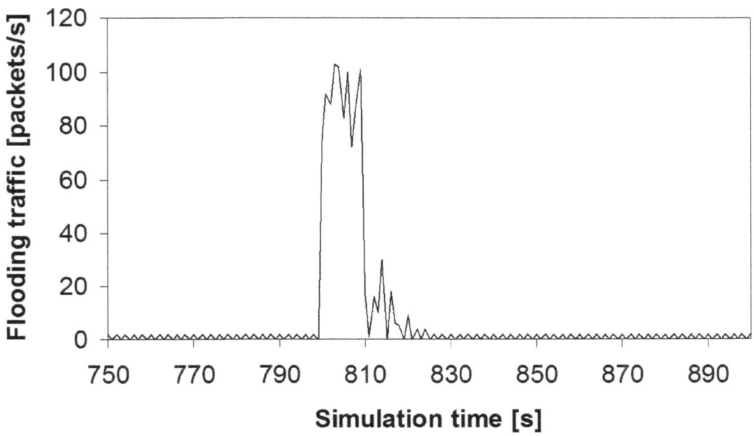

Fig. 8.25: TCP connections: normalized aggregated flooding (RSTP).

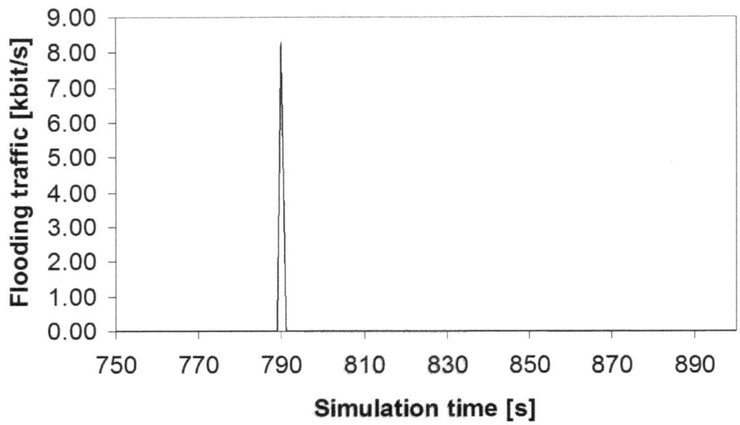

Fig. 8.26: TCP connections: normalized aggregated flooding (VLAN).

8.7 Conclusions

Over the past decades, with the emergence of many multimedia Internet applications and services, the research community has devoted a big effort in an attempt to satisfy their stringent and varied QoS requirements. A clear example of this effort is the initiative by IETF in proposing two IP QoS frameworks. These frameworks are mainly designed with terrestrial networks in mind. However, the problems of achieving QoS in networks with wireless medium such as satellite networks are much more complicated since the link is dependent on channel conditions. Hence, the resource management block is vital in realizing the IP QoS frameworks.

Standard mechanisms which operate solely in the network layer most often cannot guarantee the QoS when the end-to-end path involves satellite or wireless networks as they disregard the variability of channel conditions. This leads to the investigation of utilizing MAC layer resource management schemes or protocols to improve this situation. More recently, the idea of using cross-layer techniques further open up the potential of what can be achieved in terms of QoS provision.

Being in adjacent layers in the protocol stack, resource management (layer 2) in satellite networks is always tightly coupled with the IP QoS frameworks (layer 3). This Chapter has been dedicated to the cross-layer interactions and issues between these two layers. A review of the current state of the IP QoS frameworks in relation with the satellite network shows that DiffServ is being increasingly accepted and an example implementation of relative DiffServ is given as an illustration on how MAC layer scheduling can support the QoS provisioning. The problem of mapping between the QoS mechanisms operating at the two layers has been formulated and a measurement-based approach has been presented. The problem is also discussed in two other scenarios; namely the dual network access and Switched Ethernet over LEO satellites.

From the discussions and results presented in this Chapter, it is clear that achieving IP QoS in a satellite environment can certainly benefit from cross-layer mechanisms from layer 2. Nevertheless, caution must be observed when designing such cross-layer schemes. Uncontrolled implementation of cross-layer mechanisms may cause other problems that may not be apparent in a short period of time. Cross-layer design aimed at improving a specific performance metric may not have the entire system performance considered while cross-layer design involving multiple layers may lead to 'spaghetti design' with high number of convoluted interactions. All these aspects will increase system complexity and hence will pose problems for future innovations. Worse, system update may require complete redesign. Another example of a negative impact of uncontrolled cross-layer design is on network security issues: the increased interactions among layers may increase the channels for security attacks. In conclusion, designers must have the long-term effects in mind.

References

[1] S. Shenker, J. Wroclawski, "General Characterization Parameters for Integrated Service Network Elements", IETF RFC 2215, September 1997.

[2] R. Braden, D. Clark, S. Shenker, "Integrated Services in the Internet Architecture: an Overview", IETF RFC 1633, June 1994.

[3] R. Braden, L. Zhang, S. Berson, S. Herzog, S. Jamin, "Resource Reservation Protocol (RSVP) - Version 1 Functional Specification", IETF RFC 2205, September 1997.

[4] S. Shenker, C. Partridge, R. Guerin, "Specification of Guaranteed Quality of Service", IETF RFC 2212, September 1997.

[5] J. Wroclawski, "Specification of the Controlled-Load Network Element Service", IETF RFC 2211, September 1997.

[6] S. Blake, D. Black, M. Carlson, E. Davies, Z. Wang, W. Weiss, "An Architecture for Differentiated Services", IETF RFC 2475, December 1998.

[7] C. Dovrolis, D. Stiliadis, P. Ramanathan, "Proportional Differentiated Services: Delay Differentiation and Packet Scheduling", *IEEE/ACM Transactions on Networking*, Vol. 10, No. 1, pp. 12-26, February 2002.

[8] K. Wang. *Quality of Service Assurances in Multihop Wireless Networks.* PhD. Dissertation, University of Wisconsin-Madison, 2003.

[9] Y. Xue, K. Chen, K. Nahrstedt, "Achieving Proportional Delay Differentiation in Wireless LAN via Cross-Layer Scheduling", *Journal of Wireless Communications & Mobile Computing*, Vol. 4, No. 8, pp. 849-866, November 2004.

[10] W. K. Chai, M. Karaliopoulos, G. Pavlou, "Scheduling for Proportional Differentiated Service Provision in Geostationary Bandwidth on Demand Satellite Networks", in *Proc. of IEEE GLOBECOM 2005*, St. Louis, MO, USA, November 28 - December 2, 2005.

[11] E. Rosen, A. Viswanathan, R. Callon, "Multiprotocol Label Switching Architecture", IETF RFC 3031, January 2001.

[12] ETSI, "Digital Video Broadcasting (DVB); Interaction channel for satellite distribution systems," ETSI European Standard (Telecommunications series), EN 301 790 V1.3.1 (2003-03).

[13] G. Açar. *End-To-End Resource Management in Geostationary Satellite Networks.* PhD. Dissertation, University of London, November 2001.

[14] L. Kleinrock. *Queueing Systems.* New York, Wiley, 1976, Vol. II.

[15] ETSI, "Satellite Earth Stations and Systems (SES). Broadband Satellite Multimedia, Services and Architectures", *ETSI Technical Report*, TR 101 984 V1.1.1, November 2002.
[16] ETSI, "Satellite Earth Stations and Systems (SES). Broadband Satellite Multimedia, IP over Satellite", *ETSI Technical Report*, TR 101 985 V1.1.2, November 2002.
[17] N. Iuoras, T. Le-Ngoc, "Dynamic Capacity Allocation for Quality-Of-Service Support in IP-Based Satellite Networks", *IEEE Wireless Communications Magazine*, Vol. 12, No. 5, pp. 14-20, October 2005.
[18] N. Iuoras, T. Le-Ngoc, M. Ashour, T. Elshabrawy, "An IP-Based Satellite Communication System Architecture for Interactive Multimedia Services", *International Journal of Satellite Communications and Networking*, Vol. 21, No. 4-5, pp. 401-426, July-October 2003.
[19] T. Le-Ngoc, V. Leung, P. Takats, P. Garland, "Interactive Multimedia Satellite Access Communications", *IEEE Communications Magazine*, Vol. 41, No. 7, pp. 78-85, July 2003.
[20] S. Combes, L. Goegebeur, M. Fitch, G. Hernandez, A. Iuoras, S. Pirio, "Integrated Resources and QoS Management in DVB-RCS Networks", in *Proc. of the 22nd AIAA International Communications Satellite Systems Conference & Exhibit 2004* (ICSSC), Monterey, CA, May 2004.
[21] M. Marchese, M. Mongelli, "Rate Control Optimization for Bandwidth Provision over Satellite Independent Service Access Points", in *Proc. of IEEE Globecom 2005*, St. Louis, MO, USA, pp. 3237-3241, November 28 - December 2, 2005.
[22] M. Marchese, M. Mongelli, "On-Line Bandwidth Control for Quality of Service Mapping over Satellite Independent Service Access Points", *Computer Networks*, Vol. 50, No. 12, pp. 1885-2126, August 2006.
[23] M. Marchese, M. Mongelli, "Real-Time Bandwidth Control for QoS Mapping of Loss and Delay Constraints over Satellite Independent Service Access Points", *submitted to IEEE Transactions on Wireless Communications.*
[24] C. G. Cassandras, G. Sun, C. G. Panayiotou, Y. Wardi, "Perturbation Analysis and Control of Two-Class Stochastic Fluid Models for Communication Networks", *IEEE Transactions on Automatic Control*, Vol. 48, No. 5, pp. 770-782, May 2003.
[25] N. Celandroni, F. Davoli, E. Ferro, "Static and Dynamic Resource Allocation in a Multiservice Satellite Network with Fading", *International Journal of Satellite Communications and Networking*, Vol. 21, No. 4-5, pp. 469-487, July-October 2003.
[26] Y. Wardi, B. Melamed, C. G. Cassandras, C. G. Panayiotou, "Online IPA Gradient Estimators in Stochastic Continuous Fluid Models", *Journal of Optimization Theory and Applications*, Vol. 115, No. 2, pp. 369-405, November 2002.
[27] R. Guérin, H. Ahmadi, M. Naghshineh, "Equivalent Capacity and its Application to Bandwidth Allocation in High-Speed Networks", *IEEE Journal on Selected Areas in Communications*, Vol. 9, No. 7, pp. 968-981, September 1991.
[28] Web site with URL:
http://www-tkn.ee.tu-berlin.de/research/trace/trace.html.

[29] L. Rizzo, L. Vicisano, "RMDP: an FEC-Based Reliable Multicast Protocol for Wireless Environments", *Mobile Computer and Communication Review*, Vol. 2, No. 2, pp. 23-31, April 1998.

[30] M. Zorzi, "Performance of FEC and ARQ Error Control in Bursty Under Delay Constraints", in *Proc. of VTC '98*, Ottawa, Canada, May 1998.

[31] P. Barsocchi, A. Gotta, F. Potortì, F. J. González-Castaño, F. Gil-Castiñeira, J. I. Moreno, A. Cuevas, "Experimental Results with Forward Erasure Correction and Real Video Streaming in Hybrid Wireless Networks", in *Proc. of IEEE International Symposium on Wireless Communication Systems 2005* (ISWCS 2005), ISBN 0-7803-9206-X, Siena, Italy, September 5-9, 2005.

[32] E. Feltrin, E. Weller, E. Martin, K. Zamani, "Implementation of a Satellite Network Based on Skyplex Technology in Ka Band", in *Proc. of* 9th *Ka and Broadband Communications Conference*, Ischia, Italy, November 2003.

[33] Web site with URL: http://votos.isti.cnr.it.

[34] L. Rizzo, "Effective Erasure Codes for Reliable Computer Communication Protocols", *ACM Computer Communication Review*, Vol. 27, No. 2, pp. 24-36, April 1997.

[35] ANSI, "American National Standard for Telecommunications - Digital Transport of One-Way Video Signals - Parameters for Objective Performance Assessment", T1.801.03, 1996.

[36] ATIS, "Objective Video Quality Measurement Using a Peak Signal-to-Noise Ratio (PSNR) Full Reference Technique", Alliance for Telecommunications Industry Solutions, 1200 G Street, NW, Suite 500, Washington, DC 20005, Technical Report T1.TR.74, October 2001.

[37] A. Cuevas *et al.* "Usability and Evaluation of a Deployed 4G Network Prototype", *Journal of Communications and Networks* (ISSN: 1229-2370), Vol. 7, No. 2, pp. 222-230, June 2005.

[38] M. Liebsch *et al.*, "Candidate Access Router Discovery (CARD)", IETF RFC 4066, July 2005.

[39] ANSI/IEEE Std 802.2, 1998 Edition, "Part2: Logical Link Control".

[40] ANSI/IEEE Std 802.1D, 1998 Edition, "Part 3: Media Access Control (MAC) Bridges".

[41] IEEE, "802.1Q - IEEE Standards for Local and Metropolitan Area Networks - Virtual Bridged Local Area Networks", Std 802.1QT, 2003.

[42] IEEE, "IEEE Standard for Local and Metropolitan Area Networks - Common specifications Part 3: Media Access Control (MAC) Bridges - Amendment 2: Rapid Reconfiguration", Std 802.1w-2001.

[43] IEEE, "802.1s - IEEE Standards for Local and Metropolitan Area Networks - Virtual Bridged Local Area Networks - Amendment 3: Multiple Spanning Trees", Std 802.1sT-2002.

[44] V. Mancuso, G. Bianchi, N. Blefari Melazzi, U. Birnbacher, "Switched Ethernet Networking over LEO Satellite", in *Proc. of IEEE International Symposium on Wireless Communication Systems 2005* (ISWCS 2005), ISBN 0-7803-9206-X, Siena, Italy, September 5-9, 2005.

[45] The official OPNET Internet site with URL: http://www.opnet.com.

9

RESOURCE MANAGEMENT AND TRANSPORT LAYER

Editors: Gorry Fairhurst[1], Michele Luglio[2], Cesare Roseti[2]

Contributors: Nedo Celandroni[3], Paolo Chini[4], Gorry Fairhurst[1], Giovanni Giambene[4], Michele Luglio[2], Francesco Potortì[3], Cesare Roseti[2]

[1]UoA - University of Aberdeen, UK

[2]UToV - University of Rome "Tor Vergata", Italy

[3]CNR-ISTI - Research Area of Pisa, Italy

[4]CNIT - University of Siena, Italy

9.1 Introduction

The main challenges to be faced by transport protocols using a satellite link are the variability of the channel, due to weather conditions and the large propagation delay. Adaptive network management and control algorithms are therefore desirable to guarantee the *Quality of Service* (QoS) to data flows over *Additive White Gaussian Noise* (AWGN) channels (a baseline assumption in the following analysis).

Many popular Internet applications including email, file transfer, remote access, and Web browsing require a reliable data delivery service. End-to-end reliability for Internet traffic is guaranteed by the *Transmission Control Protocol* (TCP) at the transport layer. TCP specification covers a wide family of implementations, some of them having traditionally very poor performance over satellite links [1],[2]. Furthermore, TCP performance actually depends on the adopted *Radio Resource Management* (RRM) techniques. The DVB-RCS standard [3],[4] (see Scenario 2 for GEO-based communications, detailed in Chapter 1, Section 1.4) defines a set of *Demand Assignment Multiple Access* (DAMA) schemes, based on control loops with time constants similar to those used by TCP. Specific design choices and circumstances can lead to unwanted interactions between the link layer, implementing DAMA, and TCP, with a significant degradation of the overall performance. In particular, the so-called *access delay* (DAMA loop) becomes an important factor that affects TCP performance (see the following Section 9.4 for further details).

The variability of the satellite channel conditions also impacts Internet flows that do not utilize TCP (e.g., streaming multimedia utilizing *User Datagram Protocol*, UDP). This traffic typically is less demanding in terms of reliability, but more demanding in terms of jitter. Loss and/or corruption of multimedia packet payloads can be efficiently handled by the multimedia *codec* at the receiver and does not imply the need for retransmissions (as in TCP). The effect of propagation delay is also much less important, except for interactive services requiring low round trip delay (e.g., Internet gaming, which is not suited for satellite use).

Adaptive RRM procedures can be optimized for multimedia traffic. In particular, cross-layer optimization approaches are becoming widespread for wireless networks in general, and their application to satellite links needs to be studied both from a performance and an architectural viewpoint. This Chapter addresses the design and the implementation of cross-layer interactions between RRM (layer 2) and transport layer (layer 4) protocols in satellite environments. In particular, Section 9.2 offers an overview of TCP over GEO satellite links; Section 9.3 proposes an interaction between TCP and lower layers aiming at maximizing the TCP connection throughput; Section 9.4 describes the design of a cross-layer interaction between TCP and *Medium Access Control* (MAC) layer; Section 9.5 focuses on UDP and the performance of multimedia applications over satellite links, and, finally, Section 9.6 provides conclusions.

9.2 Overview of TCP over satellite

TCP is the primary transport protocol in the TCP/IP suite, designed to provide good performance over congested packet-switched networks [5],[6]. Similarly to other window-based protocols, TCP aims at guaranteeing a reliable and fair data exchange, despite congestion events (i.e., segment losses).

Satellite links present characteristics that significantly differ from those of wired links; moreover, TCP protocol mechanisms can be impacted by the significant delay and the presence of transmission errors [1]. In the GEO scenario, satellites are located at an altitude of approximately 36,000 km above the Earth at the equator latitude. In such a scenario, the *Round Trip propagation Delay* (RTD) between two ground stations at the Equator is on the order of 500 ms; this latency will be increased by other factors in the network (e.g., the transmission and propagation time of other links in the terrestrial path, and queuing delays in gateways and routers). Hence, the high RTD value entails a slow TCP *congestion window* (*cwnd*) increase that significantly affects the end-to-end transfer rate [7].

Much research has been directed to improve the TCP mechanism efficiency over satellite links. This Section provides a complete survey of different solutions proposed.

9.2.1 TCP standard mechanisms

TCP is implemented on top of the *Internet Protocol* (IP) and provides a reliable, byte-streaming and bi-directional connection, allowing applications to receive all segments in the correct order (a sequence number is associated to each TCP segment). Two concepts can be identified as the basis of the protocol: the *acknowledgement* (ACK) and the *sliding window.* ACKs are short packets generated by the receiver, when a TCP segment is received. They report to the sender the sequence number of the next segment the receiver expects. The *sliding window* defines the amount of unacknowledged data that can be sent at any given time [6].

TCP implements two basic mechanisms: *flow control* and *congestion control.* The *flow control* scheme allows an adequate exchange of data between two TCP nodes using a *sliding window* protocol, while the *congestion control* scheme is based on two algorithms called *Slow Start* (SS) and *Congestion Avoidance* (CA) [8]. These algorithms typically use two variables: *cwnd* and *slow start threshold* (*ssthresh*). In particular, each sending end-system probes the network congestion by gradually increasing the window of transmitted segments (outstanding in the network) until the network becomes congested and drops segments (or marks them as having experienced congestion).

Initially, the *cwnd* increase is exponential during the SS phase. The SS algorithm is quite simple and based on the amount of data sent in each *Round Trip Time* (RTT). Note that RTT is meant as the delay perceived at the transport layer, that is the RTD plus extra delays due to queuing, transmission time, processing, etc. At the beginning, the source sends one (or a small number, e.g., 2) TCP segment and waits for the relative ACK. Then, for each received ACK, it sends two segments. Therefore, each time the sender receives the ACKs for the window of data it has sent (*cwnd*), it doubles the amount of packets sent in the next RTT. When the *cwnd* size reaches the *sstresh* value, the increase becomes linear (i.e., one extra segment for each RTT period), thus allowing for a gentler probing of the available capacity; this is the CA phase.

TCP loss recovery

The default loss recovery mechanism of TCP is the *Retransmission TimeOut* (RTO) [6]. TCP performs a retransmission if RTO expires before the related segment is acknowledged. When RTO is over, TCP re-enters the SS phase by resetting *cwnd* and by re-transmitting the first unacknowledged segment. Basically, TCP uses the ACKs reception time to estimate the RTT and a smoothed average of RTT is used to set the RTO timer.

In modern implementations, TCP RTO is the final fall-back method and the timer rarely expires, hence, other methods are used for loss recovery. In particular the *Fast Retransmit* algorithm [8],[9] allows a sender to re-transmit a lost segment before RTO expires by exploiting the reception of *duplicate ACKs* (*dupACKs*) generated when segments are received out of order. In fact, TCP interprets the reception of a small number of *dupACKs* (usually 3) as an indication that a segment has been lost and retransmits it without waiting for the RTO expiration. Furthermore, TCP considers the loss as a congestion signal and reduces its transmission rate. To this purpose, the *Fast Recovery* algorithm halves *cwnd* as described in [9] and TCP enters the CA phase. Most TCP implementations, used within the Internet, add a method called *Selective ACK* (SACK) [10], to further optimize loss recovery when multiple losses occur in the same window of data. This option allows a sender to recover quickly from multiple lost segments. SACK also permits to achieve better performance with respect to multiple fast retransmissions. When SACK is used, a sender is generally able to determine which segments need to be retransmitted in the first RTT following loss detection. This avoids a slow start period (especially costly for high-delay links) following multiple segment losses and permits the sender to continue to transmit segments (retransmissions and new segments, if appropriate) at a suitable rate.

9.2.2 Criticalities of TCP on satellite links

TCP standard mechanisms, optimized to work correctly in wired (congested) networks, suffer from a certain number of factors when used over satellite links [1]. In particular, the high latency, the large *Bandwidth-Delay Product* (BDP), link asymmetry, and channel errors can negatively impact TCP performance.

TCP problems are experienced both while increasing the sliding window and in the steady-state. The former is due to the dependence of the TCP congestion control algorithms on the experienced RTT that in a satellite link is one or two order of magnitude greater than that in terrestrial networks. In fact, both SS and CA algorithms increase *cwnd* by using RTT as time parameter. The formulas below describe the time spent in SS and CA phases at the beginning of a TCP connection:

$$\begin{cases} slow\_start\_time = RTT \cdot \log_2(ssthresh) \\ congestion\_avoidance\_time = RTT \cdot (W - ssthresh) \end{cases} \tag{9.1}$$

where W denotes a suitable *cwnd* value reached in the CA phase (the ideal one corresponding to BDP).

Hence, the time needed to reach a given W is proportional to RTT. Furthermore, packet losses are interpreted by TCP as possible indication of congestion, leading to a *cwnd* reduction, thus slowing down the growth process.

In the latter type of problems, the maximum achievable throughput and then TCP steady-state performance are limited by the following formula:

$$Max\_throughput = \frac{TCP\_receiver\_window}{RTT} \quad (9.2)$$

where *TCP_receiver_window* indicates the maximum amount of data the receiver can store in its buffer at every time. In many systems, this value is advertised using a 16-bit field in the TCP header (maximum *TCP_receiver_window* = 65535 bytes).

Despite widespread support for much larger *TCP_receiver_windows* within the current deployed protocol stacks, these are rarely enabled by default; therefore, in GEO satellite links (with RTT about equal to 540 ms), the upper bound for the throughput is around 1 Mbit/s.

9.2.3 Survey of proposed solutions

Many solutions can be adopted to improve the TCP efficiency over satellite links; some of them are specifically proposed for the satellite environment while others for more general cases. Such solutions can be classified as follows:

- *Enhancements of the standard*
  - Loss recovery enhancements (i.e., SACK option [10]);
  - Large initial window;
  - Delayed acknowledgements to reduce the ACK flow;
  - Byte counting.
- *Modified algorithms*

  Experimental implementations of modified flow control mechanisms and options (i.e., TCP Vegas [11], TCP Peach [12], TCP Hybla [13], TCP Westwood [14]).
- *Modified architecture*

  Performance Enhancing Proxies (PEPs) [15] are often employed to improve the TCP performance. Many deployed satellite systems use PEPs to improve the performance of the TCP protocol to compensate for effects such as: delay, appreciable packet loss, variable bandwidth, asymmetry, mobility or other effects. A range of PEP techniques can be and are used; there is no "standard PEP" that satisfies all needs, and the most appropriate method will depend upon the service requirements (whether IPv6, mobility, IPsec, etc. are used), the link characteristics and the degree of complexity that users can tolerate in a middlebox. One common PEP method is to modify the end-to-end architecture at the transport protocol level (i.e., splitting the path, terminating connections, acknowledging packet receptions) by using one of the following approaches:

- "Enhanced" TCP version;
- Optimized protocols (i.e., XTP [16], SCPS-TP [17], etc.)

9.3 Cross-layer interaction between TCP and physical layer

It is possible to optimize the TCP performance over satellite links without changing the TCP behavior by operating on transmission parameters. Let us focus here on how TCP performance can be improved by trading packet loss rate for bottleneck link bandwidth. This procedure does not interfere in any way with the normal behavior of the TCP stack, but requires the capability of tuning link parameters at the physical level.

For a given available radio spectrum, antenna size and maximum transmission power, there are many choices for the selection of modulation scheme, symbol rate and *Forward Error Correction* (FEC) type. Commonly used wireless systems make such design choices in a static mode, permitting the user to change manually some of them, or in some cases to switch dynamically among a limited number of preset parameters. For each possible set of parameters, we define *Information Bit Rate* (IBR) as the link speed seen by the network layer, that is, the product of the symbol rate, the number of bits per symbol and the FEC rate, as detailed in what follows. Even in those cases where a wide range of IBR values is available, the criterion to switch among them is only dependent on the perceived channel quality, i.e., on the performance measured at the physical layer [18]. For terrestrial wireless environments, an example of physical layer with multiple choices is provided by the IEEE 802.11 standard, where it is possible to change dynamically modulation and coding schemes [19].

The hopping among different sets of physical layer parameters is due to the highly variable physical characteristics of most wireless links; in particular, all satellite links are subject to variable atmospheric attenuation of the signal; *Low* and *Medium Earth Orbit* (LEO and MEO) satellite constellations additionally suffer from variable signal attenuation (due to changing slant path), blocking (due to obstacles in the *Line of Sight*, LoS), multipath fading, and changing satellite distance.

Internet has been conceived assuming that *Packet Error Rate* (PER) should be as low as possible for a good TCP performance [2]: a common rule of thumb is to engineer the link so that PER due to link errors is negligible with respect to the loss rate caused by congestion. However, this could not be the appropriate choice; in [20] in fact, it is shown that a good rule of thumb is to set the ratio between PER and congestion loss to a value equal to the number of TCP connections. This finding opens the possibility for defining adaptive algorithms that dynamically choose the optimal channel parameters (e.g., modulation and coding) with the aim of maximizing the TCP performance. Implementing such algorithms requires a cross-layer approach, because physical layer parameters such as modulation and coding need to be tuned depending on information available at the transport layer.

The following analysis of the TCP performance highlights the cross-layer issues and interactions with the physical layer by relating PER with the TCP throughput. In 1997, a simple and elegant formula, relating the steady-state performance of TCP to its segment loss rate, was discovered [21]. This formula connects the maximum throughput on an unlimited bandwidth channel with the *Maximum Segment Size* (MSS), RTT, and the packet (segment) loss rate, PER:

$$throughput = K\frac{MSS}{RTT\sqrt{PER}} \qquad for\ PER < 1\% \tag{9.3}$$

where K is a constant equal to 1.31 in the case of random segment losses without delayed ACKs.

Subsequently, (9.3) was modified to take into account the TCP behavior in the presence of timeouts [22], thus allowing for a greater accuracy at higher PER, resulting in:

$$throughput = \frac{MSS}{RTT\sqrt{\frac{2bq}{3}} + RTO\min\left(1, 3\sqrt{\frac{3bq}{8}}\right)q\left(1+32q^2\right)} \tag{9.4}$$

where b is the number of segments acknowledged by each ACK and q is the PER.

Finally, in [23], a method for computing the TCP throughput in band-limited channels was proposed.

The above described relationship between PER and TCP throughput was used in [24] to optimize single-connection TCP performance, and in [20] for the case of multiple connections. In fact, for a given wireless transmission system, it is realistic to assume that some of the parameters are dynamically tunable, so that the channel PER can be traded for IBR in order to find the optimal configuration that maximizes the TCP throughput computed at the end-user (i.e., *goodput*). For example, it is possible to change the modulation scheme, thus reducing the channel bit-rate, in order to obtain higher bit energy to one-sided noise spectral density ratio, E_b/N_0. The modulation scheme can be changed together with the FEC characteristics to have a wider range of choices and to exploit as much as possible the available radio spectrum. Most modern transmission systems provide for variable bit-rates by changing the used FEC redundancy, some of them for each individual packet. It is even possible to seamlessly change the FEC, while maintaining bit time synchronization of the data stream, by using rate-compatible punctured convolutional codes [25].

Concerning the modulation scheme, let us consider a satellite carrier modulated at a rate of S symbols/s. We envisage widely diffused M-ary modulations such as *Amplitude and Phase Shift Keying* (APSK) or *Quadrature Amplitude Modulation* (QAM) types. In these schemes, M is the number of points, in the phase-amplitude space, relative to the constellation of the modulated symbols. Typical values of M are 2, 4, 8, 16, 32, and 64; *Binary Phase Shift Keying* (BPSK) and *Quadrature Phase Shift Keying* (QPSK) schemes correspond to M values of 2 and 4, respectively.

FEC types come in a multitude of modes, often concatenated between them. We define the coding rate r as the inverse of the coding redundancy. For example, in the case of a (255, 223) Reed-Solomon code concatenated with a 1/2 convolutional code, the resulting r is $223/(255 \times 2) = 0.437$.

The IBR (i.e., the TCP bottleneck rate) is given by

$$IBR = Sr \log_2 M. \tag{9.5}$$

For a given channel condition C/N_0, a modulation scheme with M points and a coding rate r, the operating E_b/N_0 expressed in dB is given by

$$E_b/N_0 = C/N_0 - 10 \log_{10} IBR \quad \text{[dB]}. \tag{9.6}$$

Diagrams similar to the one shown in Figure 9.1 [20] are used to estimate PER, given the channel parameters.

For each C/N_0, PER is a function of M and r. Since the TCP goodput depends on IBR and PER, it is possible to find M and r that maximize the goodput for each given C/N_0. The way the goodput depends on IBR and PER can be evaluated using either experimental measurements, or simulation, or an analytical expression, such as (9.3) or (9.4). As an example, Figure 9.2 shows the goodput achievable by considering two modulation schemes, namely BPSK ($M = 2$) and QPSK ($M = 4$) and four coding schemes, all based on the standard NASA convolutional coding with constraint length 7, with code rates $r = 1/2$ (base code), $r = 3/4$ (punctured base code), $r = 7/8$ (punctured base code), $r = 1$ (no coding) [26]. Four curves are depicted, each for a different C/N_0. Each curve represents how the TCP goodput changes as a function of IBR (i.e., the bottleneck rate). For each channel condition (i.e., C/N_0 value), an optimum combination of parameters exists which gives the maximum TCP goodput. Notice that the curves may exhibit notches due to the discontinuous parameter space. In Figure 9.2, for example, this happens for the case of C/N_0 = 67 dB, where BPSK at 7/8 coding rate has both a smaller IBR and a worse PER performance than QPSK at 1/2 coding rate.

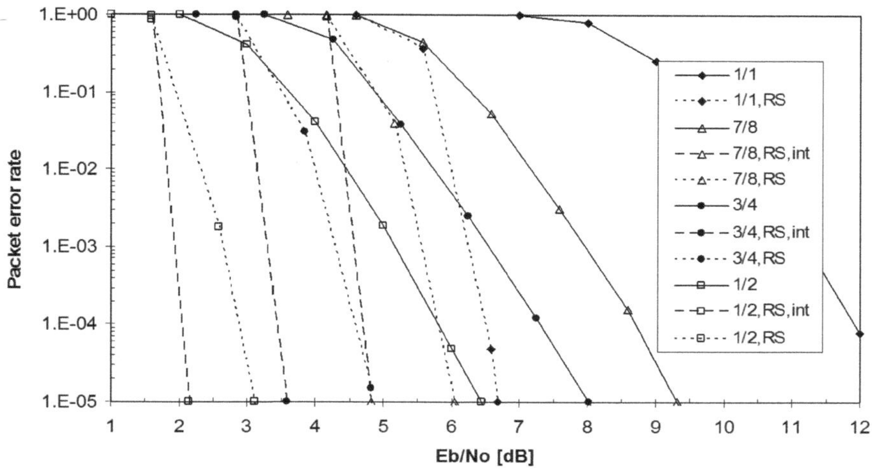

Fig. 9.1: PER as a function of E_b/N_0 for different combinations of concatenated codes. The "int" in the legend means ideal interleaving. See reference [20]. Copyright ©2006 IEEE.

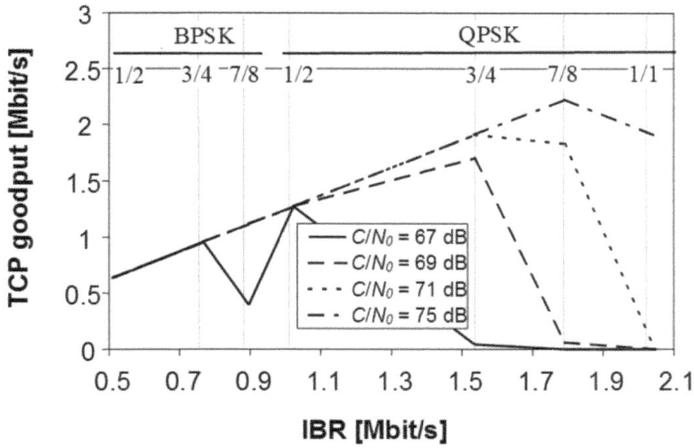

Fig. 9.2: Goodput of a single TCP NewReno connection versus the available IBR for different C/N_0 values. Packet size = 1 kB, S = 1.024 Msymbol/s, bottleneck buffer size = BDP of a GEO link. Labels indicate modulation schemes and FEC rates.

The above discussion leads to the conclusion that satellite systems could benefit from adaptive algorithms for choosing the transmission parameters by means of cross-layer interactions between transport and physical layers. An additional possibility is that MAC and physical layers interact by inserting a link-layer erasure code [20],[27] just above MAC layer, which could be an all-software solution, independent of the underlying hardware characteristics.

The recent DVB-S2 standard [28] considers very powerful error-correcting codes. For ideal AWGN channel conditions, an optimization based on channel coding would be useless because the curves that give PER versus E_b/N_0 are very steep [29], causing a sort of on-off behavior of the physical channel: either PER is negligible, or it is so high that it collapses TCP performance. However, optimizing channel parameters makes sense in non-ideal channel conditions, and, in general, on the satellite return channel [3].

9.4 Cross-layer interaction between TCP and MAC

The interaction between TCP and MAC protocols in a shared network can greatly improve the efficiency of satellite systems. MAC protocols play a fundamental role to guarantee good performance to higher-level protocols by managing the arbitration of uplink access. Two cases must be distinguished: (*i*) when TCP operates end-to-end (as the general Internet standard, or when an end-to-end IPsec protection scheme is used); (*ii*) a PEP scheme violates the end-to-end semantics. Without loss of generality, hereafter we will consider the latter case, where, referring to a DVB-RCS network (see Chapter 1, Section 1.4), the Gateway acts as a PEP (i.e., it is a local TCP receiver -from remote RCSTs-, located in the Earth station).

Satellite networks employing a DAMA scheme introduce an additional contribution to the end-to-end delay, called the *access delay* that can significantly impact the end-to-end performance of TCP flows. In a DVB-RCS-like network, the *Network Control Center* (NCC) assigns return link capacity in response to explicit requests received from RCSTs [3]. This capacity negotiation requires a signaling exchange that regulates the data flow. Therefore, when TCP is used as transport protocol, two nested control loops exist with the same time constants (i.e., RTT):

- At MAC layer: resource request - resource assignment loop;
- At TCP layer: TCP segment - acknowledgement loop.

The consequence of this interaction is an increase in the latency perceived by the end-systems. To mitigate this effect it is possible to reduce the access delay with a preventive allocation scheme driven by a cross-layer interaction between MAC and TCP layers. The idea is to use the TCP parameters, such as *cwnd* and *ssthresh*, to estimate in advance the resources needed by a given TCP flow [30],[31]. In fact, from the comparison of these two quantities, it is possible to determine the TCP congestion control status (i.e., SS or CA).

Consequently, the MAC layer can know the law according to which *cwnd* is enlarged on an RTT basis and can predict with a very good accuracy the necessary resource allocation needed by each TCP flow. In this way, it is expected to reduce significantly queuing delay, while also achieving an efficient utilization of the satellite shared capacity. More details on this approach are provided in the following sub-Sections.

9.4.1 A novel TCP-driven dynamic resource allocation scheme

The implementation of a dynamic access scheme allows optimizing the resource sharing. The DVB-RCS standard defines the following set of capacity request methods (see for more details Chapter 1, sub-Section 1.4.3): CRA, RBDC, VBDC, AVBDC, and FCA.

In particular, VBDC performs a capacity request, as long as new data arrive in the RCST queue. The amount of capacity per frame, a generic RCST requests at the k-th super-frame, can be expressed by using the formula defined in [32]:

$$r(k) = \left\lceil \frac{q(k) - n_s \cdot a(k) - n_s \cdot \sum_{j=1}^{L-1} r(k - L + j) - n_s \cdot w(k)}{n_s} \right\rceil \tag{9.7}$$

where:

- $\lceil . \rceil$ denotes rounding to the upper positive integer;
- $q(k)$ = amount of queued data;
- n_s = number of frames per super-frame;
- $n_s \cdot a(k)$ = capacity assigned in the k-th super-frame;
- L = *system response time* expressed in super-frames (also indicated as *allocation period*); it represents the time elapsed from a capacity request transmission to the actual assignment of the requested capacity;
- $n_s \cdot \sum_{j=1}^{L-1} r(k - L + j)$ = resources requested in the previous super-frames, but not yet assigned;
- $n_s \cdot w(k)$ = resources requested in the previous allocation periods and not yet assigned.

Unfortunately, the VBDC allocation method leads to a huge increase in the end-to-end delay perceived by the systems where TCP applications are running. In fact, the above mentioned access delay involves in this case the following contributions:

- *Reservation delay*: since requests are sent at a fixed rate in dedicated slots, a time interval occurs between the arrival of data in the MAC buffer and the transmission of the corresponding capacity request;
- *RTD contribution*: sum of the time to propagate the capacity request from the RCST to the NCC and the time to deliver the *Terminal Burst Time Plan* (TBTP) in the opposite direction;

- *Processing (and synchronization) delay*: time spent by the DAMA controller (in the NCC) to transmit the TBTP message with the capacity assignment;
- *Forwarding delay*: time between the reception of the TBTP by the RCST and the actual transmission of data.

On the basis of the above delay contributions, the RTT values corresponding to the VBDC case can be of the order of 1.6 s ([1]) in a standard GEO bent-pipe system [33].

The DVB-RCS standard also supports an RBDC capacity request method. In this case, resources are allocated on the basis of the rate at which an RCST wishes to transmit (usually based on monitoring the arrival rate at its layer 2 queue). This method reduces the access delay.

Most RCS systems provide a wide range of *Bandwidth on Demand* (BoD) schemes based on a combination of both methods (VBDC and RBDC). As already anticipated in Section 9.4, our interest here is in reducing the access delay, keeping optimal network efficiency, by using TCP status information to predict the amount of data that will feed the RCST queue in the future. In order to exchange cross-layer signaling between layer 2 and layer 4, dedicated local messages [31] are generated each time that TCP parameters (e.g., *cwnd*) go beyond a certain threshold; this is according to an explicit cross-layer method.

Let us assume a *system response time* greater than the physical RTD ([2]), in computing the $r(k)$ request. Such assumption allows to the proposed algorithm predicting the further data that will be present in the RCST queue when the resources will be allocated, according to both the amount of data transmitted in the k-th super-frame and the TCP phase (SS or CA):

$$Q'(k) = \begin{cases} 2 \cdot n_s \cdot a(k) & Slow\_Start \\ n_s \cdot a(k) \cdot \left(1 + \frac{1}{cwnd}\right) & Congestion\_Avoidance \end{cases} . \tag{9.8}$$

Therefore, in our TCP-driven RRM a new term is added to (9.7) and, therefore, the amount of resources per frame requested at the k-th super-frame, $r(k)$ is:

$$r(k) = \left\lceil \frac{q(k) - n_s \cdot a(k) - n_s \cdot \sum_{j=1}^{L-1} r(k-L+j) - n_s \cdot w(k)}{n_s} + \frac{Q'(k)}{n_s} \right\rceil . \tag{9.9}$$

[1] The value of RTT ≈ 3 RTD is due to the use, for the simulations, of an architecture where NCC is separated from the Gateway.

[2] This assumption is appropriate to current DVB-RCS systems when the TCP flow is not encrypted, especially when PEP mechanisms are used at the satellite Gateway to end TCP connections within the satellite segment.

Finally, in addition to $r(k)$, also the TCP phases will be communicated by the RCST to the NCC in the capacity request message by setting the following flag (*TCP phase flag*):

- 1 $\longrightarrow$ Slow Start;
- 0 $\longrightarrow$ Congestion Avoidance.

On the other side, the NCC serves all incoming requests by considering two priority levels: a *High priority level* associated to requests with the *TCP phase flag* set to 1, and a *Low priority level* associated to requests with the *TCP phase flag* set to 0. Our aim is to prioritize connections in the SS phase with respect to those operating in the CA one to favor both short transfers and just started connections. In each queue (i.e., the queue for requests in the SS phase and the queue for requests in the CA phase), requests are satisfied according to *Maximum Legal Increment* (MLI) algorithm [34] to guarantee a fair allocation among the different competing flows.

If the amount of needed resources exceeds those available in a super-frame, the NCC creates a "*waiting list*" to assign the resources in the next super-frames and stops the *cwnd* growth of all the connections coming from the RCSTs that have not obtained the requested resources. In particular, the proposed allocation scheme at the NCC performs the following two tasks:

- Assure that resources are fairly shared among all the active TCP connections;
- Provide a further cross-layer action that sets a new variable, named *cwnd**, in order to modify the current *cwnd* value used by the TCP source in the RCST as follows: *cwnd* $\longleftarrow$ *cwnd**. Note that the NCC (acting like a PEP) sends back the *cwnd** value by using a field for TCP options (layer 4 ACKs) in the headers. The rationale of this modification on the TCP protocols is to avoid internal congestion on the RCST side and, then, the possibility of layer 2 buffer overflows.

The main expected effects of the proposed cross-layer-based access scheme are:

- *Reduction of the access delay*: since the request algorithm predicts also the amount of data that will feed the RCST queue due to the TCP congestion control mechanism, the access delay will be reduced of an RTD;
- *Avoidance of internal congestions at the RCSTs*: the cross-layer interaction between RRM and TCP layers permits to prevent layer 2 buffer overflows due to satellite network congestion;
- *Efficient and dynamic resource allocation*: resources are dynamically assigned on a super-frame basis according to explicit requests, thus allowing a better utilization of the available capacity.

Analysis of the allocation process

A simulator has been implemented using ns-2 (release 2.27) [35], in order to evaluate the performance of the cross-layer allocation process and the resulting performance. In particular, the ns-2 extensions that reproduce a traditional GEO satellite network have been modified to simulate a centralized *Multi Frequency - Time Division Multiple Access* (MF-TDMA) scheme and the NCC functionalities.

The interaction between the TCP *cwnd* trend and the corresponding allocation process has been analyzed by means of the average resources assigned (in slots) as a function of time; such parameter has been monitored for one or more TCP connections sharing the return link of a communication network compliant to Scenario 2 described in Chapter 1, sub-Section 1.4.5. The main simulation parameters are detailed in Table 9.1.

Physical parameters	
Physical RTT (RTD)	$\sim$ 515 ms
Return link bandwidth	2048 kbit/s
Maximum number of RCSTs	32
Frame parameters	
Super-frame duration	96 ms
Number of slots per frame	32
Protocols	
Transport Protocol	TCP NewReno
Application Protocol	FTP
TCP parameters	
TCP packet size	1500 bytes
PER	Variable, from 0 to 0.0001

Table 9.1: Main simulation parameter values.

In particular, by considering a file transfer (where the application layer is achieved by means of the *File Transfer Protocol*, FTP) from an RCST to the NCC, Figure 9.3 highlights how the allocated resources (continuous line) are strictly correlated to the *cwnd* trend (dotted line) with our scheme. In particular, three different phases can be recognized in the allocation process according to the following sequence:

1. An initial exponential growth corresponding to the TCP SS phase;
2. A clear reduction of the allocated resources (approximately one half) when the Fast Recovery mechanism is invoked as reaction to the detection of a loss;
3. A linear growth corresponding to the TCP CA phase.

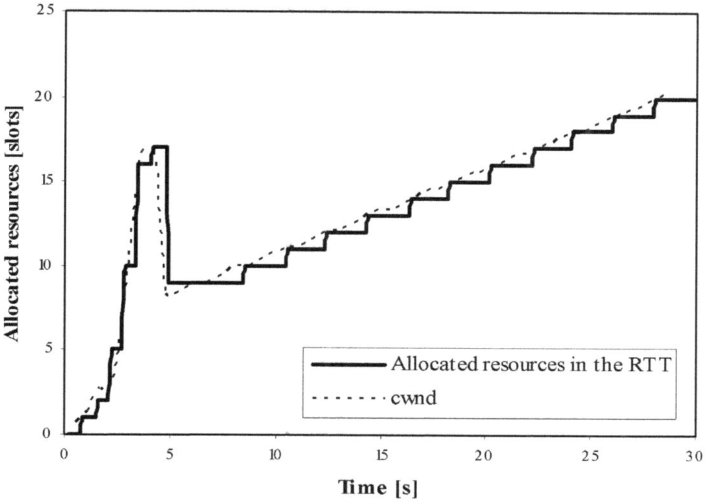

Fig. 9.3: Comparison between allocated resources and *cwnd* trend versus time (1 TCP connection, PER = 10^{-4}).

Referring to our TCP-driven RRM scheme, Figure 9.4 focuses on the fair resource sharing between two TCP connections, when losses occur. At the beginning, the capacity is saturated (i.e., the NCC stops the *cwnd* growth of both the connections in order to prevent congestion and losses): the overall capacity is perfectly divided between the two connections. When a connection is affected by a transmission error (loss), with consequent *cwnd* reduction, the NCC re-assigns temporarily the unused capacity to the other connection in order to optimize the utilization of resources.

Performance evaluation

The TCP performance strictly depends on the perceived latency at the end-systems, as shown by (9.1) and (9.2). Therefore, RTT can represent a valid parameter to evaluate the TCP performance. Hence, we have compared our TCP-driven RRM scheme with the classical CRA and VBDC capacity allocation techniques [3]. The main simulation parameters, compliant to Scenario 2, are those provided in the previous Table 9.1.

Figure 9.5 shows the average perceived RTT for the three considered access schemes. In particular, the obtained results allow the following considerations:

- VBDC presents the higher delay equivalent to about three times the physical RTD (see Chapter 1 for RTD characteristics) [33]: 1 RTD for the capacity request (on the basis of new data in the layer 2 queue, RCST side) and notification exchange; 1 RTD for the TCP segment and ACK

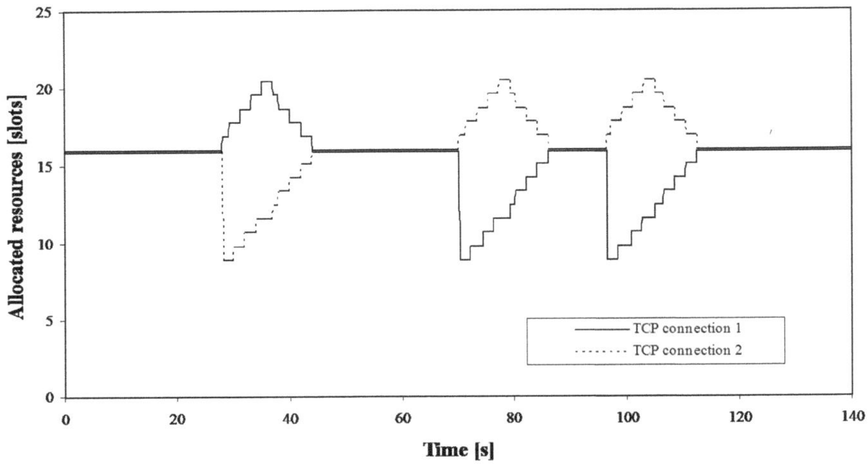

Fig. 9.4: Comparison among allocated resources in the RTT versus time (2 TCP connections, PER = 10^{-4}).

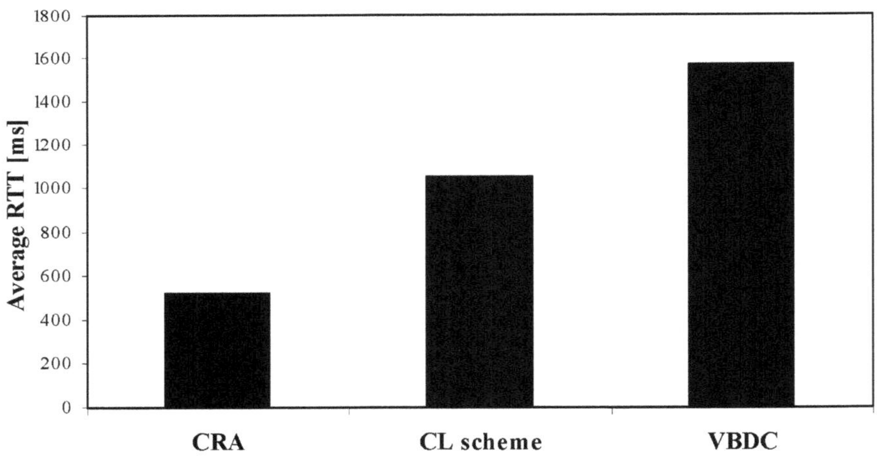

Fig. 9.5: Comparison among average RTT values obtained with the following techniques: VBDC, CRA and cross-layer scheme.

exchange; 1 RTD for the capacity allocation for the availability of the channel for ACK transmissions (Gateway side).
- In the CRA case, RTT is only affected by the physical delay RTD, since the capacity is not negotiated, but permanently assigned in the set-up phase of a connection;
- The proposed TCP-driven RRM scheme (also simply called "cross-layer scheme" in what follows) reduces the overall VBDC delay by almost 1 RTD, trying to predict the amount of data that will feed the RCST queue.

Then, by evaluating only the end-to-end performance in terms of RTT for a single TCP connection, the proposed cross-layer technique represents a good trade-off solution between VBDC and CRA.

The principle of assigning capacity on the basis of the real needs of data sources leads to significant improvements in terms of both end-to-end performance and network utilization when multiple TCP connections compete for the overall capacity.

The following simulations have been performed considering: 20 FTP transfers coming from different RCSTs and with start time instants spaced of 5 s; 10 Mbytes files have to be uploaded to a remote system through the satellite Gateway. As a reference, a fixed allocation scheme (i.e., CRA) is considered where the capacity is equally divided at the beginning among the RCSTs in a static manner. The average file transfer time has been measured for different PER values and then compared with the mean transfer time of the proposed cross-layer scheme. The results, shown in Figure 9.6, highlight that the TCP-driven RRM scheme with cross-layer information allows a reduction of the mean transfer time ranging from 12.3% (PER = 0) to 26.5% (PER = 0.01).

Finally, Figure 9.7 highlights the benefits derived from the use of the proposed cross-layer scheme with respect to CRA in terms of channel utilization. In fact, the continuous line indicates the percentage of the average utilization increase, for the cross-layer scheme with respect to CRA, when 5 FTP transfers (10 Mbytes) are running at instants spaced of 5 seconds with PER = 10^{-3}. This figure also shows the curve representing the instantaneous channel utilization when the cross-layer scheme is used (dashed line), in order to show the optimal values constantly achieved.

9.5 Overview of UDP-based multimedia over satellite

This Section focuses on multimedia transport in satellite networks, with a specific reference to Scenario 2 described in Chapter 1, sub-Section 1.4.5. Cross-layer methods offer new opportunities for satellite systems to adapt RRM to the needs of multimedia traffic. The challenge is the design of cross-layer mechanisms that can optimize the overall end-to-end multimedia application performance over satellite links, while minimizing the utilized

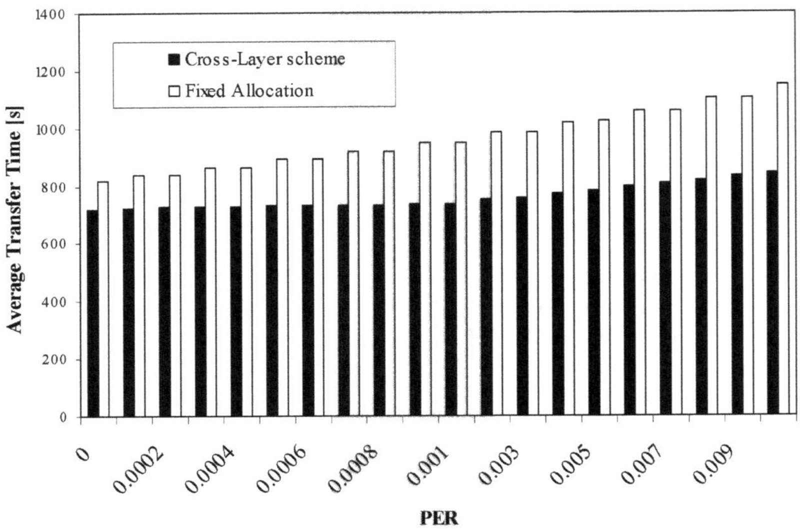

Fig. 9.6: Average file transfer time versus PER (20 FTP transfers starting at instants spaced of 5 s).

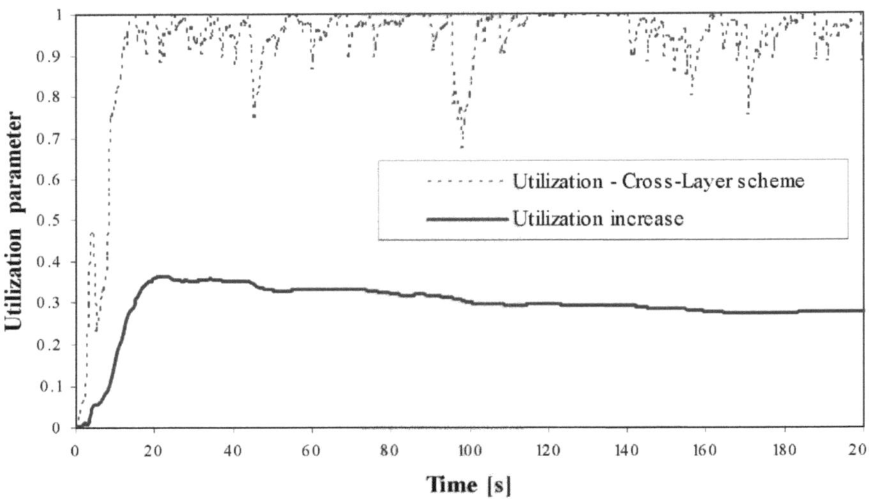

Fig. 9.7: Cross-layer access scheme: utilization and percentage of the average utilization increase with respect to the CRA scheme (5 FTP transfers starting at instants spaced of 5 s, PER = 10^{-3}).

radio resource. This topic requires a combination of expertise in propagation analysis, channel modeling, coding and modulation, jointly with consideration of link framing design and transport protocol design/evaluation. Analysis can be performed by combining physical simulation (based on propagation models) with packet-level protocol simulation (including application modeling).

9.5.1 Cross-layer methods for UDP

Examples of multimedia cross-layer methods include adapting transport protocols and application mechanisms to make them more robust to changes in the link quality conditions [36].

A first type of cross-layer method uses RRM and QoS techniques to tailor lower layer parameters to the characteristics of particular multimedia flows (as proposed for TCP in earlier Section 9.4). The requirements for multimedia traffic can differ from application to application. This kind of cross-layer communication also implies some form of signaling exchange between different protocol layers.

Recognizing the emergence of error-tolerant codecs, IETF has recently standardized a new multimedia transport protocol, named UDP-Lite [37], allowing an application to specify the required level of payload protection, while maintaining end-to-end delivery checks. In order to benefit from using UDP-Lite, the changes at the transport layer must be reflected in the design of satellite link and physical layers. Hence, it is important to tune the characteristics of lower layers in terms of modulation and coding (trading BER for IBR).

Cross-layer signaling may also be valuable to indicate the prevailing system performance to transport entities (in PEPs or end-hosts); this could also permit multimedia applications to adjust their choice of media codec in response to increased delay or reduced capacity. Hence, the use of cross-layer methods can provide increased information to the transport layer and applications concerning the quality and characteristic of the channel they are using. This new flexibility gives opportunity to higher-layer protocols to react in appropriate ways.

The success of multimedia cross-layer approaches relies not only on the development of suitable techniques, but also on the selection of appropriate signaling methods, and on the adoption of design methodologies that will permit cross-layer systems to inter-work and to evolve.

9.6 Conclusions

This Chapter provides an overview of the key issues that concern transport protocol performance over paths that include a GEO satellite segment. In particular, it gives a detailed survey of several approaches that permit a better interaction of transport layer protocols with RRM and physical layers.

Adaptive resource management can both guarantee efficient network utilization and satisfy QoS requirements, using satellite links that are affected by variable weather conditions and large propagation delays. An approach tuning the satellite link parameters at the physical or link layer and trading bottleneck bandwidth with segment error rate can permit to improve TCP performance. Moreover, where it is possible to evaluate the channel conditions in real-time, further sophisticated cross-layer interactions could be exploited for an adaptive selection of the physical layer parameters.

DAMA schemes may be used to achieve an efficient resource allocation, but they degrade the TCP performance by adding an access delay that increases the whole end-to-end delay. Then, explicit cross-layer approaches can also mitigate the interactions between TCP and DAMA (MAC layer). The rationale is to use TCP information to estimate in advance the amount of resources needed for a given TCP source. This should permit MAC layer to perform capacity requests based on both the volume of queued data and the predicted TCP traffic behavior. Simulations show that this TCP-driven RRM scheme represents a good trade-off solution with respect to both fixed access schemes, optimizing TCP performance, and classical dynamic access schemes, optimizing network efficiency.

Transport protocols not based on TCP can also benefit from cross-layer methods. Multimedia traffic flows using UDP and UDP-Lite are also expected to benefit from improved communication between protocol layers; in fact, application performance can be tuned to link and physical layer conditions, in order to achieve a system optimization.

References

[1] H. Kruse, "Performance of Common Data Communications Protocols over Long Delay Links: an Experimental Examination", in *Proc. of the* 3rd *International Conference on Telecommunication Systems Modeling and Design*, 1995.

[2] C. Partridge, T. J. Shepard, "TCP/IP Performance over Satellite Links", *IEEE Network*, Vol. 11, No. 5, pp. 44-49, September/October 1997.

[3] ETSI, "Digital Video Broadcasting (DVB); Interaction Channel for Satellite Distribution Systems", EN 301 790, V1.3.1, 2003.

[4] ETSI, "Digital Video Broadcasting (DVB); Interaction Channel for Satellite Distribution Systems; Guidelines for the use of EN 301 790", TR 101 790, V1.2.1, 2003.

[5] J. Postel, "Transmission Control Protocol", IETF RFC 793, September 1981.

[6] W. Stevens. *TCP/IP Illustrated.* Vol. 1, Ed. Addison Wesley, 1994.

[7] M. Luglio, C. Roseti, M. Gerla, "The Impact of Efficient Flow Control and OS Features on TCP Performance over Satellite Links", *ASSI Satellite Communication Letter (Sat-Comm Letter),* 9th *edition, special issue on Multimedia Satellite Communication*, Vol. 3, No. 1, pp. 1-9, June 2004.

[8] W. Stevens, "TCP Slow Start, Congestion Avoidance, Fast Retransmit and Fast Recovery Algorithms", IETF RFC 2001, January 1997.

[9] V. Jacobson, M. J. Karels, "Congestion Avoidance and Control", in *Proc. of ACM SIGCOMM*, 1988.

[10] M. Mathis, J. Mahdavi, S. Floyd, A. Romanow, "TCP Selective Acknowledgement Options", IETF RFC 2018, April 1996.

[11] L. S. Brakmo, L. L. Peterson, "TCP Vegas: End-to-end Congestion Avoidance on a global internet", *IEEE J. Select. Areas Commun.*, Vol. 13, No. 8, pp. 1465-1480, October 1995.

[12] I. F. Akyldiz, G. Morabito, S. Palazzo, "TCP-Peach: a New Congestion Control Scheme for Satellite IP Networks", *IEEE/ACM Trans. on Networking*, Vol. 9, No. 3, pp. 307-321, June 2001.

[13] C. Caini, R. Firrincieli, "TCP Hybla: a TCP Enhancement for Heterogeneous Networks", *International Journal of Satellite Communications and Networking*, Vol. 22, No. 5, pp. 547-566, September 2004.

[14] C. Casetti, M. Gerla, S. Mascolo, M. Y. Sanadidi, R. Wang, "TCP Westwood: End-to-End Congestion Control for Wired/Wireless Networks", *Wireless Networks Journal*, Vol. 8, No. 5, pp. 467-479, September 2002.

[15] J. Border, M. Kojo, J. Griner, G. Montenegro, Z. Shelby, "Performance Enhancing Proxies Intended to Mitigate Link-Related Degradations", IETF RFC 3135, 2001.
[16] "XTP Protocol Definition Revision 3.6", Protocol Engines Incorporated, PEI 92-10, Mountain View, CA, January 11, 1992.
[17] "Space Communications Protocol Specification-Transport Protocol (SCPS-TP)", CCSDS 714.0-B-1, http://www.scps.org.
[18] C. B. Cox, T. A. Coney, "Advanced Communications Technology Satellite (ACTS) Fade Compensation Protocol Impact on Very Small-Aperture Terminal Bit Error Rate Performance", *IEEE Journal on Selected Areas in Communications*, Vol. 17, No. 2, pp. 173-179, February 1999.
[19] ISO/IEC 8802-11; ANSI/IEEE Std 802.11, 1999 edn Information technology - Telecommunications and information exchange between systems - Local and metropolitan area networks - Specific requirements. Part 11: Wireless LAN Medium Access Control (MAC) and Physical Layer (PHY) specifications.
[20] N. Celandroni, "Comparison of FEC Types with Regard to the Efficiency of TCP Connections over AWGN Satellite Channels", *IEEE Trans. on Wireless Communications*, Vol. 5, No. 7, pp. 1735-1745, July 2006.
[21] M. Mathis, J. Semke, J. Mahdavi, T. Ott, "The Macroscopic Behavior of the TCP Congestion Avoidance Algorithm", *Computer Communications Review*, Vol. 27, No. 3, pp. 67-82, July 1997.
[22] J. Padhye, V. Firoiu, D. F. Towsley, J. F. Kurose, "Modeling TCP Reno Performance: a Simple Model and its Empirical Validation", *IEEE/ACM Trans. Networking*, Vol. 8, No. 2, pp. 133-145, April 2000.
[23] T. V. Lakshman, U. Madhow, "The Performance of TCP/IP for Networks with high Bandwidth-delay Products and Random Loss", *IEEE/ACM Transactions on Networking*, Vol. 5, No. 3, pp. 336-350, June 1997.
[24] N. Celandroni, F. Potortì, "Maximising Single Connection TCP Goodput by Trading Bandwidth for BER", *International Journal of Communication Systems*, Vol. 16, No. 1, pp. 63-79, February 2003.
[25] J. Hagenauer, "Rate-Compatible Punctured Convolutional Codes (RCP Codes) and their Applications", *IEEE Transactions on Communications*, Vol. 36, No. 4, pp. 389-400, April 1988.
[26] C. Caini, R. Firrincieli, M. Marchese, T. de Cola, M. Luglio, C. Roseti, N. Celandroni, F. Potortì, "Transport Layer Protocols and Architectures for Satellite Networks", *International Journal of Satellite Communications and Networking*, Vol. 25, No. 1, pp. 1-26, January/February 2007.
[27] C. Barakat, E. Altman, "Bandwidth Trade-off between TCP and Link-Level FEC", *Computer Networks*, Vol. 39, No. 2, pp. 133-150, June 2002.
[28] ETSI, "Digital video broadcasting (DVB); second generation framing structure, channel coding and modulation systems for broadcasting, interactive services, news gathering and other broadband satellite applications", EN 302 307, June 2004.
[29] G. Albertazzi, S. Cioni, G. E. Corazza, N. De Laurentiis, M. Neri, P. Salmi and A. Vanelli-Coralli, "Adaptive Coding and Modulation Techniques for Future Ka Band Satellite Systems-Part I: Forward Link", in *Proc. of the* 10th *Ka and Broadband Communications Conference*, Vicenza, Italy, pp. 329-335, September 2004.
[30] P. Chini, G. Giambene, D. Bartolini, M. Luglio, C. Roseti, "Dynamic Resource Allocation based on a TCP-MAC Cross-Layer Approach for Interactive

Satellite Networks", in *Proc. of IEEE International Symposium on Wireless Communication Systems 2005* (ISWCS 2005), ISBN 0-7803-9206-X, Siena, Italy, September 5-9, 2005.

[31] P. Chini, G. Giambene, D. Bartolini, M. Luglio, C. Roseti, "Cross-Layer Management of Radio Resources in an Interactive DVB-RCS-based Satellite Network", in *Proc. of the* 20th *International Symposium on Computer and Information Sciences* (ISCIS2005), pp. 124-135, October 2005.

[32] M. Karaliopoulos, R. Tafazolli, B. G. Evans, "Providing Differentiated Service to TCP Flows over Bandwidth on Demand Geostationary Satellite Networks", *IEEE Journal on Select Areas in Communications*, Vol. 22, No. 2, pp. 333-347, February 2004.

[33] M. Sooriyabandara, G. Fairhurst, "Dynamics of TCP over BoD satellite Networks", *International Journal of Satellite Communications and Networking*, Vol. 21, No. 4-5, pp. 427-449, July 2005.

[34] G. Açar, C. Rosenberg, "Algorithms to Compute for Bandwidth on Demand Requests in a Satellite Access Unit", in *Proc. of* 5th *Ka-band Utilization Conference*, pp. 353-360, 1999.

[35] NS-2 Network Simulator (Vers. 2.27), URL: http://www.isi.edu/nsnam/ns/nsbuild.html.

[36] G. Fairhurst, M. Berioli, G. Renker, "Cross-Layer Control of Adaptive Coding and Modulation for Satellite Internet Multimedia", *International Journal of Satellite Communications and Networking, special issue on Cross-Layer Protocols for Satellite Communication Networks*, Vol. 24, No. 6, pp. 471-491, November/December 2006.

[37] L. A. Larzon, M. Degermark, S. Pink, L. E. Jonsson, G. Fairhurst, "The Lightweight User Datagram Protocol (UDP-Lite)", IETF RFC 3828, 2004.

10

CROSS-LAYER METHODS AND STANDARDIZATION ISSUES

Editors: Gorry Fairhurst[1], María Ángeles Vázquez Castro[2], Giovanni Giambene[3]

Contributors: Gorry Fairhurst[1], Giovanni Giambene[3], Gonzalo Seco Granados[2], Alessandro Vanelli-Coralli[4], María Ángeles Vázquez Castro[2], Fausto Vieira[2]

[1]UoA - University of Aberdeen, UK

[2]UAB - Universitat Autónoma de Barcelona, Spain

[3]CNIT - University of Siena, Italy

[4]UoB - University of Bologna, Italy

10.1 Introduction

This Chapter describes a number of different techniques, approaches and architectures for cross-layer design. It also seeks to position the work presented throughout this book with respect to current and anticipated standards, indicating opportunities for future standardization.

The challenge to be faced is the design of cross-layer mechanisms that can optimize the overall end-to-end application performance over satellite links, while minimizing the utilized radio resources. This optimization can also require additional signaling between the protocol layers. This new area of work is consistent with the end-to-end argument [1], provided that system-level implications are understood [2]. Suitable methods are expected to improve significantly the performance of applications in the next generation of satellite systems, but will require changes to the design of protocols and systems, with implications on the related standards.

The discussion in this Chapter utilizes some basic ideas introduced in the previous Chapters 1 and 4.

10.2 Cross-layer design and Internet protocol stack

The current Internet protocol stack in Chapter 4 is used as the reference architecture for discussion of cross-layer design throughout this Chapter. Design principles categorize and define the placement and operation of functions within a given system. These design principles impose a structure on the design *area*, rather than solving a particular design *difficulty*. This structure provides a basis for discussion and analysis of trade-offs, and suggests a strong rationale to justify design choices.

The various standardization bodies define protocols that may be used by a system to exchange information typically specifying a protocol at a single layer of the system architecture. A cross-layer design goes beyond this structure in two ways: by increasing the awareness between layers or by implicitly conveying information between layers.

- The first case usually entails an exchange of information between protocols to enable them to work jointly towards a specific goal.
- The second case requires a redesign of the system architecture. This redesign allows layers to exchange implicitly information by, for example, mapping the functionality of one layer into a queue of an adjacent layer, without the need for cross-layer signaling. There is no actual exchange of information between layers: the traffic passing through a queue provides sufficient in-band information for the cross-layer method.

There are many mechanisms that display these properties and which have already been standardized, although these were not considered cross-layer approaches, since the term was not then defined. One possible example of cross-layer design is *Random Early Detection* (RED) that was initially proposed in 1993 [3]. The on-going standardization of cross-layer design will allow a better understanding of current schemes and "cleaner" approaches for future systems.

10.3 Cross-layer methodologies for satellite systems

The following sub-Sections provide a classification of cross-layer methodologies, based on a review of current literature and the work that has been presented in the previous Chapters.

10.3.1 Implicit and explicit cross-layer design methodologies

An important aspect for differentiating cross-layer methods, that was highlighted in Chapter 1, was the presence or absence of signaling of the internal protocol state between protocol layers. This may be used as a basis to differentiate between *implicit* and *explicit* cross-layer design/techniques, as summarized below.

In an *implicit* cross-layer design, cross-layer interactions are taken into consideration during the design phase, but there is no exchange of control information among different layers during operation. Layers are designed to complement each other and unnecessary duplicated functions can be eliminated. For example, the objective may be to prevent MAC-level collisions in the case of network flooding, or to apply a retransmission policy at the link-layer that is aware of the requirements and behavior of the transport layer protocols.

An *explicit* cross-layer design requires the exchange of additional control information between different layers during operation. This method can be used to tailor dynamically the operation and/or performance of the various protocol entities, for example to signal periods of outages to higher layers, or expected link capacity requirements to the lower layers.

10.3.2 Cross-layer techniques categorized in terms of the direction of information flow

Another cross-layer classification method considers the direction of the cross-layer information flow [4]. This approach is appropriate to an explicit cross-layer design and primarily focuses on optimizing the information flow. Such an approach should allow efficient *ad hoc* optimizations for each layer and/or protocol, compatible with future versions of current protocols. Moreover, it could provide an optimized cross-layer mechanism that could be used for different kinds of optimization, rather than defining isolated cases that are optimized for a particular communication system.

Developing an integrated cross-layer framework may be important to the satellite community, since it not only leads to improved multimedia performance over existing networks, but could also provide valuable insights into the design of next-generation algorithms and protocols for satellite multimedia systems.

This approach does not follow a traditional layered design. History has shown that devoting time to build a solid framework (like the OSI reference model) failed, when the more integrated TCP/IP protocol stack has succeeded. However, if the Internet continues its current gradual evolution, this may be too slow to be able to satisfy the immediate needs for cross-layer satellite optimizations.

One criterion for the evaluation of cross-layer methods is the *efficiency*, i.e., a flow of information would be considered more efficient if a maximum of information is available to other layers when passing a minimum set of parameters or signaling. Another criterion is the evaluation of the chosen *ad hoc performance parameter* that benefits from the information flow.

The impact on system design is a key constraint when designing to achieve efficient information flow. A cross-layer approach does not necessarily require a re-design of existing protocols, and can be performed by selecting and jointly optimizing the upper layers and the strategies available at the lower layers, such as admission control, resource management, scheduling, error protection, power control, etc.

Bottom-up approach

A *bottom-up* method seeks to design an efficient information flow among layers from the lowest layers up to the application. This approach can be appropriate to a satellite system, implementing *Adaptive Coding and Modulation* (ACM), since it would enable upper layers to be informed of the dynamics and adaptation that is taking place at the physical layer. This cross-layer solution may be less optimal for multimedia transmissions over terrestrial wireless systems, due to the delays incurred with respect to the fast variations of the radio channel conditions. However, in broadband satellite communications, slow channel variations can allow a bottom-up method to signal upper layers in time for them to react.

This approach requires defining general per-layer parameters that could be useful to the upper layers. Moreover, protocols operating at each layer should be reviewed assuming that all cross-layer parameters flowing up from lower layers are (instantaneously) available.

One serious issue is that of finding appropriate parameters that application developers will wish to utilize in the design of their applications. The wide variety of different environments in which modern Internet applications operate makes it unattractive to tune applications to specific types of networks (WiFi, WiMAX, satellite, fiber-channel, etc.), although one could envisage the communication of common transmission path characteristics (e.g., an indication of path change, of QoS change, etc.) in the same way that network stacks currently respond to indications of congestion or network-reachability information.

Top-down approach

A *top-down* approach designs an efficient information flow among layers from the application layer down to the physical layer. This can be seen as an application-centric approach: applications indicate their expectations of required network behavior, and lower-layers can then use this information to optimize lower layer parameters.

There are drawbacks with this approach. One problem is that applications are frequently unaware of the network paths over which they operate. They are therefore unable to express their requirements in a way that maps easily to the capabilities of specific lower-layers. Moreover, applications typically operate over longer time-scales with coarser data granularities (multimedia flows or blocks of data) than those used at lower layers (operating on bits or frames). It is therefore non-trivial to perform adaptive source-channel coding tradeoffs, given the time-varying channel conditions and the fact that multimedia applications cannot be expected to adapt instantaneously their behavior to achieve an optimal performance.

While lower layers can benefit from notifications of requirements (capacity estimates, delay bounds, FEC/ARQ needs, priority, etc.) this does not provide a complete solution. For example, it has limited benefit for a satellite system implementing ACM, since the upper layers may not be able to influence usefully the behavior of lower layers, rather, the channel dynamics require upper layers to adapt themselves.

Hybrid approach

There are cases in which system level constraints are refined in a top-down fashion, while the target architecture performance is abstracted in a bottom-up fashion and a "meet in the middle" approach decides the final optimization. In this case, strategies are determined by exhaustively trying/combining all the possible techniques of both the top-down type and the bottom-up one; the aim is to achieve the best performance. This presents the highest flexibility in design choices.

However, this *hybrid* approach can have draw-backs. Constraints on the design will often prevent an exhaustive analysis of all the possible strategies (and their parameters) to choose an optimized composite strategy that would lead to the best possible performance. When designing a cross-layer methodology, general software architecture principles, such as information hiding, modularization, and separation of concerns should be considered. A hybrid approach also poses challenges to design.

10.4 Potential cross-layer optimizations for satellite systems

This Section provides a summary of the set of cross-layer optimizations for the satellite systems presented throughout the previous Chapters of this book. The summary is presented ordered by scenario.

10.4.1 Optimizations aiming at QoS harmonization across layers

This sub-Section addresses aspects of QoS harmonization across layers for multimedia traffic in IP networks that contain a GEO satellite node. Two approaches have been investigated, as summarized below:

- **MAC resource utilization optimization to support IP QoS** (see Section 8.3). Current IP QoS frameworks are considered (i.e., IntServ and DiffServ) to investigate how to manage the available resources (layer 2) in an IP-based satellite network. The aim is to be as compliant as possible with a predefined QoS specification. The envisaged system is based on Scenario 2 defined in Chapter 1, Section 1.4.
- **Optimization of resource sharing mechanisms at transport layer** (see sub-Sections 3.4.1 and 3.4.2). In this case, the optimization is performed at the transport layer referring to the delayed real-time service (streaming services), an interesting application, employing buffers to add an artificial delay at the beginning of the play-out, so that a recovery procedure can be started when data is lost. The aim is to enhance the legacy satellite broadcast service with a specific multicast recovery algorithm. The envisaged scenario employs GEO satellites (Scenario 2 in Chapter 1, Section 1.4).

10.4.2 Optimization of the Radio Resource Management

Radio Resource Management (RRM) optimization normally involves the physical and MAC layers. However, the selection of RRM techniques (and the consequent optimization techniques) strongly depend on the envisaged scenario. The following techniques were presented in previous Chapters for a GEO scenario with fixed users (i.e., Scenario 2 in Chapter 1, Section 1.4). Results include:

- **Parametric optimization of bandwidth allocation strategies for TCP connections** (see Section 9.3). This study addresses bandwidth allocation to TCP connections sharing a satellite bottleneck based on the cross-layer adaptation of bit-rate and coding rate. A cross-layer method is used to coordinate the actions at the satellite link physical layer (where the fade countermeasure technique is applied) as well as at the MAC layer (where the satellite bandwidth is allocated) to optimize the TCP goodput; long-lived TCP connections are considered.
- **Bandwidth allocation strategies and dynamic bandwidth segmentation algorithms to maximize fairness and the net satellite return capacity** (see sub-Section 7.3.4). This study assumes that the total available resource is defined as a region in the time-frequency plane, i.e., a MF-TDMA frame. ACM and/or *Dynamic Rate Adaptation* are assumed. It seeks to design jointly the bandwidth segmentation, the time slots duration and the bandwidth allocation to users in a way that provides maximum fairness and efficiency in the use of the return link (DVB-RCS system).

The following technique was presented in a previous Chapter with results for a GEO mobile scenario (i.e., Scenario 1 in Chapter 1, Section 1.4):

- **Scheduling strategies for satellite HSDPA transmissions** (see sub-Section 5.3.2). The aim of this study is to investigate the applicability of the HSDPA air interface in the GEO satellite case, characterized by high propagation delays, but also by slow channel variations. Suitable scheduling techniques are investigated to guarantee the QoS support for multimedia traffic flows sent to mobile users.

The following technique has been proposed in Chapter 6 along with results for a LEO mobile scenario (i.e., Scenario 3 in Chapter 1, Section 1.4):

- **Handover and Call Admission Control** (see sub-Section 6.4). An inter-satellite scenario with satellite diversity in a multimedia LEO satellite system has been investigated. The interest is in identifying handover techniques that are able to fulfill the requirements in terms of both call blocking probability (new calls) and call dropping probability (calls in progress).

10.4.3 Optimizations combining higher and lower layers

A set of approaches summarized below have been investigated, that yield cross-layer methodologies and optimization involving MAC and IP and/or transport layers.

- **Bandwidth allocation taking into account TCP behavior** (see Section 9.4). TCP results in a bandwidth request that is time-dependent, following the slow start and congestion avoidance mechanisms. A fixed RRM allocation strategy can lead to either wastage of resources, if dimensioned to the maximum, or inability to satisfy transient requirements. This study has proposed a TCP-driven RRM scheme that allocates resources by taking into account the behavior of the TCP congestion window (internal state of the TCP protocol) for each flow. In this case, Scenario 2 has been considered. The results obtained highlight that the proposed cross-layer RRM scheme can improve the performance at the TCP level.
- **Protocol integration between Ethernet-like layer-2 and satellite-specific MAC** (see Section 8.6). This optimization focuses on the encapsulation of MAC frames defined in the IEEE 802 project into satellite MAC frames for a LEO-based satellite system. Additionally, the reuse of bridging concepts as defined in IEEE 802.1 and the use of an extended LLC sub-layer (derived from IEEE 802.2) is considered to harmonize the satellite-dependent levels and Ethernet technologies.
- **Optimization of the bandwidth provision at the SD layer considering QoS management issues arising at the SI-SAP interface** (see Section 8.4). This optimization refers to a general satellite network with multiple traffic classes. This study has provided a technique that defines a resource allocation at layer 2 on the basis of the requirements of IP-based traffic (layer 3 queue are also considered) in terms of IP packet loss rate and IP packet average delay.

- **Optimization of higher layer coding for an efficient delivery of multimedia contents across hybrid wireless networks** (see Section 8.5). This work is related to the optimization of the packet loss performance across multiple wireless access networks (i.e., satellite, and wireless networks) that are characterized by different radio channel conditions and, hence, by different packet loss patterns. Erasure codes (FZC) are proposed for implementation between the transport and network layers. The introduction of FZC allows a packet loss rate reduction without introducing additional delays.

A summary of the above cross-layer optimization proposals is provided in Table 10.1 below.

Scenario	**Users**	**Above Section**	**Layers Involved**	**Main Optimization**	**Requirements**
2	FIXED	10.4.1	2,3	Efficiency	DiffServ, IntServ
2	FIXED	10.4.1	3,4	QoS	DiffServ, IntServ
2	FIXED	10.4.2	1,2,4	TCP goodput	Parametric optimization ACM
2	FIXED	10.4.2	1,2	Throughput, bandwidth segmentation	Parametric optimization ACM
1	MOB	10.4.2	1,2,3	Transport format, layer 3 QoS	ACM of HSDPA type
3	MOB	10.4.2	2	Call blocking and call dropping	LEO multimedia sat. system
1	FIXED	10.4.3	2,4	TCP goodput	DVB-RCS
3	FIXED	10.4.3	2	service disruption due to LEO network topology change	IEEE 802, LLC layer extended to a satellite scenario
General sat. syst.	FIXED	10.4.3	2,3	Resource allocation optimization	SI-SAP
1-like	MOB	10.4.3	1,2,4 & above	Packet loss	FZC codes

Table 10.1: Summary of cross-layer approaches for satellite systems addressed in this book, in the order of appearance in the previous sub-Sections 10.4.1 - 10.4.3.

10.5 Cross-layer signaling for satellite systems

A number of signaling methods have been proposed that may carry cross-layer information between layers. According to [5], four different techniques can be considered (see Figure 10.1), as described below.

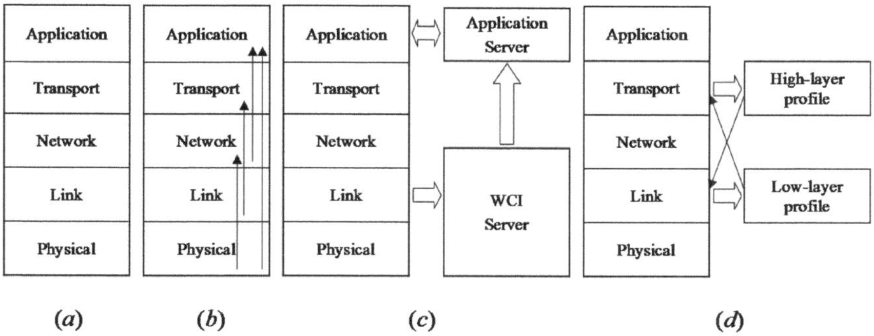

Fig. 10.1: Signaling (*a*) based on packet headers, (*b*) based on ICMP, (*c*) based on a network service, (*d*) based on local profiles.

Method based on packet headers

A packet header method uses IP data packets as in-band message carriers with no need to use a dedicated internal signaling protocol [5]. An IP packet normally can only be processed layer-by-layer, and it is not easy for higher layers to access the IP-level header. This method can be visualized as a "signaling pipe" (see Figure 10.1*a*).

Method based on *Internet Control Message Protocol* (ICMP)

ICMP is a widely-deployed signaling protocol in IP-based networks. In comparison with the "signaling pipe" previously described, this method tries to "punch holes in the protocol stack" and propagates information across layers by using ICMP messages (see Figure 10.1*b*). In this system, desired information is abstracted to parameters, measured by corresponding layers. A new ICMP message is generated only when a parameter exceeds a suitable threshold. Cross-layer communications are provided through selected "holes" and not through a general "pipe". For this reason, this method seems more flexible and efficient; moreover, it is more mature since it has been already implemented in the Linux operating system with suitably-developed *Application Programming Interfaces* (APIs). However, an ICMP message is always encapsulated in an IP packet and, hence, such message has to pass by the network layer even if the signaling is only desired between link layer and application. Utilizing ICMP messages generated within a network also requires extreme care, since this creates a vulnerability to denial-of-service attacks, and can also lead to confusing indications in cases where not all packets follow the same path through a network.

Method based on a network service

In this scheme, channel and link states from the physical layer and link layer are collected, abstracted and managed by third parties, i.e., distributed servers (e.g., *Wireless Channel Information*, WCI, server, see Figure 10.1*c*). Interested applications access the servers for their required parameters from the lowest two layers. Even if there is not a cross-layer signaling scheme within a terminal, this is a complementary solution to the two above presented schemes. Nevertheless, intensive use of this method could introduce considerable signaling overhead and delay over a radio access network.

Method based on local profiles

In this approach, local profiles are used on end-hosts to store periodically updating information: cross-layer information is abstracted from each necessary layer and stored in separate profiles. Other interested layer(s) can select the profile(s) to obtain the desired information as shown in Figure 10.1*d*.

10.6 Standardization issues

Some of the mechanisms summarized in Section 10.5 may be ready for standardization in the short-term (a few years). Initial discussions are already proceeding in many organizations, but a generalized framework and higher-layer interactions will require more substantial research before standardization may start.

A wide range of organizations participating in standardization of components of the satellite systems has been identified. In particular, we can refer here to the *Internet Engineering Task Force* (IETF) [6] at the transport layer (e.g., TCP, UDP, RTP, QoS), to organizations such as the *European Telecommunications Standards Institute* (ETSI) [7] and the *International Telecommunication Union - Telecommunication sector* (ITU-T) [8] that deal with the network layer and QoS, to satellite/broadcast/mobile *fora* and organizations that focus on lower layer functions (e.g., DVB, SatLabs, 3GPP).

Below the network layer, mobile and broadband systems have traditionally been standardized by different organizations (e.g., 3GPP, DVB-Forum) or by different areas within the same organization (e.g., ETSI *Broadband Satellite Multimedia* -BSM-, ETSI *Satellite-UMTS*, S-UMTS). This is likely to continue for cross-layer methods. It is therefore important to disseminate information about key issues, available options and requirements to these groups in preparation for future standardization work. Cross-fertilization of ideas and results may be of benefit to both mobile and broadband communities, allowing them to converge (perhaps using the common traffic classes established for QoS interworking).

Above the network layer, there has been little attention paid to the demands and benefits of cross-layer optimization. Recent work within the *Internet Architecture Board* (IAB) suggests that after much exploration of the issues, IETF is starting to understand the architectural issues, and standardization within appropriate IETF working groups could follow in the longer-term.

The multi-disciplinary nature of cross-layer approaches, not only complicates the analysis of the system, but is expected to pose also practical problems to standardization. To be successful, standards for cross-layer mechanisms and interfaces will require close liaisons and information flow between these organizations on both technical and architectural issues. Such close liaisons are complicated by differing terminology and by different standardization processes employed, and are not the current norm. Standardization of cross-layer methods will therefore pose its own challenges. A first example of standardization of cross-layer methods can be represented by encapsulation allowing adaptive coding in DVB-S2, a work recently started within *DVB-Return Channel via Satellite* (DVB-RCS), *DVB-Global Broadcast Service* (DVB-GBS) and IETF working groups.

The following Sections introduce the key standardization bodies and groups relevant to the eventual standardization of cross-layer methods for satellite systems [9].

10.6.1 Standardization bodies and groups

The groups of interest in connection with standardization activities that could be related to cross-layer issues are listed in Table 10.2.

10.6.2 European Conference of Postal and Telecommunications Administrations

The *European Conference of Postal and Telecommunications Administrations* (CEPT), through its permanent *European Radiocommunication Office* (ERO), is a body of policy-makers and regulators with 44 country members covering almost the entire geographical area of Europe. Within CEPT, the *Electronic Communications Committee* (ECC) is responsible amongst others for developing policies on electronic communication activities in a European context and for harmonizing within Europe the efficient use of the radio spectrum.

10.6.3 ETSI

The objective of ETSI [7] is to produce and perform the maintenance of the technical standards and other deliverables which are necessary to achieve a large, unified European market for telecommunications and related areas. ETSI is an independent, non-profit organization, based in Sophia Antipolis (France). The principal role of ETSI is technical pre-standardization and standardization at the European level in the following fields:

Organization	Working group	Sub-group	Layers
IETF	-	-	L2-L7
3GPP	-	-	L1-L2
CEPT	-	-	L1
CEN/CENELEC	-	-	L1-L2
ETSI	TC-SES	S-UMTS	L1-L3
ETSI	TC-SES	SDR	L1-L2
ETSI	TC-SES	BSM	L2-L3
DVB	DVB-CM	-	L1-L2
DVB	DVB-TM	-	L1-L2
DVB	DVB-RCS	-	L1-L2
DVB	DVB-S2	-	L1
ITU-R	SG4	WP 4 A	L1
ITU-R	SG4	WP 4 B	L1
ITU-R	SG6	-	L1
ITU-R	SG8	WP 8 D	L1
ITU-R	SG8	WP 8 F	L1
WorldDAB	-	-	L1-L3
GBSI-ITSO	Standards and Regulatory Groups	-	L1-L3

Table 10.2: Standardization *fora* of interest for cross-layer issues.

- Telecommunications.
- Information and communication technology in co-ordination with the *European Committee for Standardization* (CEN) and the *European Committee for Electro-technical Standardization* (CENELEC).
- Areas common to telecommunications and broadcasting (especially audio-visual and multi-media matters) in co-ordination with CEN, CENELEC and the *European Broadcasting Union* (EBU).

The ETSI *Technical Committee for Satellite Earth Stations and Systems* (TC-SES) is responsible for all types of satellite communication services (including mobile and broadcasting) and for all types of Earth station equipment (especially the radio frequency interfaces and network and/or user interfaces). It maintains an internal liaison with the ETSI *EMC and Radio spectrum Matters* (ERM) working group (for electromagnetic compatibility issues and radio spectrum matters), with the ETSI *Special Mobile Group*, SMG (for GSM and S-UMTS), and with the working group TM4 of the ETSI Technical Committee *Transmission & Multiplexing*, TM (for fixed radio links). TC-SES also maintains external liaisons with other bodies, including: ITU-R (SG4 on *Fixed Satellite Services*, JWP10-11S on satellite broadcasting, WP 8 D on *Mobile Satellite Services*, TG8/1 on IMT-2000), CEPT-ERO and the *European Co-operation on Space Standardization* (ECSS). Many of the standards produced by the TC-SES are relevant to mobile satellite systems, broadcasting satellite systems and hybrid networks, comprising satellite and terrestrial infrastructures.

The ETSI TC-SES S-UMTS working group oversees the *Satellite* component of the UMTS as part of the *International Mobile Telecommunications* (IMT-2000) standard. It is the ETSI focal point for liaising with the other bodies for the development of standards on S-UMTS/IMT2000. S-UMTS systems will complement *Terrestrial UMTS* (T-UMTS) and interwork with other IMT-2000 family members through the UMTS core network. S-UMTS mobile satellite services will be delivered utilizing either *Low* or *Medium Earth Orbit* (LEO, MEO), or *Geostationary* (GEO) satellite(s). One of the main objectives of the S-UMTS working group is to enforce a significant level of compatibility with T-UMTS in order to minimize user terminal modifications required to receive S-UMTS mobile satellite services. Three main directions are currently explored:

- The adaptation of the 3GPP W-CDMA specifications to satellite;
- The adaptation of the 3GPP *Multimedia Broadcast Multicast Service* (MBMS) specifications to satellite;
- The analysis of the feasibility of an OFDM air interface for mobile satellite networks.

The S-UMTS Family G specification set aims at achieving the satellite air interface fully compatible with W-CDMA-based systems. However, due to the differences between terrestrial and satellite channel characteristics, not all the T-UMTS specifications are directly applicable, but some of them need modifications. Family G has been released as a multipart standard consisting of the following six documents ([1]) specific to the satellite air interface:

- Part 1, "Physical channels and mapping of transport channels into physical channels (S-UMTS-A 25.211)", defines transport channels and physical channels [11];
- Part 2, "Multiplexing and channel coding (S-UMTS-A 25.212)", describes multiplexing and channel coding [12];
- Part 3, "Spreading and modulation (S-UMTS-A 25.213)", specifies spreading and modulation [13];
- Part 4, "Physical layer procedures (S-UMTS-A 25.214)", describes physical layer procedures [14];
- Part 5, "UE Radio Transmission and Reception", establishes the minimum RF characteristics for the user equipment [15];
- Part 6, "Ground stations and space segment radio transmission and reception", describes the space segment RF characteristics [16].

The TC-SES S-UMTS technical activity related to *Satellite MBMS* (S-MBMS) is based on the design of a *Satellite Digital Multimedia Broadcasting*

[1] Part 1 through part 4 are based on their counterparts developed within 3GPP for terrestrial UMTS in frequency division duplexing mode.

(S-DMB) system [17]. This offers a unidirectional point-to-multipoint bearer service from a single source entity (satellite) to multiple recipients using broadcast or multicast mode. S-MBMS is defined by six specifications currently under approval within TC-SES S-UMTS.

Finally, TC-SES S-UMTS is studying OFDM as a possible satellite air interface. OFDM techniques are being used by several digital broadcast terrestrial systems and are characterized by high spectral efficiency. These techniques have been considered for 3G air interfaces, but entail technical challenges due to both the rather high peak-to-average power ratio and the non-linear distortion induced by the on-board *High Power Amplifier* (HPA). The TC-SES S-UMTS studies show that with *ad hoc* pre-distortion techniques and turbo coding the effect of the HPA non-linear distortion drastically reduces, thus allowing for the adoption of OFDM on satellite air interfaces [18].

Although current specifications and study items do not explicitly deal with cross-layer aspects, the need for TC-SES S-UMTS to maintain compatibility with the T-UMTS system evolution will indeed require opening new work items related to the development of capacity-improving techniques, such as interference mitigation, multi-user detection, and macro- and micro-diversity algorithms. TC-SES S-UMTS is a crucial working group regarding interests for cross-layer activities on S-UMTS. Its liaisons with organizations outside ETSI include: 3GPP, ITU-R SG8 WP 8 D, and ITU-R SG8 WP 8 F (for IMT-2000 and systems beyond).

10.6.4 DVB

The DVB Project was initiated in 1992 [19] and has subsequently implemented an approach of pre-competitive co-operation in the development of open digital TV standards that can be freely adopted worldwide. The motivation was to promote a common, standard, European platform for digital TV broadcasting, and the idea was supported by all players (i.e., broadcasters, operators, standardization bodies, media groups and industry). Today, DVB has 220 members from more than 30 countries worldwide. By incorporating both commercial and technical bodies within the organization, DVB has succeeded in delivering transmission standards for television systems operating over a range of media, including DVB-S, DVB-C and DVB-T standards. The advent of interactive networks stimulated the standardization of *Return Channels for Cable* (i.e., DVB-RCC), *Satellite* (i.e., DVB-RCS), *Local Multipoint Distribution System*, LMDS (i.e., *DVB-Return Channel for LMDS*, DVB-RCL), and *Terrestrial* (i.e., DVB-RCT) systems.

The work in the DVB technical area is organized in *ad hoc* groups. Each of them works on commercial requirement documents provided by the *Commercial Module*. This is a set of user requirements that outline market parameters, such as user functions, timescales and price range. A DVB specification is developed in the *Technical Module* and its working groups, where technological implications of user requirements are examined and available technologies are explored. Once the Technical Module reaches consensus on the resulting specification, and the Commercial Module's support for it has been ensured, the Steering Board is solicited to give the final approval. It is then offered for standardization to ETSI or CENELEC through the EBU/ETSI/CENELEC *Joint Technical Committee* as well as sometimes to ITU-T or ITU-R.

The main DVB standards that are relevant to satellite communications are considered below.

DVB-RCS

The *Digital Video Broadcasting-Technical Module* (DVB-TM) created an *ad hoc* group early 1999, called DVB-RCS, which lead to specification ETSI EN 301 790 [20]. This document specifies a satellite terminal known as a *Satellite interactive Terminal* (ST) or *Return Channel Satellite Terminal* (RCST) that supports a two-way DVB satellite system.

The return link in DVB-RCS uses an MF-TDMA air interface where STs have allocated capacity in slots within a certain time-frequency structure. The entire system is controlled by a *Network Control Center* (NCC) (e.g., at the Gateway side of the satellite) controlling the ST behavior. The NCC is responsible for synchronization of the system, via the *Network Clock Reference* (NCR), and sends out a number of specific system tables in order to give the STs all the information needed for receiving and transmitting in the system. This includes, in addition to the tables already existing in the DVB-S system, tables informing on frame composition, capacity allocation, regulation of ST timing and frequency offsets, etc.

The DVB-RCS standard adopts the DVB-S standard for the forward link, that uses a *Time Division Multiplexing* (TDM) carrier, usually with practical data rates on present Ku band transponders ranging up to tens of Mbit/s. The second generation of DVB-S, DVB-S2 (see later) is also compatible with the DVB-RCS standard.

The DVB-RCS standard does not currently include any specific features for mobility management, and this issue is a current research topic and a standardization target. To achieve this, it is important to devise a robust variation of the DVB-RCS return link to support high-speed mobility as maintaining synchronization after acquisition. More details are provided in a further sub-Section.

Finally, the interoperability issues between DVB-RCS terminals and networks are addressed by the SatLabs Group, an international, non-profit association whose members are interested in promoting two-way satellite networks based on the DVB-RCS open standard. The SatLab Web site [21] contains a wide collection of documents (some of them have a public access) dealing with specifications, recommendations and technical issues. The SatLabs qualification programme was defined to achieve DVB-RCS interoperability testing and certification. The SatLabs Group is lead by ESA with the participation of many manufacturers, operators and service providers in the field of satellite communications.

DVB-S2

The DVB-S2 [22] standard for satellite transmission supports ACM, which enables high data-throughput efficiency. ACM is applicable in networks where a return channel allows transmission of information concerning the reception quality from the satellite receiver to the satellite uplink station. The standard defines the reception quality parameter and its binary coding. The transport of this parameter back to the uplink station is not in the scope of the standard and is specified separately for the different return channel systems. This has been done already for satellite return channel in the current release of the DVB-RCS standard [20].

Another potential application for ACM in DVB-S2 is hybrid satellite-terrestrial networks for high-speed Internet access. In this kind of networks, a user terminal receives data over satellite and transmits data over a terrestrial dial-up connection. The more efficient use of satellite capacity could make such hybrid networks more attractive and therefore enable a larger market for DVB-S2 receiver chips with the interactive services profile implemented.

Applicability of DVB-S2-like ACM as a countermeasure to fading due to terminal mobility is also a possibility. ACM does not help against the fast fading that occurs in land mobile scenarios due to multipath and, further, against typically short shadowing and blocking events. The adoption of ACM in DVB-S2 is intended to counteract rain fading; therefore, it is important to investigate how terminal mobility changes the time variability of rain fade events and, hence, the efficiency of ACM, e.g., when a car or a high-speed train travel through a rain cell.

DVB-S2 extension for mobile usage

Current expectations of users are to access the Internet and to receive multimedia contents while on the move. This is the reason way there is interest in evolving the DVB-S2/-RCS standard to allow the mobile usage (possible scenarios are: users on plains, trains and in land masses). This extension need to address many challenging issues such as [23]: stringent frequency regulations (Ku band), Doppler effect, frequent handovers, and impairments in synchronization acquisition and maintenance. In addition to this, the railway scenario is affected by shadowing, fast fading (due to mobility, there are deep and frequent fades caused by the poles of the electrified lines) and long blockages (presence of tunnels and large train stations with non-LoS propagation conditions to the satellite). The new standard should address important issues that are outlined below.

- *Spectrum spreading techniques*: the stringent regulations for Ku band mobile terminals require a careful study for the possible use of spreading techniques, especially in the return link for terminals with small antennas. The adoption of spectrum spreading is a possible solution to reduce the EIRP, while preserving the required SNR, at the expenses of reduced spectral efficiency. In the forward link, the introduction of spreading requires the design of a new DVB-S2 receiver. In the return link, each terminal could in principle implement direct spreading within the assigned time and frequency slot (MF-TDMA approach).

- *Fading countermeasures*: the more challenging propagation conditions of the non-LoS scenario can be mitigated by adopting advanced techniques such as diversity and higher layer FEC schemes. Moreover, new synchronization acquisition and maintenance procedures need to be employed to cope better with frequent fades.
- *Resource management techniques*: efficient RRM schemes need to be adopted to account for mobility, such as: impact of spreading on the MF-TDMA allocation process (DVB-RCS); support of handover requests with suitable protocols; interworking with terrestrial networks in shadowed areas (e.g., tunnels, cities, etc.) where gap fillers can be used; adaptive schedulin techniques for the forward link that are aware of the physical layer behavior.

All these innovative aspects require a cross-layer system design aiming at optimizing the choices made at different layers. The DVB-TM is now working to specify the modifications that are needed for the mobile extension of the DVB-S2 standard [23]. The SatNEx II project [24] is actively involved in this standardization process.

DVB-H

The broadcast of digital television signals was originally targeted to fixed reception, although mobile reception is also feasible with current digital television standards (DVB-T, DVB-S2). The Commercial Module of DVB decided to launch commercial requirements for the production of *ad hoc* specifications able to provide broadcasting to one specific niche of the mobile receivers: handheld terminals. This is the aim of the *DVB-Handheld* (DVB-H) standard.

Conditional access is important in all broadcast radio/satellite networks to prevent unauthorized access to the broadcast content by eaves-dropping. In DVB-H, an *IP-based Conditional Access System* (IP-CAS) can provide link-layer encryption (scrambling) for DVB-H services. CAS messages are delivered over IP and may take advantage of time-slicing to save power at a receiver. The DVB common scrambling algorithm on Transport Stream packets is also employed (DVB-CAS): it uses *entitlement control messages* to send keys to receivers and *entitlement management mode* messages to deliver management messages.

10.6.5 International Telecommunication Union

ITU is an international organization of the United Nations where governments and industries coordinate global telecom networks and services. ITU is divided in three sectors: ITU-T that aims at the definition of high-quality standards covering all fields of telecommunications; ITU-R that plays a fundamental role in the management of the radio-frequency spectrum, physical layer issues, and satellite orbits; and ITU-D, dealing with *Telecommunications Developments.*

ITU-R is charged with determining the technical characteristics and operational procedures for a huge and growing range of wireless services. This Sector also plays a vital role in the management of the radio-frequency spectrum, a finite natural resource that is increasingly in demand due to the rapid development of new radio-based services and the enormous popularity of mobile communication technologies.

In its role as global spectrum coordinator, ITU-R develops and adopts the *Radio Regulations*, a voluminous set of rules that serve as a binding international treaty governing the use of the radio spectrum for different services around the world. ITU-R also acts, through its Bureau, as a central registrar of international frequency use, recording and maintaining the *Master International Frequency Register*, which currently includes around 1,265,000 terrestrial frequency assignments, 325,000 assignments servicing 1,400 satellite networks, and another 4,265 assignments related to satellite Earth stations. Moreover, ITU-R is responsible for coordinating efforts to ensure that communication, broadcasting and meteorological satellites in the world's increasingly crowded skies can co-exist without causing harmful interference each other. The Union facilitates agreements between both operators and governments, and provides practical tools and services to help frequency spectrum managers.

The portion of the radio-frequency spectrum suitable for communications is divided into 'blocks', the size of them varying according to individual services and their requirements. These blocks are called 'frequency bands' and are allocated to services on an exclusive or shared basis. The full list of services and frequency bands allocated in different regions forms the *Table of Frequency Allocations*, which is a part of the radio regulations.

10.7 Conclusions

A range of cross-layer optimization techniques have been proposed and evaluated in this book for three different scenarios (i.e., DVB-S/DVB-RCS via GEO bent-pipe satellite, S-UMTS via GEO bent-pipe satellite, and LEO constellation with regenerating satellites). The most significant techniques have been summarized in this Chapter to provide final guidelines for both standardization efforts and further research directions.

Cross-layer methods have been categorized, considering: (*i*) either explicit signaling or an implicit scheme with a joint optimization of different protocol layers; (*ii*) the definition at higher layers of requirements to be used for appropriate settings at lower layers or, vice-versa, the lower layers progressively determining the requirements at higher layers. As for explicit cross-layer, we have described different mechanisms for the exchange of internal protocol state information between non-adjacent protocol layers, thus violating the classical ISO/OSI layered philosophy.

We have proved that the cross-layer techniques can improve the overall end-to-end quality of service, while optimizing the efficiency in utilizing the scarce satellite radio resources. However, standardization *fora* have not yet significantly addressed cross-layer issues. To this aim, there is a need for a new framework, as well as the strong cooperation of different standardization bodies. One of the aims of this book has been to provide some useful insights that may promote new standardization activities on cross-layer air interface design for satellite communication networks.

References

[1] J. H. Saltzer, D. P. Reed, D. D. Clark, "End-to-End Arguments in System Design", *ACM Transactions in Computers Systems*, Vol. 2, No. 4, pp. 277-288, November 1984.

[2] P. Karn, C. Bormann, G. Fairhurst, D. Grossman, R. Ludwig, J. Mahdavi, G. Montenegro, J. Touch, L. Wood, "Advice for Internet Subnetwork Designers", BCP 89, IETF RFC 3819, July 2004.

[3] S. Floyd, V. Jacobson, "Random Early Detection Gateways for Congestion Avoidance", *IEEE/ACM Transactions on Networking*, Vol. 1, No. 4, pp. 397-413, August 1993.

[4] M. van der Schaar, S. Shankar, "Cross-Layer Wireless Multimedia Transmission: Challenges, Principles, and New Paradigms", *IEEE Wireless Communications Magazine*, Vol. 12, No. 4, pp. 50-58, August 2005.

[5] Q. Wang, M. A. Abu-Rgheff, "Cross-Layer Signalling for Next-Generation Wireless Systems", in *Proc. of IEEE Wireless Communications and Networking Conference 2003* (IEEE WCNC 2003), New Orleans, USA, pp. 1084-1089, March 2003.

[6] The Internet Engineering Task Force (IETF); Web page with URL: http://www.ietf.org.

[7] European Telecommunications Standards Institute (ETSI); Web page with URL: http://www.etsi.org.

[8] International Telecommunication Union; Web page with ULR: http://www.itu.int/home/index.html.

[9] MoSSA, Advanced Satellite Mobile Systems-Task Force Specific Support Action, Project IST-507557, Deliverable "Survey on Standardization and Regulatory Activities"; Web site with URL: http://asms1.wss.bcentral.com/mossa/default.htm.

[10] ETSI TC-SES working group; Web page with URL: http://portal.etsi.org/ses/.

[11] ETSI, "*Satellite Earth Stations and Systems* (SES); Satellite Component of UMTS/IMT2000; G-family; Part 1: Physical channels and mapping of transport channels into physical channels (S-UMTS-A 25.211)", TS 101 851-1.

[12] ETSI, "*Satellite Earth Stations and Systems* (SES); Satellite Component of UMTS/IMT2000; G-family; Part 2: Multiplexing and channel coding (S-UMTS-A 25.212)", TS 101 851-2.

[13] ETSI, "*Satellite Earth Stations and Systems* (SES); Satellite Component of UMTS/IMT2000; G-family; Part 3: Spreading and modulation (S-UMTS-A 25.213)", TS 101 851-3.
[14] ETSI, "*Satellite Earth Stations and Systems* (SES); Satellite Component of UMTS/IMT2000; G-family; Part 4: Physical layer procedures (S-UMTS-A 25.214)", TS 101 851-4.
[15] ETSI, "*Satellite Earth Stations and Systems* (SES); Satellite Component of UMTS/IMT2000; G-family; Part 5: UE Radio Transmission and Reception (S-UMTS-A 25.101)", TS 101 851-5.
[16] ETSI, "*Satellite Earth Stations and Systems* (SES); Satellite Component of UMTS/IMT2000; G-family; Part 6: Space Segment Radio Transmission and Reception (S-UMTS-A 25.104)", TS 101 851-6.
[17] IST-MAESTRO project, "Mobile Applications & sErvices based on Satellite & Terrestrial inteRwOrking"; Web site with URL: http://ist-maestro.dyndns.org, 2006.
[18] ETSI, "Evaluation of the OFDM as a Satellite Radio Interface Satellite Earth Stations and Systems (SES); Satellite Component of UMTS/IMT-2000", TR 102 433, 2006.
[19] Digital Video Broadcasting (DVB) Project; Web page with URL: http://www.dvb.org.
[20] ETSI, "Digital Video Broadcasting (DVB); Interaction channel for Satellite Distribution Systems", EN 301 790.
[21] SatLabs official Web site with URL: http://www.satlabs.org/.
[22] ETSI, "Digital Video Broadcasting (DVB); Second Generation Framing Structure, Channel Coding and Modulation Systems for Broadcasting, Interactive Services, News Gathering and other Broadband Satellite Applications", EN 302 307.
[23] S. Scalise, G. E. Corazza, C. Párraga Niebla, P. Chan, G. Giambene, F. Hu, A. Vanelli-Coralli, M. A. Vázquez Castro, "Towards the Revision of DVB-S2/RCS Standard for the Full Support of Mobility", *SSC Newsletter*, Vol. 17, No. 2, November 2006.
[24] SatNEx II Web site with URL: http://www.satnex.org.

Index

A

Access protocol, 120, 132
Adaptive algorithms, 209, 295
Adaptive coding and modulation, 16, 24, 106, 139, 208, 316
Asynchronous transfer mode, 109

B

Broadband satellite multimedia, 28, 31, 69, 98, 256
Broadcast and multicast services, 5, 80, 152, 160

C

CAC, 45, 51, 100, 110, 177, 179, 184, 189, 199, 257
  Complete partitioning, 52, 179
  Complete sharing, 52, 179
Call handover, 53, 189, 195, 214, 233
  Inter-satellite handover, 54, 190, 194, 214
  Intra-satellite handover, 54, 190, 191, 214
CDMA, 14
Channel quality indicator, 138
Channel utilization, 162, 305
Combined free/demand assignment multiple access, 48, 101, 256
Commercial solutions, 82, 89
Communications architecture, 314, 320
  Cross-layer signaling, 320
Congestion control, 164, 186
Control-theoretic approach, 213
Cross-layer approach, 34, 156, 164, 256, 314
  Bottom-up approach, 316
  Hybrid approach, 317
  Top-down approach, 316
Cross-layer design, 35, 36, 45, 96, 105, 214, 221, 256, 270, 313, 314
  Explicit cross-layer, 35, 36, 133, 145, 156, 164, 300, 315
  Implicit cross-layer, 35, 36, 45, 95, 217, 270, 290, 315, 317

D

Delayed real-time services, 83
Demand assignment multiple access, 18, 290, 298, 300
  Access delay, 290, 298
  Rate-based dynamic capacity, 20, 211, 221, 249, 299
  Volume-based dynamic capacity, 20, 211, 221, 249, 299, 303
DiffServ, 36, 77, 107, 183, 246
DVB-S, 16, 80, 105, 187, 326
  DVB-RCS, 17, 81, 186, 211, 249, 289, 298, 323, 327
    Adaptive coding, 218
    Implementation issues, 218, 228
    MF-TDMA scheme, 15, 17, 184, 211, 251, 302, 327

E

ETSI TC SES S-UMTS working group, 15, 121, 325

Explicit congestion notification, 107

F

FDMA, 13

G

GEO satellite systems, 4, 10, 68, 71, 125, 131, 141, 184, 209, 265, 290

H

Handover algorithms, 51
  Handover queuing, 53
  Predictive resource reservation, 54
HSDPA, 16, 108, 138, 139, 141, 144, 148
Hybrid satellite networks, 265
  Erasure codes, 265
  QoS, 266
  WiFi networks, 267

I

Infinitesimal perturbation analysis, 99, 216, 257
IntServ, 36, 77, 107, 183, 244

L

LEO satellite systems, 4, 10, 54, 68, 71, 132, 141, 189, 192, 195

M

MAC, 18, 97, 98, 105, 110, 119, 139, 248, 256, 298
MEO satellite systems, 4, 10, 48, 71, 141
Modeling and simulation, 54

N

NCC, 17, 178, 208, 298, 327
Network layer, 243
Node-B, 139

O

OSI model, 34, 102

P

Packet scheduler, 134, 137, 140, 152, 155, 164
Performance enhancing proxies, 29, 99, 293
Power allocation and control, 50
  Closed loop, 50
  Feedback loop, 50
  Open loop, 50
Proactive algorithms, 210

Q

QoS classes, 156, 165
QoS for multimedia services, 68
  Background services, 76
  Conversational services, 70
  Interactive services, 73
  Performance requirements, 70, 73, 74, 76
  QoS based IP models, 76
  Streaming services, 74
QoS mapping, 98, 256, 260

R

Radio resource management, 43, 54, 96, 101, 119, 177, 289, 303, 318
  Cross-layer approach, 45, 60, 96, 99, 101, 104, 214, 295, 303, 305
    Joint optimization, 95, 97–100
    MAC-centric approach, 105
  Dynamic allocation, 20, 47, 49, 55, 99, 101, 110, 191, 198, 208, 211, 213, 214, 218, 233, 248, 251, 256, 299
  Fairness, 44, 217, 232
Reactive algorithms, 210
Receding horizon controller, 214
Resource allocation, 23, 46, 99, 121, 138, 158, 179, 184, 208, 225, 249, 299
  Frequency allocation, 46
  Space allocation, 46
  Time allocation, 46

S

S-UMTS, 15, 58, 108, 121, 131, 152, 325
Satellite constellations/orbits, 4, 10, 275
Satellite digital multimedia broadcasting, 325
Satellite IP networks, 31, 69, 76, 109, 248

IP QoS, 76, 109, 183, 244, 248
Proportional DiffServ, 249
Scheduling scheme, 134
Channel-aware scheduling, 135
Exponential Rule scheduler, 138
Maximum C/I scheduler, 138
Proportional Fair scheduler, 138, 147
Service level agreement, 96, 222, 246
SI-SAP, 31, 98, 256
Smith predictor controller, 214
Standardization, 322
Static algorithms, 209

T

TCP over satellite, 290
Cross-layer interactions, 294, 298
MODCOD optimization, 294
TDMA, 13
Transport layer, 99, 289
Congestion control, 291
TCP, 99, 273, 290, 294, 298
UDP, 273, 290, 305

U

UMTS, 15, 58, 121

V

VLANs for LEO constellations, 270
Voice over IP, 50, 70, 262

W

W-CDMA, 16, 46, 137